Disordered Materials

Advanced Texts in Physics

This program of advanced texts covers a broad spectrum of topics which are of current and emerging interest in physics. Each book provides a comprehensive and yet accessible introduction to a field at the forefront of modern research. As such, these texts are intended for senior undergraduate and graduate students at the MS and PhD level; however, research scientists seeking an introduction to particular areas of physics will also benefit from the titles in this collection.

Springer
Berlin
Heidelberg
New York
Hong Kong
London
Milan
Paris
Tokyo

Paolo M. Ossi

Disordered Materials

An Introduction

With 189 Figures

Paolo M. Ossi
INFM and Dipartimento di Ingegneria Nucleare
Politecnico di Milano
via Ponzio, 34/3
20133 Milano, Italy
e-mail: paolo.ossi@polimi.it

Library of Congress Cataloging-in-Publication Data: Ossi, Paolo M., 1953–. Disordered materials: an introduction/ Paolo M. Ossi. p. cm. – (Advanced texts in physics, ISSN 1439-2674) Includes bibliographical references. ISBN 3540413286 (acid-free paper) 1. Order-disorder models. 2. Amorphous substances. I. Title. II. Series. QC173.4.O73 O74 2002 530.4'13–dc21 2002030318

ISSN 1439-2674

ISBN 3-540-41328-6 Springer-Verlag Berlin Heidelberg New York

Springer-Verlag Berlin Heidelberg New York
a member of BertelsmannSpringer Science+Business Media GmbH

http://www.springer.de

Printed in Germany

Typesetting: Data prepared by the author using a Springer TEX macro package
Final page layout: Verlagsservice Ascheron, Mannheim
Cover design: *design & production* GmbH, Heidelberg

Printed on acid-free paper SPIN 10765262 57/3141/ba 5 4 3 2 1 0

Teach your tongue to say
"I don't know"
for fear
of inventing things
and then being ensnared.

Berachot, 4a

Preface

The teaching of solid state physics essentially concerns focusing on crystals and their properties. We study crystals and their properties because of the simple and neat results obtained from the analysis of a spatially periodic system; this is why the analysis can be made considering a small set of atoms that represent the whole system of many particles.

In contrast to the formal neat approach to crystals, the study of structurally disordered condensed systems is somewhat complicated and often leads to relatively imprecise results, not to mention the experimental and computational effort involved. As such, almost all university textbooks, including the advanced course books, only briefly touch on the physics of amorphous systems.

In any case, both the fundamental aspect and the ever wider industrial applications have given structurally disordered matter a role that should not be overlooked. The study of amorphous solids and their structure, stability and properties is a vibrant research branch; it is difficult to imagine how any physicist, chemist or engineer who has to deal with materials could possibly ignore this class of systems.

The author of Disordered Matter – an Introduction uses this course book at the Politecnico in Milan, Italy. Collecting the material for the course proved no mean task, leading him to have to prepare ad hoc didactic material. The continual exchange between teacher and student has led to the current version of the book.

The goal in preparing this book was to supply a selected range of topics in a way that would allow the reader to understand the various aspects of a highly complicated and many facetted problem, such as the investigation and modelling of the structure of a structurally disordered condensed system.

To this aim Chapter 1 of the book briefly examines the geometrical and symmetrical properties of the platonic solids, taken as constructive elements that can be used to represent atomic structures, even highly complicated ones, by matching them together in various ways.

In Chapter 2 cellular and topological disorders are defined and the entropy approach to describing these kinds of disorder is introduced.

The glass transition, with an emphasis on the kinetic and thermodynamic features, and an examination of the material parameters that may inhibit the formation of a crystal are then discussed in Chapter 3.

Chapter 4 contains an in-depth look at the characteristics and limits of the main experimental techniques used in structural investigations; the various modelling strategies are then examined. Much effort is placed on the ability to geometrically represent the elementary units that, when bonded together, give rise to a structure with local order that is partially similar to the order obtained from experiments on real disordered systems. It is little wonder that an amorphous solid and its corresponding crystal should have a number of structural elements in common. As such we try to study the structure of disordered systems using the very same elements that define the ordered system.

Attention is then placed on how the structure of a system evolves with changing its dimensions. Chapter 5 examines how the transition from an atom-molecule to a solidoccurs; atomic clusters allow for exploring the role of the surface in the stability of a system and clearly indicate that the structure of "small" systems of atoms is non-crystalline. The analysis of noble-gas clusters offers direct proof of the problems arising when we realise an extended packing of elementary structural building blocks. These blocks are made up of only a few atoms and are also locally the most energetically favourable. However, it is impossible to endlessly juxtapose them without introducing defects into the structure, which would be destabilised. On the other hand, the evolution of alkali-metal clusters provides a good example of the role of electrons in stabilising the structure of the system. The family of Carbon clusters with a closed-cage structure, Fullerene C_{60} being the prototype, is then described. An introduction to the cluster-assembled nanocrystalline materials that have recently been synthesised, and that are at the heart of much research activity, is a logical development to cluster physics and is the final step along the path from the atom to the solid.

The quasicrystals are dealt with in Chapter 6: they are a paradigm of extended orientational order but do not have the translational order that is so distinctive of usual periodic crystals. The study of these systems, which constitute an intermediate phase between the crystalline phase and topologically disordered matter, leads to a new definition of the concept of the crystal itself.

Although all topics are treated self consistently, this book is directed at a reader with a reasonable undergraduate background in the physics of crystals. Each chapter is fully supported with a set of figures and completed with references, both general and specialised. Given the plan of the text it is addressed to third, fourth and fifth-year undergraduate students majoring in physics, materials sciences, chemistry, materials, chemical and electronic engineering. Since the book is the outcome of research work, including very recent research results, it can be used as a reference book for researchers working in the field of structurally disordered condensed systems.

I would like to extend my gratitude to several colleagues who helped me with their constructive comments and valuable suggestions at various stages of the preparation of the manuscript. Particular thanks go to Rosanna Pastorelli who critically read some parts of the draft. I am indebted to my students, who listened to the lectures, made useful comments and pointed out many mistakes; all remaining errors are my responsibility alone. Giorgio Benedek was the first person to suggest I should publish an English version of the book. Timothy Dass made a painstaking translation and provided many suggestions to make reading the book easier. I am especially grateful to Evelina Kaneclin for her skill in word processing the book, Mirko Verona for the careful preparation of the figures and Angela Lahee, my editor at Springer, for her patience and assistance. Finally I wish to thank the publishers of the journals carrying the papers I refer to throughout the book.

Milano
May, 2002

P. M. Ossi

Contents

1. Solids: Geometrical and Symmetrical Properties

It does seem somewhat difficult and artificial to talk about condensed matter with irregular structure without first making some reference to its "non pathological" counterpart, those crystals that made up the entire world of solid state physics up to a few years ago.

When we consider the amorphous metals and the amorphous solidified noble gases we see that both the position of the interstices after high density packing of hard sphere atoms and the structure resulting from the disposition of partially deformable spheres, relaxed in a suitable potential, consist mostly of tetrahedra and octahedra, all distorted to some extent or another. Tetrahedra and octahedra are the very same structural units we find in closely packed crystals. Moreover, the way such simple structural units, with a well defined geometrical shape and composition, are packed lies at the heart of those structural models that have been developed both for covalent glasses and for metals whose stoichiometry depends on chemical factors and not just on geometric constraints.

As such we shall examine the geometrical and symmetrical properties of certain highly regular solids in order to help us model amorphous solids (see Chap. 4).

A brief introduction will be given to hypersolids in more than three dimensions space and to the techniques to project them onto usual three-dimensional space. This is particularly relevant in order to study the quasicrystals (see Chap. 5).

Lastly the essential concepts of "classic" crystallography will be recalled; these can be traced back to the recognised crystal perfection and symmetry and supply a simple framework to describe the properties of crystals. Indeed, to study disordered solids special reference will be made initially to their corresponding crystals.

1.1 The Platonic Solids and Their Duals

The five platonic solids, tetrahedron, cube, octahedron, dodecahedron and icosahedron, are perfect examples of highly regular and symmetrical structures. Each has the same kind of regular convex polygon faces, whether they

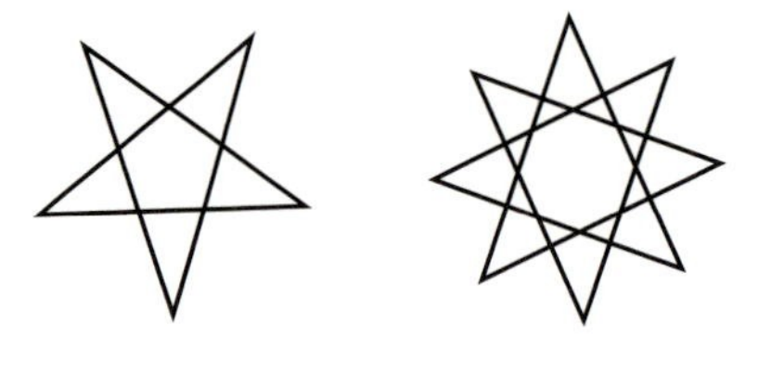

Fig. 1.1. Stellated polygons according to Kepler

be triangle, square or pentagon, and the vertices are all alike. The term polygon means a closed flat region bounded by n straight lines simply connected together. The polygon can be contracted as much as you like and a closed line originally drawn within it will contact accordingly and still within the polygon.

The vertices of a regular polygon, that is equilateral and equiangular, are equally distanced from the centre. As such it is easy to draw a circle around the polygon. The sides are also the same distance from the centre, which means a circle can be drawn inside the polygon that touches each side. A regular polygon $\{n\}$ has n sides and angles.

Apart from studying the regular or primary polygons whose sides do not intersect Kepler was the first to study in a systematic way stellated polygons whose sides are obtained by extending non-adjacent sides of a primary polygon until they intersect, provided that the perimeter is a single line (Fig. 1.1).

A convex polyhedron is said to be regular if its faces are all regular, i.e. the polygons that make up the polyhedron and its solid angles are all regular. For simplicity we shall refer to the term vertex figure rather than solid angle. So, let's consider a vertex V of a polygon: the vertex figure connected to it is the segment enclosed by the centres of the sides that meet at vertex V. For polygon $\{n\}$ whose sides are L length, the length of the vertex figure is given by

$$L_V = L \cos \frac{\pi}{n} . \tag{1.1}$$

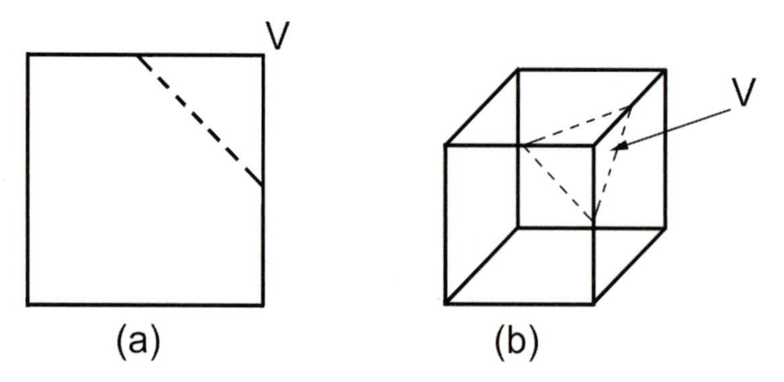

Fig. 1.2. Vertex figure relative to vertex V of a square (**a**) and (dashed-line triangle) at vertex V of a cube (**b**)

Figure 1.2 shows (part a) the vertex figure of a vertex V in a square. The vertex figure of vertex V in a polyhedron is a polygon whose sides are the vertex figures of the faces for V. Figure 1.2 shows (part b) the vertex figure (triangle) of a vertex V of a cube.

The sum of the external angles that make up a polygon is given by 2π, thus each external angle of $\{n\}$ is $2\pi/n$, whereas the corresponding supplementary internal angle is given by $\pi\,(1-2/n)$. Furthermore, the sum of the internal angles is given by $(2n-4)\pi/2$. In a regular convex polyhedron the solid angle of any vertex has r face angles, each of which is $\pi\,(1-2/n)$. The sum of these angles must be less than 2π otherwise we would get re-entrant vertices and not all the vertices would be equal; thus the following would stand:

$$\left(1-\frac{2}{n}\right)<\left(\frac{2}{r}\right) \quad \text{or} \quad \frac{1}{n}+\frac{1}{r}>\frac{1}{2}, \quad \text{thus } (n-2)(r-2)<4\,. \tag{1.2}$$

A platonic solid whose vertex is surrounded by r faces, and each face having $\{n\}$ sides, is called a regular polyhedron and is given as $\{n,r\}$.

From (1.2) we can deduce that the only platonic solids are reported in Table 1.1.

$\{n,r\}$	Faces	Vertices	Edges	Touching faces at each vertex	Name
$\{3,3\}$	4	4	6	3	Tetrahedron
$\{3,4\}$	8	6	12	4	Octahedron
$\{4,3\}$	6	8	12	3	Cube
$\{3,5\}$	20	12	30	5	Icosahedron
$\{5,3\}$	12	20	30	3	Dodecahedron

Table 1.1. Features of the platonic solids

It is now possible to build onto this result. In fact as at least three faces have to meet at each vertex the smallest face angle is $2\pi/3$. This is the value of the angles of a regular hexagon. However, as can be seen from Fig. 1.3, three regular hexagons with a common vertex lie on the same plane and as the value of an angle of a regular polygon increases with the number of sides, the faces of a regular polyhedron can only be a triangle, a square or a pentagon. As the angles of a square are right angles there can be no more than three at any vertex since the sum of the face angles that meet at the vertex is less than 2π; thus the cube is obtained. By the same analogy there cannot be more than three pentagons at any vertex of a regular polyhedron (dodecahedron); four or more cannot be joined together without there being some overlapping.

If equilateral triangles are joined together then the rule for the sum of the face angles at each vertex is fully met, with three equilateral triangles giving a tetrahedron, four giving a tetrahedron and five giving a icosahedron, whereas if we use six faces we get a hexagon.

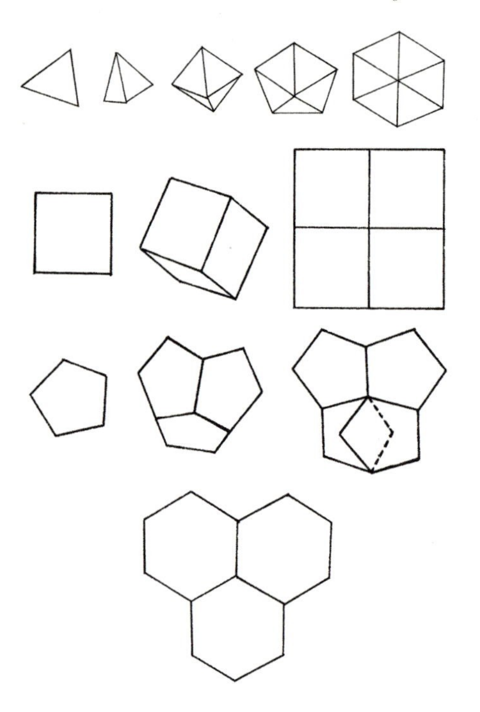

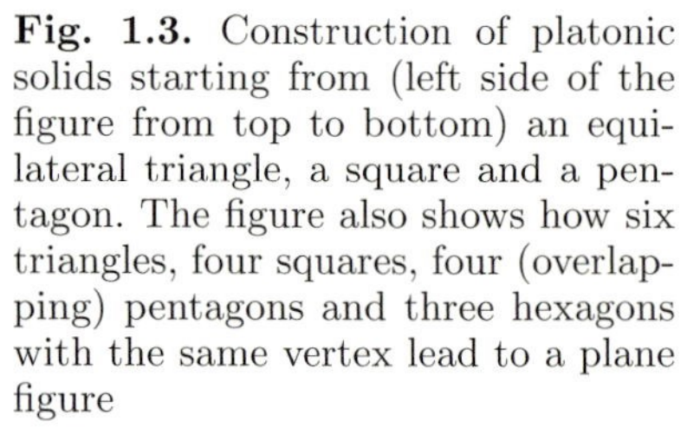

Fig. 1.3. Construction of platonic solids starting from (left side of the figure from top to bottom) an equilateral triangle, a square and a pentagon. The figure also shows how six triangles, four squares, four (overlapping) pentagons and three hexagons with the same vertex lead to a plane figure

Given that all the faces of a regular polyhedron are regular and thus all the edges are the same length L and all the vertex figures are regular, then all the faces must be the same. If, though, two adjacent faces are different, they would have the same vertex V but the vertex figure for V would have different length sides, each given as $L\cos(\pi/n)$ with different values for n. Furthermore, the dihedral angle, between two adjacent faces, must be the same since the faces that meet at a given vertex V are part of a right angled pyramid whose base is the vertex figure. Each lateral face of this pyramid is an isosceles triangle whose sides are $L/2$, $L/2$, $L\cos(\pi/n)$. The number of sides to the base is given by the value of the dihedral angle. This value, r, is the same for all the vertices and the vertex figures must all be the same, too. The regular polyhedron $\{n, r\}$ has polygon faces with n sides each of length L and the vertex figure is a polygon $\{r\}$ whose sides are all: $L\cos(\pi/n)$.

One should note that a perpendicular line passing through the centre of a face will meet the perpendicular line passing through the centre of a vertex figure at a point O, which is the centre of the circumsphere (touching all the vertices) and of the midphere, touching all the edges, as well as being the centre of the insphere which touches all the faces.

The platonic solids are nothing but simply connected polyhedra and show the property (just like simply connected polygons) that any simple closed curve drawn onto the surface of a polyhedron can be shrunk until it is a dot but it will remain on the surface of the polyhedron. Alternatively we can see that any region bounded by edges of the polyhedron is made up of one or more faces of the solid.

For those polyhedra that are simply connected, Eulero's formula stands. This formula defines the number of faces N_f, vertices N_v and edges N_e as

$$N_\mathrm{f} + N_\mathrm{v} = N_\mathrm{e} + 2\,. \tag{1.3}$$

Equation (1.3) is an important result from Eulero's research on polyhedra and was mentioned in his letter to his friend Goldbach dated November 1750. The formula expresses how the number of different elements that make up the surface of a polyhedron are related to each other. The really new aspect is the distinction between the edges (lines on the surface of the polyhedron) and the sides (lines that define the perimeter on its faces). Given that each edge has two sides, the number N_s of sides is given by

$$N_\mathrm{e} = \frac{1}{2} N_\mathrm{s} \tag{1.4}$$

where N_s is even. Furthermore, as each face of the polyhedron has at least three sides,

$$2N_\mathrm{e} \geq 3N_\mathrm{f} \tag{1.5}$$

and as at least 3 faces are required to define a solid angle, then

$$2N_\mathrm{e} \geq 3N_\mathrm{v}\,. \tag{1.6}$$

As such (1.3) is fully met both by the platonic solids (see Table 1.1) and by the pyramids and prisms. If the base of a pyramid has n sides then there are $(n+1)$ faces, the same number of solid angles and $2n$ edges. At the same time a prism with n sides to the base will have $(n+2)$ faces, $2n$ solid angles and $3n$ edges.

Using formulas (1.3), (1.5) and (1.6) we get new conditions for N_f and N_v as given by

$$2N_\mathrm{v} + 2N_\mathrm{f} = 2N_\mathrm{e} + 4 \geq 3N_\mathrm{v} + 4$$

and get

$$2N_\mathrm{f} - 4 \geq N_\mathrm{v}\,. \tag{1.7}$$

If, on the other hand, we say

$$2N_\mathrm{v} + 2N_\mathrm{f} = 2N_\mathrm{e} + 4 \geq 3N_\mathrm{f} + 4$$

from which

$$2N_\mathrm{v} - 4 \geq N_\mathrm{f}$$

we can deduce that

$$N_{\mathrm{v}} \geq \frac{1}{2} N_{\mathrm{f}} + 2 . \tag{1.8}$$

The proof of (1.3), as drawn up by Eulero, assumed that polyhedra were convex, given that at the time polyhedra and convex solids were synonyms.

Later on exceptions to the formula were found. One, for example, is a polyhedron made up of two tetrahedra meeting at one common edge or vertex. For a polyhedron to meet the requirements of Eulero's formula it must consist of a definite number of polygons where any two vertices are connected by edges and each closed curve on the surface divides the polyhedron into two parts.

The term polygon means a plane surrounded by straight segments that are topologically equivalent to a disc which, in turn, is any flat surface that is homeomorphic to a circle.

To demonstrate (1.3) we shall divide the edges of the polyhedron into two groups, each with a given colour. For simplicity let's take a cube and colour one edge red. The two vertices must thus also be red. The next edge to colour is chosen according to the rule that it must have a red vertex (there are only two options). If we continue to colour all the edges of the cube following the above rule in the end all the vertices will be coloured red and the edges, whether coloured or not, will have red vertices.

The cube will have eight red vertices and seven red edges; there will usually be one red edge $N_{\mathrm{e,r}}$ less than the number of red vertices. The number of red vertices is, though, equal to the number of vertices in the polyhedron, N_{v}. Let us now colour all the faces yellow leaving the red edges intact. You will get $N_{\mathrm{e,y}}$ yellow edges (five for the cube), one less than the number of faces (yellow) N_{f} (six in the cube). The relation between the number of edges, vertices and faces is

$$N_{\mathrm{e,r}} = N_{\mathrm{v}} - 1 \quad ; \quad N_{\mathrm{e,y}} = N_{\mathrm{f}} - 1 .$$

Since the number of edges of the polyhedron is N_{e} and

$$N_{\mathrm{e}} = N_{\mathrm{e,r}} + N_{\mathrm{e,y}} = (N_{\mathrm{v}} - 1) + (N_{\mathrm{f}} - 1) = N_{\mathrm{v}} + N_{\mathrm{f}} - 2$$

then

$$N_{\mathrm{e}} + 2 = N_{\mathrm{v}} + N_{\mathrm{f}} .$$

In this example the polyhedron is divided into two parts, red and yellow. Both parts are topologically connected and have the same contour, the line where the two colours meet. As such each part can be deformed to give a disc. The surface obtained by joining two discs together by way of the perimeter is a sphere, so the polyhedron can be deformed uniformly to obtain a sphere. Thus, for Eulero's formula to be valid the polyhedron musty be "spherical" or otherwise defined as simple.

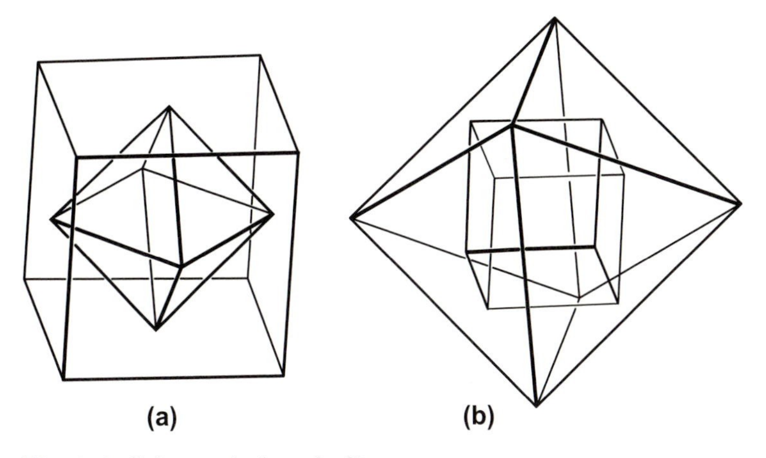

Fig. 1.4. Cube-octahedron duality

For the circumsphere of a given platonic solid, i.e. the sphere touching all the polyhedron vertices, we defined a set of non continuous rotations in space such that the vertices of the polyhedron form a set of equivalent points under the mentioned rotations and as such the polyhedron is brought into self-coincidence. The tangent planes to the circumsphere at the vertices of the polyhedron outline another regular polyhedron which, in turn, is brought into self-coincidence when the original circumsphere undergoes the same set of rotations.

This construction establishes that there is a relation between pairs of platonic solids: when applied to a octahedron, for example, it becomes a cube and vice versa the octahedron can circumscribe a cube as shown in Fig. 1.4. The two polyhedra are thus *dual* to each other. The group of rotations is called the octahedral group. There is a similar dual relation between the icosahedron and the (pentagonal) dodecahedron: the pertinent group of rotations is called the icosahedral group.

Duality has nothing to do with the fact that the platonic solids are made up of regular polygons. The principle holds because there is a relation between the number of faces and vertices for pairs of regular polyhedra. As can be seen in Table 1.1 the construction of dual solids consists in exchanging vertices and faces: where there is a face we substitute it with a vertex. The eight vertices and six faces of a cube become six vertices and eight faces of an octahedron. Likewise, the twenty vertices and the twelve faces of a dodecahedron become the twelve vertices and the twenty faces of the icosahedron. The tetrahedron, with its four vertices and four faces, is self-dual.

From the construction of dual solids we can see that the vertices of an inscribed solid coincide with the centres of the faces of its dual. If we examine pairs of dual solids whose edges are the same length we will see that in the dual construction the edge of the first solid lies perpendicular to the edge of its dual and they intersect at the centres. As such every two edges form a

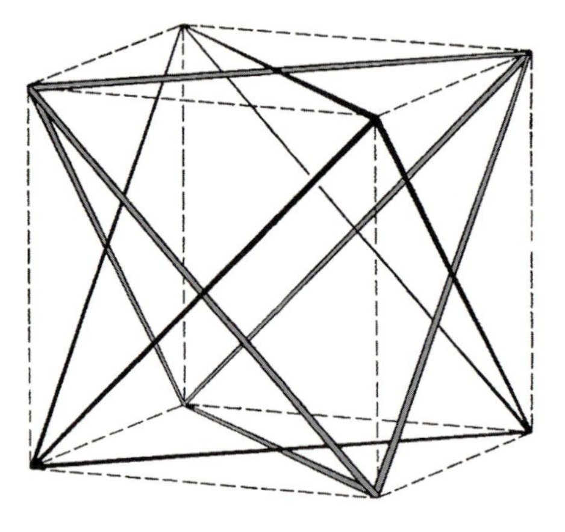

Fig. 1.5. Inscribing a tetrahedron inside a cube where the vertices of the two solids coincide. The double lines and the lines in bold show the two possible dispositions of the tetrahedron

pair of perpendicular bisectors; it is clear that one solid will have the same number of edges as its dual.

That the tetrahedron is self-dual may be related to the property, shared by all other platonic solids, that they are symmetrical to their centres; the vertices, faces and edges make up symmetrical pairs around the centre of the polyhedron. As such, for example, the straight line that connects the centre point of an edge of a cube to the cube centre intersects another edge at its centre point. However, the tetrahedron is not centrosymmetric; the straight line that connects a vertex with the centre of the polyhedron intersects the tetrahedron in the centre of a face. A tetrahedron can be inscribed into a cube in two different ways; in both cases the vertices of the two solids will coincide and the edges of the tetrahedron will make up the diagonals of faces of the cube, as shown in Fig. 1.5. This construction is possible because the tetrahedral group is a subgroup of the octahedral group. Similarly, we will see

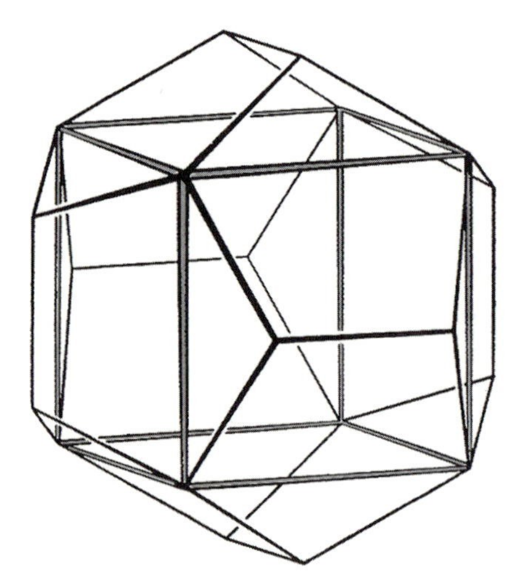

Fig. 1.6. One of the possible dispositions of a cube inscribed in a dodecahedron. Each edge of the cube lies on a face of the dodecahedron

that the octahedral group is a subgroup of the icosahedral group and as such a cube can be inscribed into a dodecahedron in the same way a tetrahedron can be inscribed into a cube. Figure 1.6 shows one of the five dispositions of a cube inscribed into a dodecahedron; each edge of the cube lies on a face of the dodecahedron, and two cubes meet at each vertex of the dodecahedron.

1.2 Elements of Symmetry in Space

The platonic solids are highly symmetrical. If we examine a cube's axes of rotational symmetry we will see (Fig. 1.7) three fourfold rotation axes, each of which intersects with the centre of two opposite faces, four threefold axes that extend from opposite vertices, and six twofold axes, each of which intersects with the centre of pairs of opposite edges. Since the octahedron is dual with the cube it has the same number of rotation axes as the cube with the three fourfold rotation axes passing through pairs of opposite vertices, the four threefold axes intersecting the centres of pairs of opposite faces, and the six twofold axes must lie as in the cube.

In the case of the dodecahedron-icosahedron dual solids (Fig. 1.8) fifteen twofold axes intersect with the centres of opposite edges, ten threefold axes intersect with the centres of opposite faces (icosahedron) and pass through pairs of opposite vertices (dodecahedron), six fivefold axes pass through pairs of opposite vertices (icosahedron) and intersect with the centres of opposite faces (dodecahedron). The tetrahedron (Fig. 1.9), on the other hand, has four threefold axes and three twofold axes. The three twofold axes intersect with the centres of pairs of opposite edges whereas each threefold axis passes through a vertex and intersects with the centre of the opposite face.

Apart from rotation the platonic solids have important reflection symmetries. If we divide a cube into two with the plane intersecting the centre of

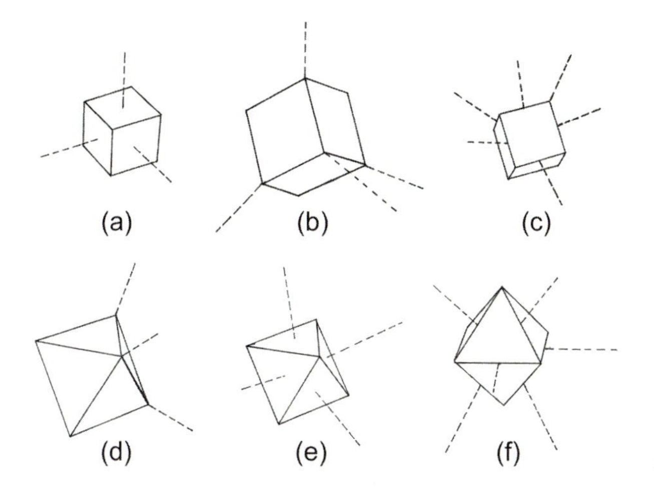

Fig. 1.7. Cube rotation axes: (**a**) fourfold; (**b**) threefold; (**c**) twofold. Octahedron rotation axes: (**d**) fourfold; (**e**) threefold; (**f**) twofold

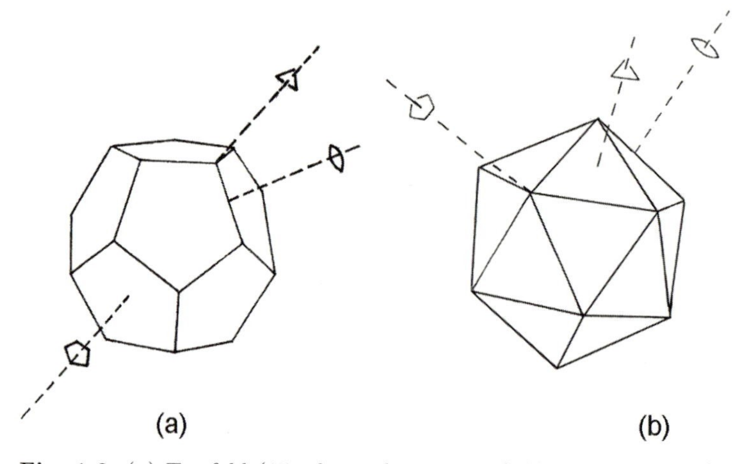

Fig. 1.8. (**a**) Twofold (15, about the centre of the opposite edges), threefold (10, about pairs of opposite vertices), fivefold (6, about centres of opposite faces) rotation axes of dodecahedron. (**b**) Analogous scheme for the icosahedron; the number of axes of each order is the same as the corresponding axes in the dodecahedron. The figure shows only one axis of each kind

two opposite faces and we imagine we have put one of the cut surfaces onto a mirror we will see a cube where one half of the cube is real and the other half is its reflection in the mirror. Whenever a plane cuts a solid into two, so that when it is put onto a mirror the reflected image gives us the original solid again, we have identified a reflection plane. All reflection planes have a common point, i.e. the centre of the platonic solid we are examining.

Since a cube has three pairs of parallel opposite faces, we have three reflection planes (Fig. 1.10) each of which lies normally to one of the three fourfold rotation axes. Furthermore, each plane, (six all together) that intersects with the diagonals of one face and its opposite face, i.e. it passes through pairs of

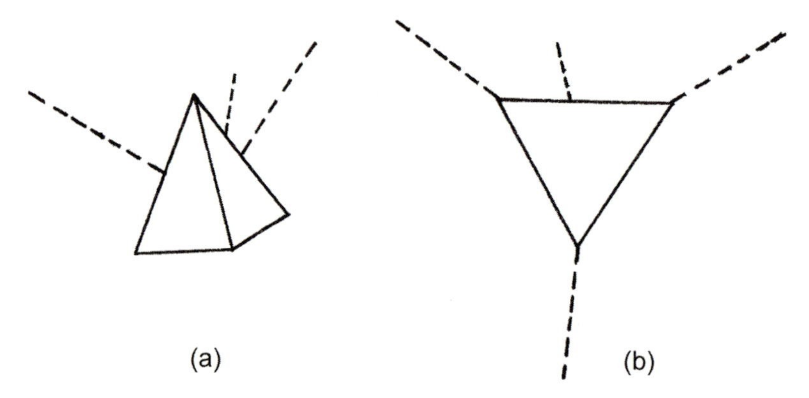

Fig. 1.9. Twofold (**a**) and threefold (**b**) rotation axes of a tetrahedron

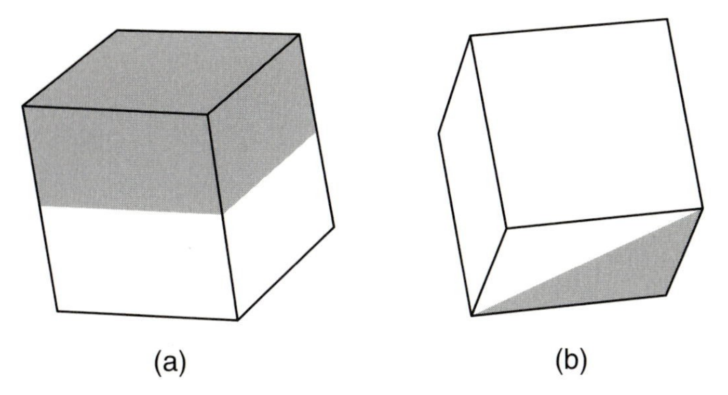

Fig. 1.10. Reflection planes lying normal to a fourfold rotation axis (**a**) and to a twofold rotation axis (**b**) of a cube

opposite edges (Fig. 1.10) is in turn a reflection plane and lies normally to a twofold rotation axis. In the end we can count nine reflection planes.

Since the octahedron is dual with the cube it too has nine reflection planes which lie the same way as the reflection planes of the cube (Fig. 1.11).

The number of reflection planes of a tetrahedron can easily be deduced, provided we remember how the tetrahedron is inscribed into a cube. The tetrahedron has six reflection planes: each plane contains an edge and cuts the opposite edge into two (Fig. 1.12). These planes coincide with the same number of reflection planes of the cube the tetrahedron is inscribed into.

It is quite easy to see the reflection planes of a dodecahedron and a icosahedron, which are dual to each other (Fig. 1.13). Each reflection plane intersects with a pair of opposite edges, cuts another two opposite edges into half and is perpendicular to one of the fifteen twofold rotation axes. There are also fifteen reflection planes.

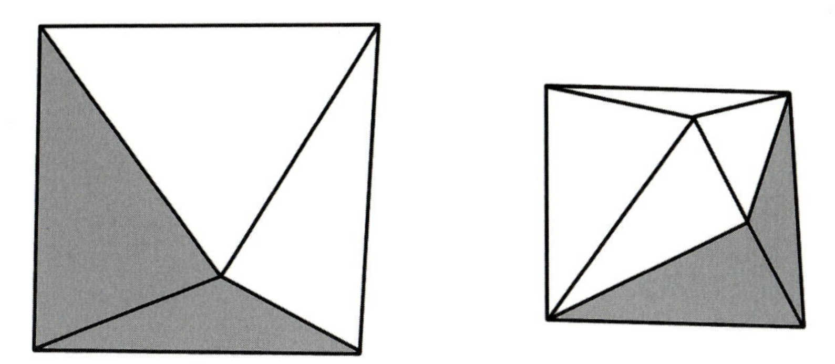

Fig. 1.11. Reflection planes of an octahedron. Notice that the disposition of the reflection planes of the cube is the same. This is due to the duality between the two solids

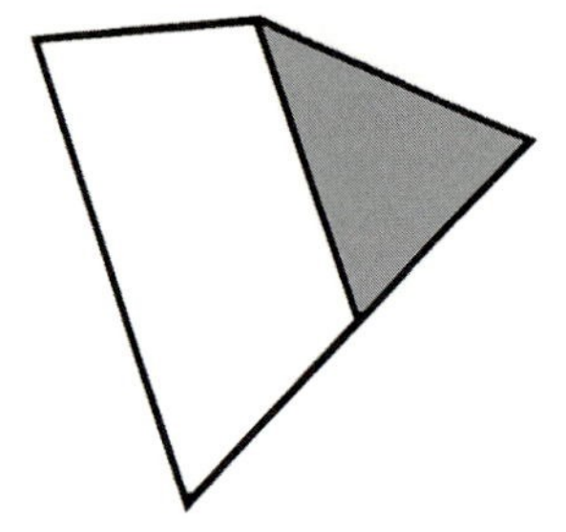

Fig. 1.12. Each reflection plane of a tetrahedron contains one edge and cuts the opposite edge in half

Let us now consider a pair of dual platonic solids: the construction method used to define duality makes the vertices of one of the two solids coincide with the centres of the same number of faces of the other solid. This way all the rotation axes and the reflection planes of the two solids coincide, and thus are also symmetry elements of the compound solid.

The method used to find the reflection planes of a tetrahedron, by way of inscribing it into a cube, where the cube is not dual to the tetrahedron, consists in making all four threefold axes of the two solids coincide with each other and then expanding or shrinking the tetrahedron until its four vertices fall on four of the cube vertices. This method can be pursued systematically every time a platonic solid can be inscribed into a different one by identifying the symmetry elements that are common to both solids. These constitute the symmetry elements of the compound solid. Where a tetrahedron is inscribed into a octahedron, which is dual to a cube, the procedure is just the same as the above explanation.

If, though, we want to inscribe a tetrahedron into a dodecahedron using the same method to inscribe it into a cube, making the vertices of the tetrahedron coincide with the same number of vertices of the dodecahedron so that there are four common threefold axes, we will see that the three twofold axes of the tetrahedron coincide with the same number of twofold axes of the dodecahedron. However, all similarity with the tetrahedron in a

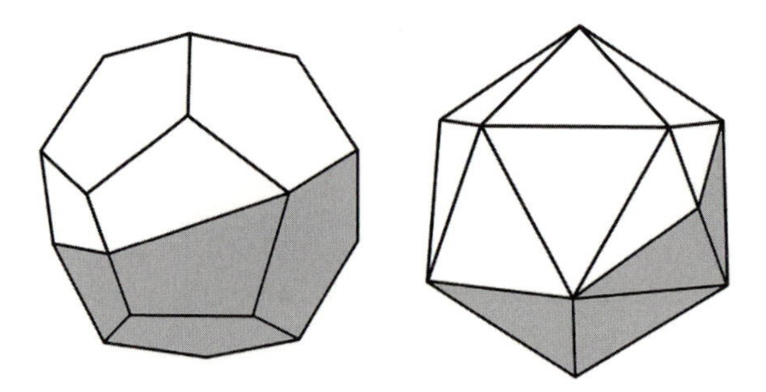

Fig. 1.13. Each reflection plane of dodecahedron (**a**) and icosahedron (**b**) passes through a pair of opposite edges, cuts another two opposite edges into half and is perpendicular to a twofold rotational axis

cube ends here because the edges of the tetrahedron do not lie on the faces of the dodecahedron and the six reflection planes of the tetrahedron do not correspond to any reflection plane of the dodecahedron. The compound solid has no reflection planes.

When a tetrahedron is inscribed into a icosahedron, which is dual to the dodecahedron, the vertices coincide with four face centres of the icosahedron, thus each of the four opposite faces of the icosahedron is parallel to one face of the tetrahedron.

As we have seen we can inscribe a cube into a dodecahedron or into its dual icosahedron. In this way if we make the threefold axes of the cube coincide with the same number of threefold axes of the icosahedron, the vertices of the cube will fall on the centre of the faces of the icosahedron parallel to those of a octahedron. However, each face of the cube will be parallel to one of the edges of the icosahedron or of the circumscribed dodecahedron. Three reflection planes of the cube and the icosahedron (dodecahedron) will coincide and make up the reflection planes of the compound solid.

1.3 Polytopes in the Four Dimensional Space and Their Projections onto the Physical Space

We can imagine the transformation of polyhedra into hyper-solids in Euclidean space having more than three dimensions by repeating the procedure by which we can transform the figure defined in a low dimension space into a figure in progressively higher dimension space.

In space of null dimension the only figure is a point P_0. In space of one dimension there can be any number of points and two of these points will bound a line-segment P_1 which can be constructed by joining point P_0 to another point. If we join P_1 to a third non collinear point we will get a triangle, the simplest polygon, P_2, on a plane; if we then join this triangle to a fourth point outside its plane we will construct a tetrahedron, the simplest form of polyhedron, P_3. If we now join a fifth point to the polyhedron outside the three dimensional space we will construct a pentatope, the simplest element P_4. This succession is exhibited in Fig. 1.14 where the equilateral triangle has the same faces as the tetrahedron.

In general any set of $(r+1)$ points that do not lie in a $(r-1)$ space are the vertices of the r-dimensional simplest figure, the so-called simplex, whose elements are simplexes formed by subsets of the $(r+1)$ points, namely the vertices themselves, $\binom{r+1}{2}$ edges, $\binom{r+1}{3}$ triangles, $\binom{r+1}{4}$ tetrahedra, $(r+1)$ *cells*: in a single relation,

$$R_l = \binom{r+1}{l+1} = \binom{r}{l+1}\binom{r}{l} . \tag{1.9}$$

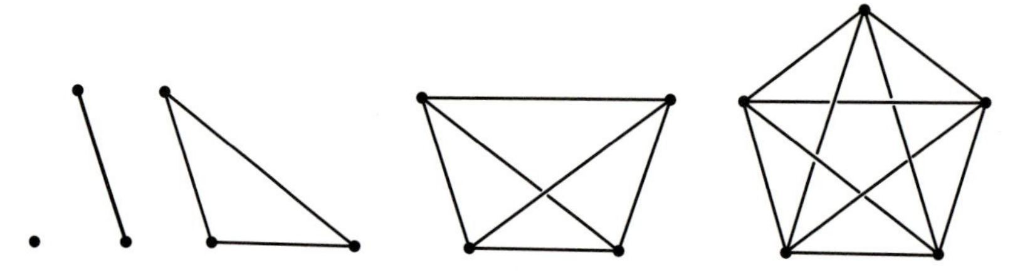

Fig. 1.14. Constructive scheme of pentatope P_4

A line-segment is bounded by two points, a triangle by three line-segments (sides), a tetrahedron by four planes (faces) and a pentatope by five three-dimensional regions (tetrahedral cells). In general a simplex is a finite region of r-dimensional space enclosed by $(r+1)$ hyper-planes. If all the $\frac{1}{2}r(r+1)$ edges are equal then we obtain a regular simplex, as shown in Fig. 1.14.

If we limit ourselves to the 4-dimensional space the regular polytopes are those whose cells are regular polyhedra. These polytopes are called r-cells if they are bounded by only r polyhedra. One fundamental property of Euclidean r-dimensional space is that we can draw r mutually perpendicular lines through any point O; r points equidistant from O along these lines will thus make the vertices of a regular simplex $(r-1)$ whereas all the lines together constitute a set of r cartesian reference axes.

If all the centre points of the edges that extend from a given vertex V of P_r lie in one single hyper-plane (e.g. if there are r edges) then these centre points are the vertices of a $(r-1)$-dimensional polytope called the vertex figure of V in P_r. Since the cells of the vertex figure are themselves the vertex figures of cells then a regular polytope P_4, whose cells are $\{n,r\}$, must have $\{r,s\}$ vertex figures where s is the number of cells that surround an edge. This polytope is $\{n,r,s\}$. Thus in four dimensions the formula

$$\sin\frac{\pi}{n}\sin\frac{\pi}{s}\geq\cos\frac{\pi}{r} \tag{1.10}$$

stands; this means that $2\pi/s$ is greater than or equal to the dihedral angle $(\pi-2\phi)$ of $\{n,r\}$. As such we can place s polyhedra $\{n,r\}$ around a common edge. The 4-dimensional Schläfli symbol $\{n,r,s\}$, gives a polytope where $\{n,r\}$ and $\{r,s\}$ must belong to the platonic solids

$$\{3,3\},\{3,4\},\{4,3\},\{3,5\},\{5,3\}.$$

The criterion in (1.10) limits the number of possible polytopes to six, as shown in Table 1.2.

The polytope $\{3,3,4\}$ is the only one with no equivalent polyhedron. Apart from being self-dual this polytope is also centrosymmetric whereas $\{3,3,3\}$, like the tetraedron, has no centre of symmetry.

From Table 1.2 duality relations can be inferred when we consider that, in four dimensional space, points correspond dually to three-dimensional lines and the lines to three-dimensional planes.

Polytope P_4	Equivalent polyhedron	Number & kind of limit polyhedra	Number of vertices	Duality
$\{3,3,3\}$; 5-cell	Tetrahedron	5 Tetrahedra	5	Self-dual
$\{4,3,3\}$; 8-cell	Cube	8 Cubes	16	Dual of $\{3,4,3\}$
$\{3,4,3\}$; 16-cell	Octahedron	16 Octahedra	8	Dual of $\{4,3,3\}$
$\{3,3,4\}$; 24-cell	-	24 Tetrahedra	24	Self-dual
$\{5,3,3\}$; 120-cell	Dodecahedron	120 Dodecahedra	600	Dual of $\{3,3,5\}$
$\{3,3,5\}$; 600-cell	Icosahedron	600 Tetrahedra	120	Dual of $\{5,3,3\}$

Table 1.2. Features of the P_4 polytopes

The study of projected polytopes P_4 into three dimensional space is simplified by first considering an analogous, but easier to visualise, case where a polyhedron is projected onto a plane. Among the various possible projections, which are based on the choice of the centre of projection and the choice of the image plane, a usual one is the parallel projection with the centre at an ideal point and whose parallel lines represent parallel lines. However, there is the disadvantage that the projected faces partially overlap. This limitation can be overcome by moving the centre of projection to a point very close to one of the polyhedron faces; we usually prefer the centre of the face and to project onto the plane of that face. The projections of the platonic solids obtained this way are shown in Fig. 1.15; this is what we see when we remove one face of the polyhedron and look at the interior through the hole.

Parallel projection is not the correct method when projecting polytopes P_4 since the polyhedra that bound the polytope are represented in three-dimensional space by polyhedra that intersect with each other and partially overlap, thus giving a distorted image. The very best results are obtained when we project from a point very close to a hyper-surface; in so doing the boundary polyhedra of the polytope P_4 are represented by sets of polyhedra in three-dimensional space, one of which has a very special role and is filled up in a simple way by the others. If we project further onto the plane we will achieve the images as seen in Fig. 1.16 for polytopes $\{3,3,3\}$, $\{4,3,3\}$, $\{3,4,3\}$ and $\{3,3,4\}$. It is noticeable that the last projection is a large octahedron filled by 23 smaller octahedra, with four different possible forms, making 24 polyhedra in all.

Projections of $\{5,3,3\}$ and $\{3,3,5\}$ are much more complicated; Figure 1.17 shows a possible projection of $\{3,3,5\}$.

Our interest on the polytopes P_4 is given by the observation that $\{3,3,5\}$ is a packing of tetrahedra where every five have a common edge. This packing represents an excellent model of the highly dense physical structures with

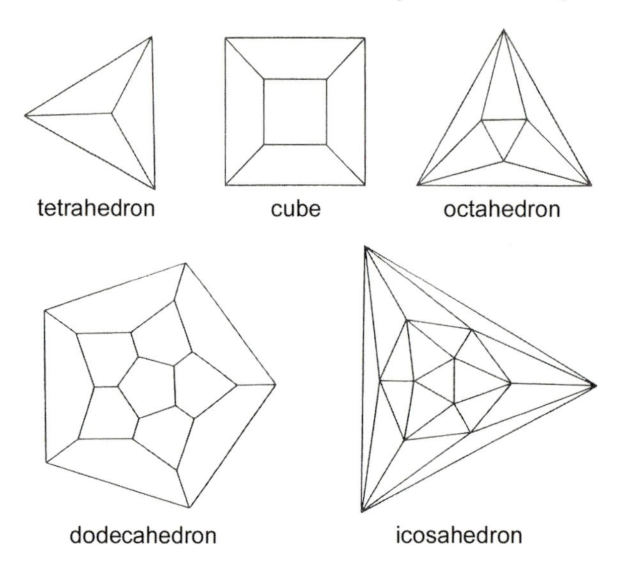

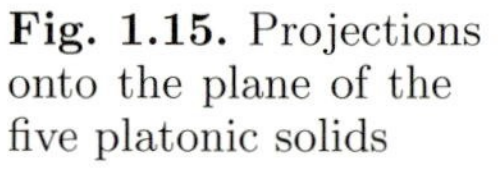
Fig. 1.15. Projections onto the plane of the five platonic solids

local icosahedral coordination such as the Frank–Kasper metallic phases and the quasicrystals. In these structures the local coordination polyhedra coincide with the icosahedral coordination polyhedron, which is typical of the atoms arranged at the vertices of the $\{3, 3, 5\}$ polytope. On the other hand the four-dimensional space has positive curvature whereas the three-dimensional space is flat. The idea of representing a physical structure (three-dimensional)

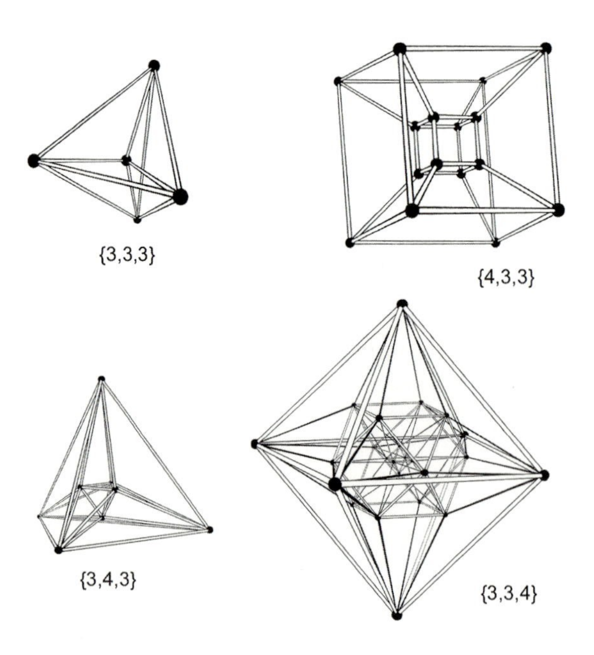

Fig. 1.16. Projections onto the plane of polytopes P_4 $\{3, 3, 3\}$, $\{4, 3, 3\}$, $\{3, 4, 3\}$, $\{3, 3, 4\}$

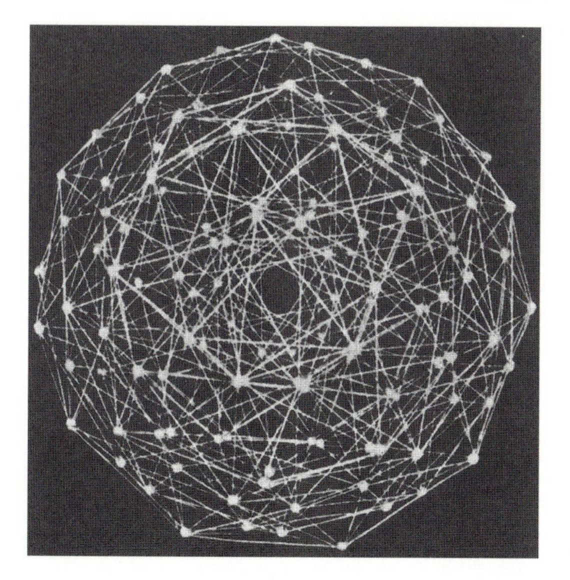

Fig. 1.17. One of the possible projections onto the plane of polytope $\{3, 3, 5\}$

using an ideal four-dimensional crystal will allow us to use the normal techniques adopted to analyse the properties of three-dimensional crystals, but it requires defects, particularly disclinations, being introduced just when the projection is made, in order to obtain, on average, a flat plane. Thus the local interaction between atoms will lead to a local icosahedral configuration which is ideal to obtain an efficient packing of metallic atoms with isotropic interaction. This configuration is however partially incompatible with the requirement to fill the three-dimensional space with matter, which is possible without altering the icosahedral coordination, only by introducing disclinations (see Chap. 5). This will be discussed in further depth in the following chapters.

1.4 Elements of Crystallography

In order to specifically study the degree of structural order of extended systems, as seen from the position in the physical space $\boldsymbol{R}$ of the atoms, or molecules, we must start by examining how regular an ideal, defect free crystal with no surfaces is. In order to discuss the regularity of atomic arrangement in a crystal we have to refer to the idea of a crystal lattice. This is a mathematical construction made of an infinite three-dimensional array of points distributed periodically in space. Each point has identical first neighbours and is equal to every other lattice point. Thus the crystal lattice is said to have perfect *translational symmetry.*

The position $\boldsymbol{r}$ of any lattice site in relation to an arbitrary origin is defined by a vector:

$$\boldsymbol{r} \equiv n_1\boldsymbol{x}_1 + n_2\boldsymbol{x}_2 + n_3\boldsymbol{x}_3 \, . \tag{1.11}$$

Equation (1.11) defines the set of lattice vectors called *lattice translations* or *primitive translations*.

A translation vector, also called lattice vector $\boldsymbol{T}$, joins equivalent points on the lattice

$$\boldsymbol{T} = \boldsymbol{r}_i - \boldsymbol{r}_j \, . \tag{1.12}$$

Equation (1.12) is true of any of the indices i, j.

In (1.11), n_i are *integers* and the non-coplanar vectors, called *basis vectors*, $\boldsymbol{x}_1$, $\boldsymbol{x}_2$, $\boldsymbol{x}_3$, are the fundamental units of translation symmetry. The volume enclosed by the three integers $(\boldsymbol{x}_1, \boldsymbol{x}_2, \boldsymbol{x}_3)$ defines what is called a *primitive cell.* Referring to a given lattice point the basis vectors cannot be chosen in a single way, as seen in Fig. 1.18 for a two-dimensional square lattice. It is usual to choose the two shortest basis vectors (top left) but each choice of $(\boldsymbol{x}_1, \boldsymbol{x}_2)$ allows us to reach any point of the lattice.

If the system is elemental, namely it is made up of a single kind of atom, then these atoms can be placed directly onto the lattice sites so that the lattice vectors coincide with the position of the atoms. More frequently we will find that each lattice site is associated with a set of atoms. In this case we speak about a lattice with a *basis* and the three vectors $(\boldsymbol{x}_1, \boldsymbol{x}_2, \boldsymbol{x}_3)$ will define a parallelepiped unit-cell (not necessarily the cell with the minimum volume). Repetition of this unit-cell in space generates the whole crystal with the associated periodicity. The unit-cell with the minimum volume is called the primitive unit cell and it contains one atom.

Each unit cell contains the same number of atoms, usually a small number, in a fixed position. Thus if the cell contains p atoms, and if we suppose that this cell is centred on the origin O, the atomic coordinates being ξ_p $(\xi = 0, 1, 2, ..., p-1)$, the distribution of the atoms in the crystal is given by the (microscopic) density $\varrho(\boldsymbol{x})$ as

$$\varrho(\boldsymbol{x}) = \sum_{\boldsymbol{r}} \sum_{\xi=0}^{p-1} \delta(\boldsymbol{x} - \boldsymbol{r} - \xi_p) \tag{1.13}$$

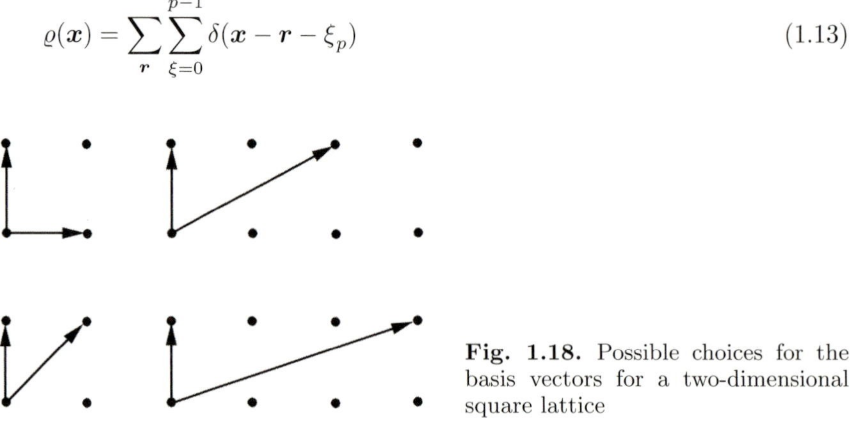

Fig. 1.18. Possible choices for the basis vectors for a two-dimensional square lattice

to which is associated the periodicity property

$$\varrho(\boldsymbol{x} + \boldsymbol{r}) = \varrho(\boldsymbol{x}) . \tag{1.14}$$

Since each basis is positioned and oriented the same way with respect to its lattice site then the resulting atomic structure is a *perfect crystal.* Some very complicated structures have even been found where the unit cells have up to a few thousand atoms. All the $\boldsymbol{r}$ vectors defined by (1.11) have greater or equal modules to the module of the shortest vector connecting two vertices of the unit primitive cell. The set of lattice translation vectors $\boldsymbol{T}$ is closed with respect to the sum and product. Thus, if we take two vectors of the lattice $\boldsymbol{T}_1$ and $\boldsymbol{T}_2$ then $\mp\boldsymbol{T}_1$, $\mp\boldsymbol{T}_2$, $(\boldsymbol{T}_1 \mp \boldsymbol{T}_2)$, $(\boldsymbol{T}_1 \mp \boldsymbol{T}_2)$ are also lattice vectors.

It is often quite useful to substitute one parallelepiped unit-cell with another unit cell, the so-called *Wigner–Seitz* cell.

Taking any point of the lattice as the centre of the cell to be constructed, we join such a centre to all its closest lattice sites, bisecting them and making the bisector planes intersect with each other. The obtained (two dimensional) regular polygon or (three-dimensional) regular polyhedron is the Wigner–Seitz cell. Like the primitive cell the Wigner–Seitz cell contains *one* lattice site; this lies at the centre of the cell whereas each of the eight lattice sites at the vertices of the primitive cell is common to eight adjacent cells. If we represent the atoms as hard spheres on the sites of the various lattices then the crystal will look like a box where the hard spheres are packed in different arrays, depending on the structure, to obtain the highest packing efficiency in the space.

The planes defining the faces of the three-dimensional Wigner–Seitz cell are given by the equation:

$$(\boldsymbol{r} + \boldsymbol{x})^2 = \boldsymbol{x}^2 . \tag{1.15}$$

This means that such planes are defined by the intersection of spheres of equal radii centred around adjacent lattice sites. This is another example of the periodicity property of functions associated with the distribution of atoms on the lattice sites.

When we examine elemental metals the most frequent structures found are the simple cubic, sc, the body-centred cubic, bcc, and the face-centred cubic, fcc (see Fig. 1.19). In the first case there are eight atoms on the vertices of the unit cubic cell, whose edge l is called the *lattice parameter.* $\frac{1}{8}$ of the total volume of each atom goes to fill volume l^3 of the cell, which thus contains one atom. The volume available for each atom is $V = \frac{4}{3}\pi\left(\frac{l}{2}\right)^3$ and the packing efficiency is given as

$$f = \frac{V}{l^3} = \frac{\pi}{6} = 52.36\% . \tag{1.16}$$

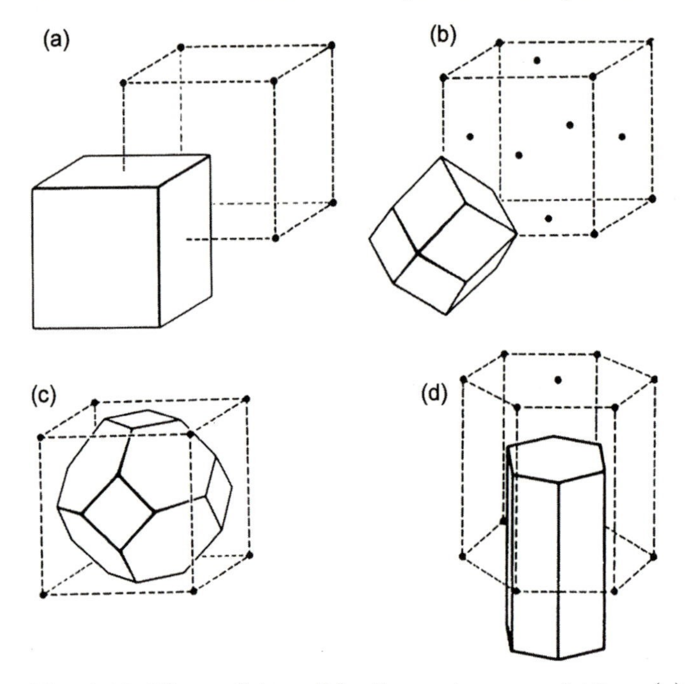

Fig. 1.19. Wigner–Seitz cell for the most common lattices: (**a**) simple cubic structure; (**b**) face-centred cubic structure; (**c**), body-centred cubic structure; (**d**), hexagonal structure

The body-centred cubic structure not only has atoms at the vertices of the cube, it also has an atom at the centre of the cube. The unit cell contains two atoms; the radius of the hard sphere is thus given as $r = l\frac{\sqrt{3}}{2}\frac{1}{2} = \frac{\sqrt{3}}{4}l$ and the packing efficiency is:

$$f = \frac{2\frac{4}{3}\pi\left(l\frac{\sqrt{3}}{4}\right)^3}{l^3} = 68.02\%\,. \tag{1.17}$$

The face-centred cubic structure, on the other hand, still has eight atoms at the cube vertices as well as an atom at the centre of each face; thus the cell contains four atoms. The radius of the hard sphere is given by $r = \frac{1}{2}\frac{l}{2}\sqrt{2} = l\frac{\sqrt{2}}{4}$ and the packing efficiency is:

$$f = \frac{4\frac{4}{3}\pi\left(l\frac{\sqrt{2}}{4}\right)^3}{l^3} = 74.05\%\,. \tag{1.18}$$

The hexagonal close-packed structure, hcp, is in turn often found in pure metals and has the same packing efficiency as the fcc. If we examine an fcc crystal and an hcp crystal from the same angle (Fig. 1.20) we will see that the

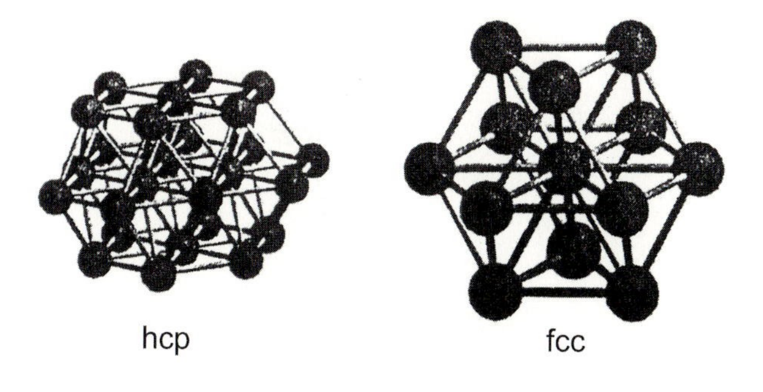

Fig. 1.20. Stereographic models of the hcp and fcc lattices seen from the same angle. In the hcp structure the plane sequence is ABAB..., whereas in the fcc structure the sequence is ABCABC...

sequence of planes in the hcp crystal is ABAB..., whereas in the fcc crystal the sequence is ABCABC... .

The geometric problem to realise the most efficient packing of hard spheres, simple in the case of elemental structures, becomes more complicated when we try to achieve highly compact structures using two different kinds of atoms of different sizes. One typical example is given by the Laves phases AB_2 where the prototype is $MgCu_2$. This is an intermetallic compound whose stability depends not only on electronic factors but especially on geometric factors such as the greatest possible symmetry, most dense space occupation and the greatest number of atoms coordinated with one given atom. Generally known as C15, the structure of $MgCu_2$ is face-centred cubic with eight chemical formulas in each unit cell which is cubic and non primitive. The arrangement of the atoms can be seen by using two sub-lattices, the cubic lattice, typical of a diamond (Fig. 1.21) with magnesium atoms, and the sub-lattice with copper atoms distributed on vertex sharing

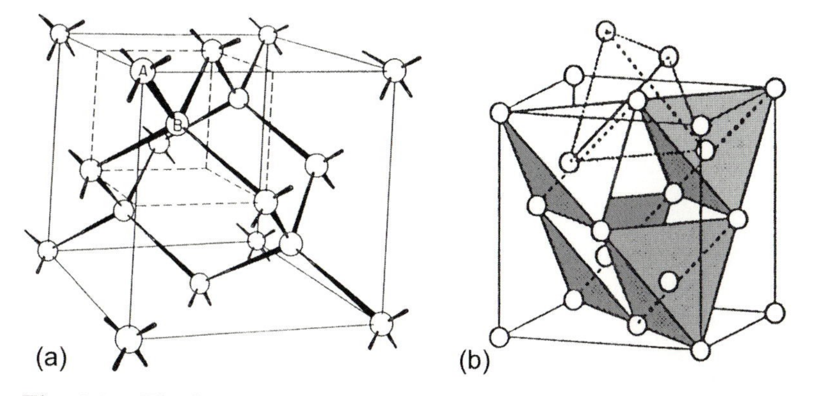

Fig. 1.21. The diamond structure: lattice (**a**) is the same as the fcc lattice though it has two atoms per unit cell. Starting with the fcc lattice, if we give one of the atoms (A) at a cube vertex the coordinates (0,0,0), an atom (B) will be inserted at a quarter of the main cube diagonal: its coordinates will be $\left(\frac{1}{4}, \frac{1}{4}, \frac{1}{4}\right)$. Each unit cube will have eight such atoms; the diamond structure consists of two interpenetrating fcc lattices. (**b**) Alternative view, highlighting the tetrahedral atom disposition

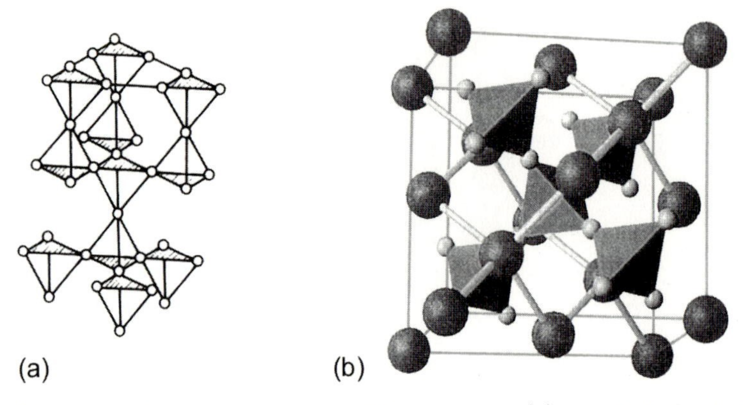

Fig. 1.22. Two pictures of the C15 structure; (**a**) tetrahedral sub-lattice of Cu atoms (in the prototype system $MgCu_2$); (**b**) small dots represent Cu atoms, arranged in tetrahedra, and large dots indicate Mg atoms

Cu_4 tetrahedra. These tetrahedra are arranged in such a way that pairs of tetrahedra overlap and are "welded" to a vertex to form a sandglass, as can be seen in Fig. 1.22. A local representation of the structure is given by the Frank–Kasper polyhedra, all of which are convex polyhedra, with triangular faces either all the same or different to each other, and at least five triangles having a common vertex. In these kinds of polyhedra the atom in the central position has 12, 14, 15 and 16 as its coordination numbers.

If the central atom of $MgCu_2$ is copper then the polyhedron is a icosahedron with coordination number 12, whereas if the centre atom is magnesium than the polyhedron has four magnesium atoms that form a tetrahedron and 12 copper atoms that make up four triangles each of which is opposite to a magnesium atom, as shown in Fig. 1.23.

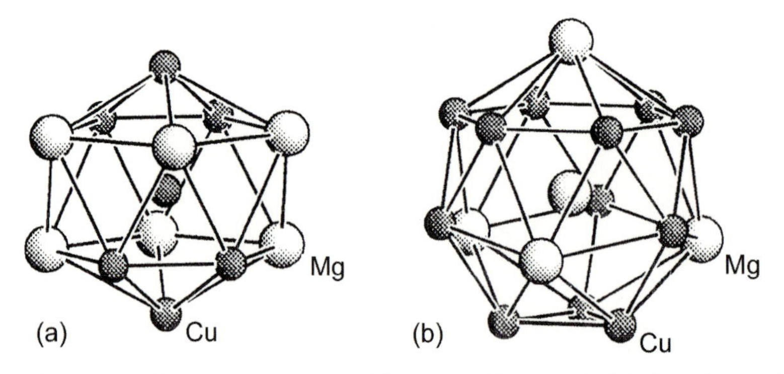

Fig. 1.23. The two polyhedra of local coordination in $MgCu_2$; (**a**) with Cu as a central atom we obtain a icosahedron: polyhedron C12; (**b**) Mg as a central atom is coordinated with four Mg atoms which form a tetrahedron and with twelve Cu atoms which form four triangles, opposite to the Mg atoms: polyhedron C16

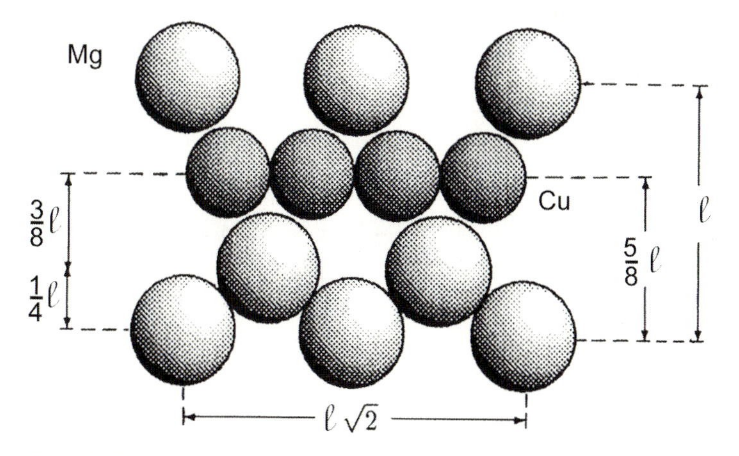

Fig. 1.24. A section of the cubic cell of $MgCu_2$ along the diagonal, to calculate packing efficiency

The packing efficiency f can be calculated looking at Fig. 1.24, which represents a section of the cubic cell along the diagonal. The four copper atoms touch each other and are aligned along the diagonal, which is $l\sqrt{2}$ long, so $r(\text{Cu}) = (1/8)l\sqrt{2}$. Two magnesium atoms on the diagonal of the unit cell and $(1/4)l\sqrt{3}$ separated from each other lead to $r(\text{Mg}) = (1/8)l\sqrt{3}$. The ideal ratio between the atomic radii in this kind of packing is thus $(r(\text{Mg})/r(\text{Cu})) = (3/2)^{1/2}$. The space filling efficiency is given by

$$f = \frac{4}{3}\pi\frac{1}{l^3}\left[8\left(\frac{1}{8}l\sqrt{3}\right)^3 + 16\left(\frac{1}{8}l\sqrt{2}\right)^3\right] = 0.710\,,$$

if we take into account that the cubic cell contains eight magnesium atoms and sixteen copper atoms. This is a little less efficient than the most dense packing of hard spheres. In the C15 structure of $MgCu_2$ the atoms of both metals touch each other which is why the crystal has two interpenetrating lattices made up of only copper atoms and magnesium atoms.

Since the translational periodicity so far examined is a common property to all crystals this property alone is insufficient to identify and classify the over twenty thousand crystal systems so far discovered. The distinctive symmetry elements of a given crystal can be determined when we consider the *rotations* about certain crystallographic axes, as centred on lattice points, the *reflections* with respect to particular planes of atoms, the *inversions* and combination among these operations and with translations. In 1848 the Russian mathematician Bravais demonstrated that first, in three dimensional crystals, lattice translations are only compatible with some kinds of rotations and, second, because of a crystal's so-called point symmetry, i.e. all the rotations, reflections, inversions and their combinations as a whole, there can

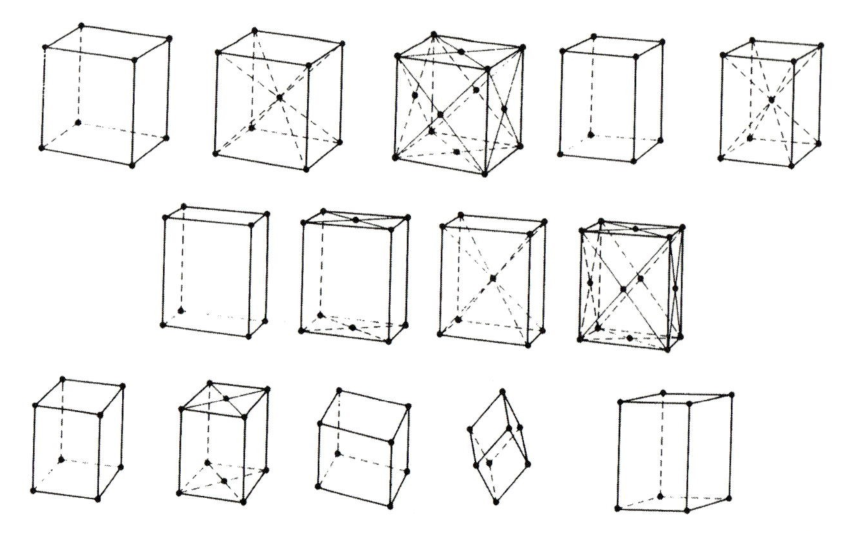

Fig. 1.25. The fourteen three-dimensional Bravais lattices. From top to bottom and from left to right: simple, body centred and face centred cubic structures, simple and body-centred tetragonal, simple, basis centred, body centred and face centred orthorhombic, simple and basis centred monoclinic; triclinic; trigonal; hexagonal

be no more than 14 different kinds of lattices. These lattices are commonly referred to as the Bravais lattices and are represented in Fig. 1.25.

In a plane the Bravais lattices can be determined by the relative length of the basis vectors $\boldsymbol{x}_1$ and $\boldsymbol{x}_2$ and by angle α between these vectors. There are five possible lattices: oblique ($\boldsymbol{x}_1 \neq \boldsymbol{x}_2$, where α is any angle), square ($\boldsymbol{x}_1 = \boldsymbol{x}_2, \alpha = 90°$), primitive rectangle ($\boldsymbol{x}_1 \neq \boldsymbol{x}_2, \alpha = 90°$), centred rectangle ($\boldsymbol{x}_1 = \boldsymbol{x}_2$, α is any angle) and hexagonal ($\boldsymbol{x}_1 = \boldsymbol{x}_2, \alpha = 60°$).

The so-called *crystallographic restriction* (Barlow theorem, 1901) requires that a lattice, when rotated around an axis of a given order, covers positions equivalent to the starting ones, unaffected by the operation: only rotations of order 1, 2, 3, 4, and 6 are allowed in three dimensional space. In two-dimensional space, which is easier to see, let us take floor tiles to illustrate the crystallographic restriction. Only if we use triangular, square or hexagonal tiles is it possible to completely and evenly cover the plane without overlapping the tiles.

The result can be explained if we consider that the sum of the inside angles of a polygon is $\pi(n-2)$ where n is the number of sides. In order to obtain regular tiling the vertex angle of the polygon $\pi(n-2)/n$ must be a divisor of 2π, namely

$$\frac{2\pi n}{\pi(n-2)} = m \qquad \text{where } m \text{ is a natural number}$$

so

$$m = \frac{2n}{n-2} = 2 + \frac{4}{n-2}.$$

As such, m can only be a natural number if 4 can be divided by $(n-2)$, and this will occur if

$$\begin{aligned} n-2 &= 1\,, \quad & n &= 3\,, \quad & &\text{triangles,} \\ n-2 &= 2\,, \quad & n &= 4\,, \quad & &\text{squares,} \\ n-2 &= 4\,, \quad & n &= 6\,, \quad & &\text{hexagons.} \end{aligned}$$

There are no other cases; these are the only polygons we can use to achieve a perfect tiling structure; and this is done by ensuring they all have the same vertex.

We could consider using a different form of tiling structure where the vertex of one polygon touches one side of a second polygon. If α is the angle between the second polygon and one of the sides that go to make up the vertex angle of the first polygon, then the following must stand:

$$m\alpha = \pi,$$

where m is a natural number; so

$$\alpha = \frac{\pi(n-2)}{n}$$

and

$$m = 1 + \frac{2}{n-2}$$

as such the condition for m can only be obtained if

$$\begin{aligned} n-2 &= 1\,, \quad & n &= 3\,, \\ n-2 &= 2\,, \quad & n &= 4\,. \end{aligned}$$

Given the above the different ways of matching tiles will not give us a different tile shape to what has already been seen.

Figure 1.26 gives us an example of the two forms of tiling obtained using triangular tiles, whereas Fig. 1.27 shows us tilings obtained from square and hexagon tiles.

Going back to three-dimensional lattices the symmetrical elements of the unit cell are usually considerably reduced by the symmetry of the atomic arrangement in the basis.

All operations that do not affect the crystal constitute the *space group* (group in the mathematical language); all the primitive translations in turn are a group, the translation group, which is a sub-group of the space group. We should note that the result of two successive translations does not depend on the order in which they are applied to a crystal, whereas the result of two rotations may depend on the order the crystal is rotated. So, while

Fig. 1.26. Two possible ways of covering the plane using triangular tiles

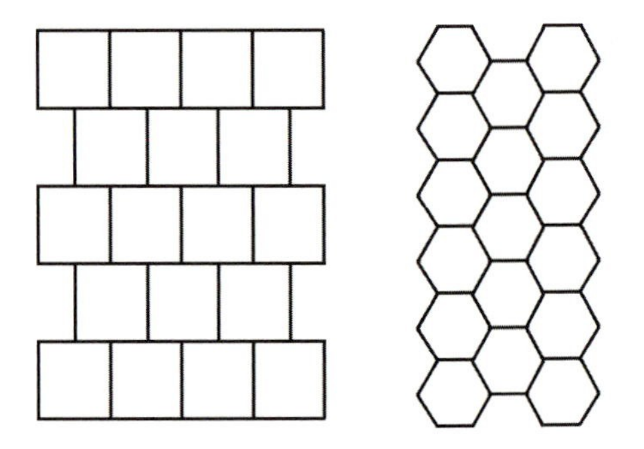

Fig. 1.27. Covering the plane using squares and hexagons

the translation group is commutative the space group generally is not. All rotations that are compatible with the symmetry of the Bravais lattice are part of the *point group* of a crystal. The symmetry elements of this group pass through a fixed point of the lattice, which we shall take as the origin of the reference axes for the crystal. The point group may not be a sub-group of the space group since the space group rotations only take place together with non-primitive translations.

The importance of the point group is that the Bravais lattice and the unit cell are invariant under whatever point group operations. This all means that the point group is the group of symmetries of the Bravais lattice or at least a sub-group of it. This limits the possible number of point groups. Particularly, in two-dimensional space there are ten groups, whereas in three dimensional space there are thirty-two and these go to make up the crystalline classes. In the end, a point group may be associated with a corresponding Bravais lattice; these, together with non-primitive translations combined with the point group elements, go to make up the space group. In two-dimensional space the five Bravais lattices and the ten point groups will give us 17 space groups; in three-dimensional space the fourteen Bravais lattices and the thirty-two point groups will give us 230 space groups. The space group of a crystal depends on the translation symmetry and on the distribution of the atoms in the unit cell, namely on the basis symmetry.

Conventional analysis of crystal directions and planes is carried out using the *Miller indices.* In a plane these indices are represented by three integers (h,k,l). In a cubic crystal the Miller indices are related to the orientation of the plane relative to the crystallographic axes where $1/h$, $1/k$ and $1/l$ are the intercepts of the plane (h,k,l) on the axes $\boldsymbol{x}_1$, $\boldsymbol{x}_2$, $\boldsymbol{x}_3$. If an intercept, for example k, has a negative value then it is shown as $\bar{k}$.

The equation for the plane is given as

$$\frac{x}{1/h}+\frac{y}{1/k}+\frac{z}{1/l}=1 \tag{1.19}$$

so

$$hx+ky+lz=1\,. \tag{1.20}$$

Thus the direction cosines of the perpendicular to the plane (h,k,l) are proportional to h, k, l, respectively.

Alternatively, if the intercepts of a plane with $\boldsymbol{x}_1$, $\boldsymbol{x}_2$, $\boldsymbol{x}_3$ axes are (l_1,l_2,l_3), then the Miller indices are the smallest integers having the ratio $(1/l_1):(1/l_2):(1/l_3)$. For example, if the intercepts on the axes are $(4,3,2)$ they define a plane with indices $(1/4):(1/3):(1/2)$, which correspond to the ratio between the integers $3:4:6$; then the Miller indices of the plane are $(3,4,6)$.

For crystals with hexagonal symmetry we use four Miller indices $(h\ k\ i\ l)$ where $i=-(h+k)$. The first three indices refer to the intercepts of the considered plane on the three axes $\boldsymbol{x}_1$, $\boldsymbol{x}_2$, $\boldsymbol{x}_3$ which define the hexagonal basal plane where the three axes are inclined at 120°, whereas l defines the intercept on the $\boldsymbol{x}'^3$ axis which is perpendicular to the basal plane.

The notation $\{h\ k\ l\}$ refers to the family of equivalent lattice planes, perpendicular to a crystal axis. For example, {100} refers to planes (100), (200), (300),... The directions in a crystal are also given by three indices (four for the hexagonal lattice). [h k l] is the perpendicular direction to a plane with Miller indices $(h\ k\ l)$. Direction $[h\ k\ l]$ defines a straight line that passes through a point with coordinates $(h\ k\ l)$.

1.5 The Reciprocal Lattice

The three Miller indices do not define a plane univocally because all parallel planes have a common perpendicular.

The notation does, however, have a simple geometric significance if we introduce a lattice, dual of the lattice in physical space $\boldsymbol{R}$, called the *reciprocal lattice.* $(h\ k\ l)$ now refers to the coordinates of a point in the reciprocal space $\boldsymbol{K}$, also called the space of wave vectors $\boldsymbol{k}$. The reciprocal lattice points are given by the relation

$$\boldsymbol{k} = m_1\boldsymbol{x}_1^* + m_2\boldsymbol{x}_2^* + m_3\boldsymbol{x}_3^* \tag{1.21}$$

where, just for (1.11), m_i are integers and $\boldsymbol{x}_1^*$, $\boldsymbol{x}_2^*$, $\boldsymbol{x}_3^*$, are the non-coplanar vectors that define primitive translations in the reciprocal lattice. These vectors are linked to the $\boldsymbol{x}_1$, $\boldsymbol{x}_2$, $\boldsymbol{x}_3$ vectors by orthogonality since each basis vector of the reciprocal space (namely $\boldsymbol{x}_1^*$) is orthogonal to the two non-homologous basis vectors of the physical space (namely $\boldsymbol{x}_2$ and $\boldsymbol{x}_3$). Given this we obtain:

$$\boldsymbol{x}_i \cdot \boldsymbol{x}_j^* = 2\pi\delta_{ij}, \qquad \text{where} \quad i,j = 1,2,3\,.$$

If this is true then

$$\boldsymbol{x}_1 = \text{ const } (\boldsymbol{x}_2 \times \boldsymbol{x}_3)$$

and since

$$\boldsymbol{x}_1 \cdot \boldsymbol{x}_1^* = \text{ const } \boldsymbol{x}_1(\boldsymbol{x}_2 \times \boldsymbol{x}_3) = 2\pi$$

we obtain

$$\boldsymbol{x}_1^* = 2\pi\frac{(\boldsymbol{x}_2 \times \boldsymbol{x}_3)}{\boldsymbol{x}_1(\boldsymbol{x}_2 \times \boldsymbol{x}_3)}\,. \tag{1.22}$$

The same goes for $\boldsymbol{x}_2^*$ and $\boldsymbol{x}_3^*$.

The basis vectors of direct space are defined in the same way as vectors of reciprocal space. For example, we get

$$\boldsymbol{x}_1^* = 2\pi\frac{(\boldsymbol{x}_2^* \times \boldsymbol{x}_3^*)}{\boldsymbol{x}_1^*(\boldsymbol{x}_2^* \times \boldsymbol{x}_3^*)}\,.$$

The relations with $\boldsymbol{x}_2$ and $\boldsymbol{x}_3$ are similar. To sum up, each Bravais lattice corresponds to one reciprocal lattice.

The common denominator in (1.22) is the volume of the primitive cell in physical space. As such the dimensions of $\boldsymbol{k}$ vectors are $[L]^{-1}$; hence the name reciprocal lattice.

The parallelepiped unit-cells given by $\boldsymbol{x}_1^*$, $\boldsymbol{x}_2^*$, $\boldsymbol{x}_3^*$, are the primitive cells of the reciprocal lattice. In the same way we can construct Wigner–Seitz cells using the same method used for direct lattices. These cells are built up around the origin of the reciprocal lattice, which is made to coincide with a reciprocal lattice point, and is called the G point. The Wigner–Seitz cells in the reciprocal lattice are called *Brillouin zones*.

The Wigner–Seitz cell and the Brillouin zone of the simple cubic lattice and the hexagonal lattice are all the same except that the hexagonal lattice

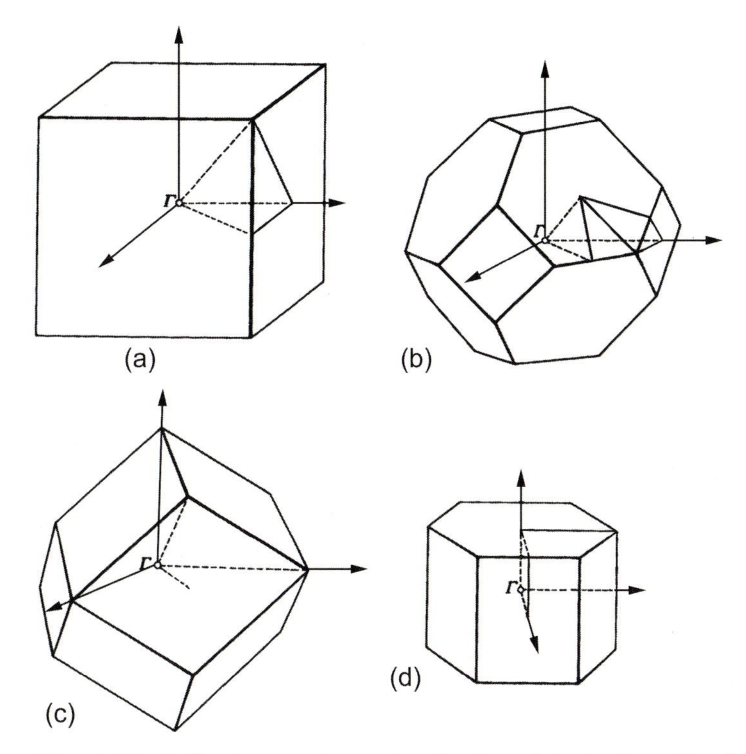

Fig. 1.28. Brillouin zone for various lattices; (**a**) simple cubic; (**b**) face-centred cubic; (**c**) body-centred cubic; (**d**) hexagonal

is rotated 30° about the $\boldsymbol{x}_3^*$ axis of the hexagonal Brillouin zone as compared to the orientation of the Wigner–Seitz cell. The shape of the Brillouin zone of the body-centred cubic lattice is the same as the Wigner–Seitz cell of the face-centred cubic lattice, and vice versa (Fig. 1.28).

The Brillouin zone is a convex polyhedron bounded by planes which meet the equation

$$(\boldsymbol{k}' + \boldsymbol{k})^2 = \boldsymbol{k}^2 \tag{1.23}$$

just as for (1.15) in direct space.

Geometrically speaking the planes that satisfy the requirements of (1.23) bisect the segments that join the origin, where $\boldsymbol{k}' = 0$, to the nearest reciprocal lattice points, so $\boldsymbol{k}' = \boldsymbol{k}$.

The relation between the points of the two spaces, namely the physical space and the reciprocal space, is

$$\boldsymbol{r} \cdot \boldsymbol{k} = 2\pi N \tag{1.24}$$

where N is an integer.

Equation (1.24) can also be expressed as

$$\exp\left[\mathrm{i}\boldsymbol{k}\cdot\boldsymbol{r}\right]=1 \tag{1.25}$$

which is of particular interest when interpreting experimental diffraction data.

All the r points that meet the requirements of (1.15) for a fixed $\boldsymbol{k}$ value lie on a plane perpendicular to $\boldsymbol{k}$. Each set of (m_1, m_2, m_3), which defines a specific $\boldsymbol{k}$ value, refers to a set of lattice planes in the $\boldsymbol{R}$ lattice. If the m_i values are the minimum compatible integers then $\boldsymbol{k}$ is the shortest primitive translation in the reciprocal lattice and the m_i's are thus the Miller indices. As such we get

$$(x,y,z)=(r_x,r_y,r_z)\equiv\boldsymbol{k}\,. \tag{1.26}$$

What's more,

$$h:k:l=k_x:k_y:k_z$$

which is equivalent to

$$(h,k,l)=\frac{1}{2\pi N}\,(k_x,k_y,k_z)\equiv\frac{\boldsymbol{k}}{2\pi N}\,. \tag{1.27}$$

From the point of view of a structural analysis of materials the most important consequence of the described symmetry, (see (1.14) and (1.25)), regarding the physical properties of a crystal is that the matter density $\varrho(\boldsymbol{x})$, is periodic and it can be Fourier transformed to

$$\varrho(\boldsymbol{x})=\sum_{\boldsymbol{k}}\varrho(\boldsymbol{k})\exp\left[\mathrm{i}\boldsymbol{k}\cdot\boldsymbol{x}\,\right]. \tag{1.28}$$

This means that the Fourier transform $\varrho(\boldsymbol{k})$ is not zero only for the vectors

$$\boldsymbol{k}=h\,\boldsymbol{x}_1^*+k\,\boldsymbol{x}_2^*+l\,\boldsymbol{x}_3^* \tag{1.29}$$

where h, k, l, are Miller indices.

2. Structural Order

2.1 Order and Disorder

Order and disorder are rather complex concepts with various and sometimes contrary exceptions and shades and this makes it somewhat difficult to give them a generally valid definition. The dictionary defines order as "the arrangement of objects in position, or of events in time" or, more generally, "the manner in which one thing succeeds another" [2.1].

Fig. 2.1. An example of ordered, but not regular, structure is given by "Path of Life II" by M.C. Escher (1958). "From the centre, representing the limit of infinitely small, four rows of fish (rays) protrude, in spiral form, swimming head to tail. The four largest examples, enclosing the square surface, change direction and colour: the white tails are still part of the school that comes from the centre while the grey heads are pointed towards the inside and are part of the grey rows pointed towards the centre." (M.C. Escher's "Path of Life II")

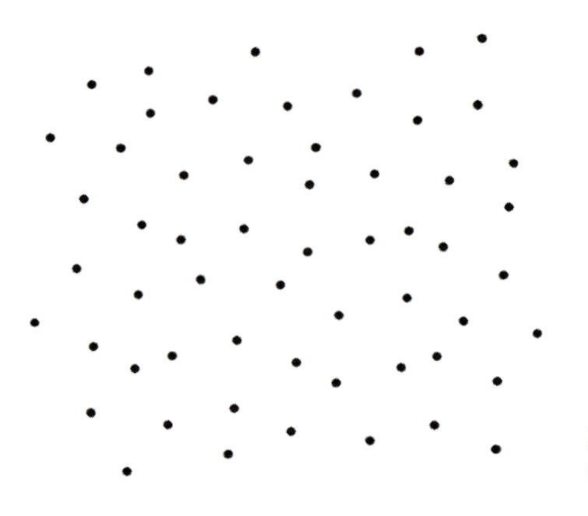

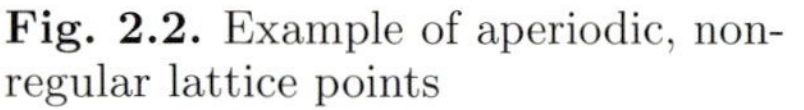

Fig. 2.2. Example of aperiodic, non-regular lattice points

Rather than the general definition of order we are more interested in a series of criteria that can be used in Physics and in particular in analysing the structure of condensed matter, and that can possibly include "objects" in Nature and in art that give the onlooker the feeling of order.

First of all it is common to consider *order* and *regularity* as equivalent, namely that they belong to the same category and refer to objects in that category. In actual fact the idea that something is regular is geometric and is based on the idea that one or more geometric elements are repeated, whereas order is probabilistic and leads to the idea of singleness in that a given macroscopic structure corresponds to a single microscopic configuration.

Structures may be ordered but not necessarily regular. An example is Escher's covering of a flat surface in his xylography called "Path of Life II" [2.2] Fig. 2.1; another example is given by the proteins which have a single configuration (ordered) with an irregular structure.

The point lattice in Fig. 2.2 is aperiodic and not regular. Figure 2.3 gives two possible ways of covering a flat surface, using different schemes, but with due consideration to the disposition of the points in Fig. 2.2. In one case the polygons are all different whereas in the other two convex regular polygons give rise to an ordered system in that the arrangement achieved by joining the polygons under certain well defined rules is unique and can be built on indefinitely.

Crystals are ordered and regular since they can be obtained by periodical translation of an elementary (geometric) cell in space. Excluding the trivial case of a vacuum, crystals constitute the highest degree of three-dimensional order that can be achieved. In a crystal an assembly of innumerable particles (atoms or molecules) is packed in a regular structure along lines in space and on planes in a geometric lattice in order to obtain the highest density possible. With exception to vibrational motion the atoms have a fixed position and are temporarily invariable. As such, from a rheological point of view a crystal is a solid. As we saw in Chap. 1 crystal structures do not change when

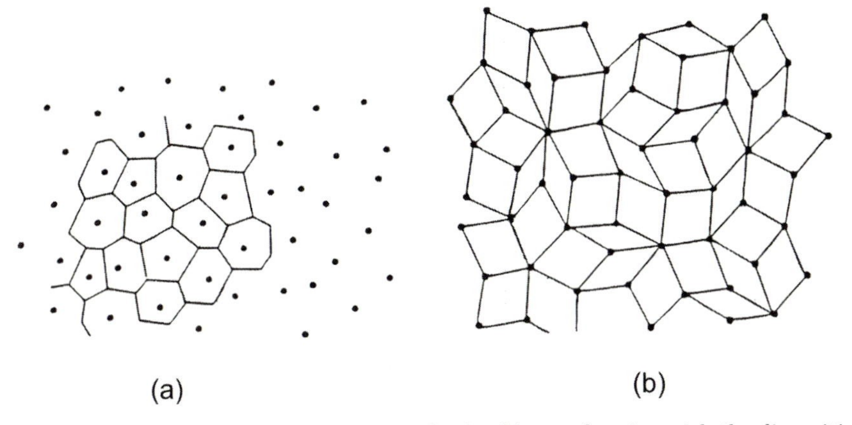

Fig. 2.3. Two possible plane coverings obtained by conforming with the disposition of the points in Fig. 2.2; (**a**) using irregular polygons; (**b**) using two regular polygons (losanges), interconnected by defined rules: the result is a Penrose tiling (aperiodic)

they undergo lattice translations, rotations about axes of different order or reflections in various planes and combinations of reflections and rotations.

The concept of disorder is intuitive and, to some extent, primitive. As such, since it is closely connected to the concept of "casual" it may be defined independently only in a specific context. It is easier to define disorder as a state of absence of, or departure from, the condition of order. The immediate consequence of this approach is that we have to give a better definition of perfect order. From this we have to be able to recognise how much more or less departure from the condition of order is achievable.

2.2 Rules of Order

The order of a set of objects is defined with respect to one particular property of the objects. An object may possess p properties, and the set may be ordered *contemporaneously* with respect to i properties where i is less than p.

In the case of a set of atoms, distributed in space, whose positions are defined by a particular point, namely the centre of mass, and if there is a rule that, when applied to all the atoms, establishes their reciprocal position, provided each atom actually occupies the position it was assigned, then the set is said to be perfectly ordered.

This rule is called the *rule of order* $\Re$.

The positional (and orientational) order in decorative motifs is associated with a particular rule of order. In the simplest of cases if these motifs are distributed along a straight line they represent a sort of "crystal" in one dimension.

The *repeated* application of the rule of order $\Re$ is essential. In fact, in a finite collection it is possible to construct a single arrangement in space according to a complex rule that defines exactly where each member of the collection is to be placed. The result, however, may be extremely irregular.

To obtain perfect order we must be able to move from one member of the set to another by repeating the same rule. This rule can thus be repeated infinitely and will give us the correct position of the objects that are progressively added to the set.

More generally, a set of objects $\{O_k\}$, each having n properties $\{p_n\}$, has perfect order in relation to property p_j provided there is a rule of order $\Re$ and that there is at least one object O_r of which we know $p_j(O_r)$. The property p_j of each of the other objects in the set $\{O_k\}$, taken singularly, can only be obtained with respect to $p_j(O_r)$ provided that the rule $\Re$ is applied repeatedly.

This means that the $\Re$ rule does not contain any factors of probability; it will give the very same result whenever it is applied to $\{O_k\}$ which, in turn, may be finite or infinite and may also consist of both physical objects (atoms, molecules) and abstract elements, whether they be geometrical (polygons, polyhedra, points,) or not. Lastly, all the properties $\{p_n\}$ may include space-time coordinates such as, in Physics, orientational coordinates, the atomic number Z, which is usually a function of the position $Z(x)$, electrical or magnetic dipole moments (atomic or molecular).

In studying condensed matter we must discuss systems that are extended in space and that include many atoms or sets of atoms. In the simple case of an elemental system we are interested in the position of like atoms.

The most important aspect of the rule of order $\Re$ is that when it is applied to the set in question sequentially we will obtain a *unique* space arrangement. It is clear that in order to achieve order the atoms only have to be the same distance from each other, though this is not a necessity. The application of the rule of order $\Re$ allows us to substantiate whether the position occupied by any object in the set is correct or not.

The rule of order $\Re$ must be exact and quantitative. In the simple case of a one dimensional "crystal" we can require that the atoms, seen as random points along a straight line, are distanced from each other by a fixed length, e.g. $L = 1$ nm. There is no sense in their being just any distance between them provided that the distance is between 1 nm and 1.5 nm; such an imprecise criterion would lead to infinite space dispositions.

In order for space order to be perfect the rule that leads to this space order must be a rule of order namely that it generates a unique space disposition when applied repeatedly.

If the objects to be ordered are not identical, e.g. if we are dealing with two different kinds of atoms, X and Y, then they can be set out in an ordered manner based on a positional (or topological) rule of order. They may then either be ordered or not with respect to any other property they have in common; among these common properties the chemical order may require that the two species of atoms occupy alternative lattice sites. This sort of order may not occur even though there may be topological order.

Multiple order occurs when a set of objects is ordered at the same time with respect to two or more properties. For example, when there is no external magnetic field, atoms of the X and Y species, with non-zero magnetic moments, arranged alternatively on simple cubic lattice sites, and having parallel or non-parallel alignment because of the atomic spin, constitute a set with multiple order, i.e. topological, chemical and magnetic order.

2.3 Order Parameters

In most cases physical systems do not exhibit perfect order, nor do they represent complete disorder. On the contrary, the non conformity with a rule of order varies greatly both as regards the degree of order and as regards such properties other than those that lead to the definition of order.

Let's now come back to structural order. Even though a crystal structure may have various kinds of defects 99% of the atoms will lie on lattice sites. The same material in the liquid state will present local order between first and sometimes second neighbours. In sufficiently diluted gases no traces of order are found. If a hypothetical observer ideally inserted an atom in the crystal system then this would more than likely be recognisable since it would occupy an interstitial site. In the case of a gas we would not be able to tell which is the additional atom, not even after a short time interval, owing to the high non-correlative atomic mobility.

On the other hand, it is more difficult to treat a liquid. Owing to the high atomic mobility we presume that the additional atom cannot be identified. However, we can see, at least after a shorter time interval than the time diffusion constant, some atomic correlation in the microscopic arrangement under observation. This is an index of specific local order due to the increased local number density. In this case of limited space order, which, however, leaves no trace over a long distance, we consider, if we start from any arbitrary point, the *radial* extension of local order by implicitly considering that the material is isotropic.

Whenever we refer to systems showing partial order (the most common situation) we must identify how they are disordered and the degree of disorder.

For each kind of phase transition where symmetry is broken we can identify an appropriate macroscopic parameter, called the order parameter η, which gives us a numeric measurement of the degree of order in the system. While we can immediately identify the order parameter in certain transitions it is very difficult to do so in others. For example, in normal liquid-crystalline solid transition the translation symmetry of the liquid phase at high temperature is broken at the transition. In fact, while the average space-time of the microscopic density of a liquid is isotropic, thus it is invariant under all the elements of the translation group, the crystal has a periodic average density and is invariant only under a sub-group of the translation group.

The order parameter, which may be a scalar, vectorial or tensorial quantity, a complex number or any other quantity, appears in the less symmetrical phase in any transition where symmetry is broken whatever the order of the transition, defined according to Ehrenfest. In this scheme the order of a phase transition is given by the lowest order derivative of Gibbs' free energy, G, which presents a discontinuous change at the transition. In liquid-crystal transition, which is first order, since the slope of the G curve is discontinuous at the transition temperature, there is an abrupt discontinuity in the state of the system, as reflected by the trend of any intensive variable (volume, density): consequently, the symmetry properties in the liquid and solid phases are not related to each other.

In the case of a continuous transition, such as in paramagnetic-ferromagnetic materials, the rotational symmetry is broken since the order parameter spontaneous magnetization defines a single direction in space. In this case the state of the system changes with continuity and the symmetry properties of the two phases are closely related; usually, though not always, the low temperature phase is the least symmetrical.

Now we consider the continuous order-disorder transition in a binary XY alloy where, by convention, in the fully ordered configuration the X atoms occupy the ξ lattice sites and the Y atoms occupy the υ lattice sites. Disorder is achieved when the X atoms move from the ξ sites to the υ sites and, at the same time, the Y atoms move from the υ sites to the ξ sites. No holes and/or interstices are allowed for. In order to quantitatively obtain the degree of order in the system we will take a crystal with N atoms where

N_X^ξ = number of X atoms on the ξ sites,
N_Y^υ = number of Y atoms on the υ sites,
N_X^υ = number of X atoms on the υ sites,
N_Y^ξ = number of Y atoms on the ξ sites.

Thus the following stand

$$\begin{cases} N = N_X + N_Y \\ N_X = N_X^\xi + N_X^\upsilon \\ N_Y = N_Y^\xi + N_Y^\upsilon \\ N_X^\upsilon = N_Y^\xi \end{cases} \tag{2.1}$$

where N, N_X and N_Y are constant during the order-disorder transition (we do not consider there may be some chemical reaction between the X and Y atoms).

$$c_X = \frac{N_X}{N} \qquad \text{and} \qquad c_Y = \frac{N_Y}{N}$$

are the atomic fractions of the X and Y atoms respectively of the crystal, and

$$f_I^i = \frac{N_I^i}{N_I}, \qquad \text{where} \quad I = X, Y; \qquad i = \xi, \upsilon$$

are the fractions of the I type of atoms that occupy (correctly) the i sites, calculated with respect to the number of I type of atoms (and not with respect to the overall number of atoms, N, in the crystal).

The *long range order parameter* S is obtained using the following relation

$$S = \frac{\left| f_X^{\xi} - c_X \right|}{1 - c_X} \tag{2.2}$$

or, in the same way, by

$$S = \frac{\left| f_Y^{\upsilon} - c_Y \right|}{1 - c_Y}. \tag{2.3}$$

S may only take on values in the interval $[0, 1]$. In particular, when the crystal has a totally ordered configuration the fraction f_X^{ξ} of the X atoms that occupy the ξ sites, calculated on the total number of X atoms, is equal to 1, thus each X atom occupies one ξ site; in this case $S = 1$.

When, though, the crystal has a totally disordered configuration, $f_X^{\xi} = c_X$ and $S = 0$; this condition corresponds to the greatest value of $N_X^{\upsilon} = N_Y^{\xi}$: the search for the greatest value is bound by the values in (2.1). The greatest disorder, thus, *does not* correspond to the condition $f_X^{\upsilon} = 1$ (all the X atoms on υ sites). For simplicity's sake let us refer to a binary alloy with equiatomic composition. This state would lead to the interchange of all the X atoms with all the Y atoms, thus back to complete order.

As an example we shall consider an equiatomic binary alloy XY lying on a two-dimensional square lattice, as shown in Fig. 2.4. The lattice sites are $N = 36$; for stoichiometric composition $N_X = N_Y = 18$, thus $c_X = c_Y = 0.5$.

Figure 2.4 shows (part a) the ξ and υ sites occupied by the X and Y atoms in a fully ordered configuration. In this case $N_X^{\xi} = 18 = N_Y^{\upsilon}$ and $N_X^{\upsilon} = N_Y^{\xi} = 0$, thus $f_X^{\xi} = 1 = f_Y^{\upsilon}$. The long range order parameter (eqs. 2.2 and 2.3) is $S = 1$.

Figure 2.4 shows (part b) partial disorder in the system shown in part (a) of Fig. 2.4. The disorder was obtained by exchanging the positions of X and Y atoms that are nearest neighbours, along any line. The scheme shows that alternative lines have a disordered atom arrangement and each position of an XY atom exchange is separated from the next by an atom in its correct position. Furthermore, the disordered arrangement between contiguous disordered lines is staggered by one atomic position. We impose cyclic boundary conditions; the lattice is thought to be closed on itself. As such we understand why we have to exchange atoms at the edge of the lowest line of the lattice. In this configuration $N_X^{\xi} = 12 = N_Y^{\upsilon}$ and $N_X^{\upsilon} = 6 = N_Y^{\xi}$, thus $f_X^{\xi} = \frac{2}{3} = f_Y^{\upsilon}$. The long range order parameter is $S = 1/3$.

Figure 2.4 shows (part c) complete crystal disorder. The lines where all the possible exchanges have taken place alternate with perfectly ordered lines.

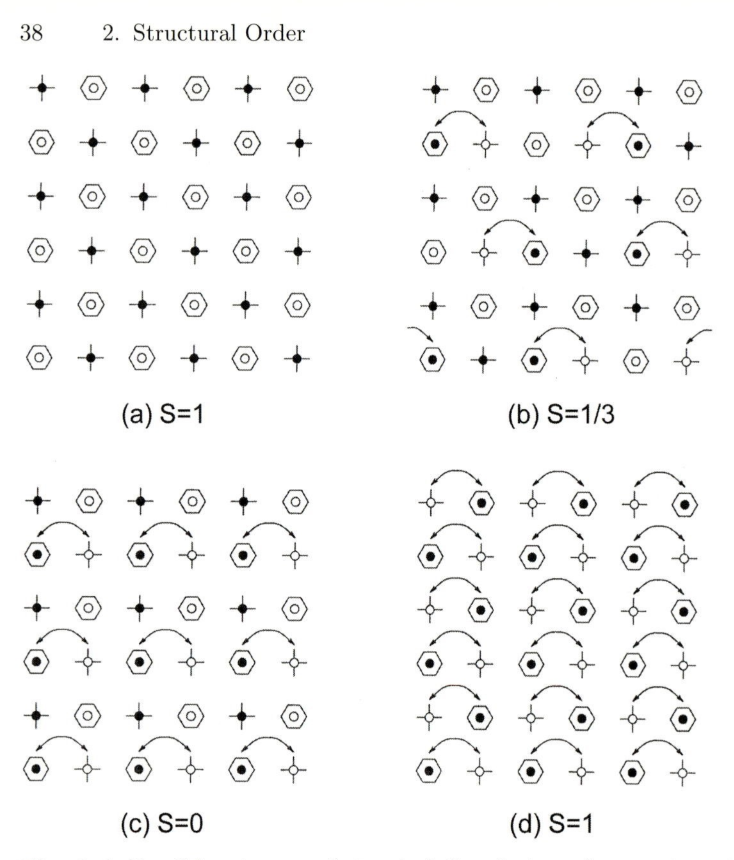

Fig. 2.4. Possible schemes of chemical disordering of an equiatomic XY binary system on a square lattice (•: X ; ∘: Y ; -¦-: ξ; ⬡: υ ; ⌒: interchange XY)

The result is thus $N_X^\xi = 9 = N_Y^\upsilon$ and $N_X^\upsilon = 9 = N_Y^\xi$; this corresponds to $f_X^\xi = 0.5 = f_Y^\upsilon$, thus S $= 0$.

Let us take any other disordering scheme, where $f_X^\xi = 9$: the long range order parameter has the same value as that of the scheme in part (c) of Fig. 2.4. However, each configuration is characterised by the short range order parameter having a different value, $a(x)$ (see Chap. 4), which gives us an index of the degree of local order around the position of a reference atom.

Figure 2.4 shows (part d) the exchange of all the X atoms with all the Y atoms. So, $N_X^\xi = 0 = N_Y^\upsilon$ and $N_X^\upsilon = 18 = N_Y^\xi$ at the same time. In this case $f_X^\xi = 0 = f_Y^\upsilon$, thus $S = 1$. The structure is hence completely ordered; only the observation point has changed.

The quantity S interpolates between the extremes of complete order and complete disorder and is an example of a scalar order parameter. By using this quantity we can give a quantitative meaning to the expression "partial order" so we can compare different systems with each other.

A *probabilistic* rule of order which does not give us a single result, leads us to partially ordered sets; perfect order, in this case, is obtained only statistically.

In actual fact, even when we consider sets, for example sets of atoms or molecules, for which the property whose degree of order is characterised by a deterministic rule of order, what mostly interests us is the degree of order in the set, considered in a thermodynamic state. Therefore, no particular meaning can be assigned to any specific microscopic configuration among the huge number of those compatible with the set of thermodynamic variables that define the state of the system; we are interested in the ensemble average of the order parameter η.

In principle the calculation for the ensemble average can be simplified provided certain conditions are true. In Statistical Mechanics we usually require that the ergodic condition is true, which can be translated into the equivalence of the ensemble averages with time averages. We can give an example of this equivalence, in general terms, as follows.

We shall start from an initial fixed instant, $t = 0$, and consider that the time evolution for two systems, **P** and **Q**, which are identical and isolated and that at the initial time are in two states, 1 and 2, of equal energy.

We will then observe what states **P** and **Q** are in at periodic intervals. For the ergodic condition to be true and having waited long enough, the **Q** system will occupy state 1 and, as of that moment, its statistical evolution will coincide with that of **P**, measured after the initial instant. When we average out very long times, the statistical behaviour of the two systems will become *identical*. The base hypothesis is that the skip frequencies V_{mn} between any two configurations m and n that are compatible with thermodynamic variables are never zero. The isolated system, with fixed constraints, must be able to move from one state to any other of equal energy (accessibility hypothesis).

The equilibrium distribution of the fluctuations of any observable physical quantity, that is its distribution averaged along a sufficiently long period of time so that it becomes independent of time, must be the same for both systems. This means that the characteristics of the equilibrium state of a system depend only on the bonds and on the energy that fix the accessible configurations, whereas any particular microscopic configuration from which system evolution begins is of no importance.

The time averages are necessary because the values of the physical variables that describe the system generally fluctuate from time to time, and thus the degree of order therein fluctuates at the same time. In the statistical set all the variables we want to determine the degree of order of, from the positions to the orientations, to the electrical and magnetic moments, ... present statistical distributions. As such it is clear that the probability that extracting a particular set of values of the variables from the distributions this set corresponds to the real degree of order is extremely low.

2.4 Cellular Disorder and Topological Disorder

We generally make a distinction between *Long Range Order* (LRO), which extends throughout the structure, and *Short Range Order* (SRO), which involves the shell of the first neighbours of any atom chosen to be the origin, depending on the spatial extension of structural order.

Since the atoms on the surface of a solid, even at thermodynamic equilibrium, undergo different forces compared to those that experience the effect of the presence of atoms in the volume, the interatomic spacing on the surface are different to those in the volume. The immediate consequence is that there is no lattice order in a conglomerate, with say 100 atoms, where a large number are on the surface. In these cases, however, we will often find short range order.

Atomic clusters are an example of this situation: the kind of structural order in them and their evolution with the dimensions, namely with the number of atoms in the cluster, will be examined in more depth later on. In glassy and liquid systems the experimentally observed short range order is imposed by the chemical nature of the constituents and by the need to achieve geometric atom packings as compact as possible. The structural arrangement among first neighbours gives rise to local order, either Chemical Short Range Order (CSRO), of chemical origin, or Topological Short Range Order (TSRO), of geometrical origin.

Medium Range Order (MRO) has progressively gained more importance in amorphous solids, initially in the analysis of elemental semiconductors, silicate and chalcogenide glasses, and later in the structural characterisation of metallic glasses. MRO typically extends on intervals of the order $0.5-2$ nm, and this is characteristic of the interaction between second neighbours.

A macroscopic structure with long range crystalline order is subject to periodic translation. Though this is the highest level of structural order possible in condensed matter these crystal structures are usually associated with various kinds of disorder, as seen in Fig. 2.5.

The most widespread sort of disorder is short-range disorder due to localised defects (point and line defects) such as vacancies, di-vacancies, interstitials, substitutional impurity atoms, F centres, dislocations. Apart from the atoms in the "core" of the defect, the first neighbour atoms of the defect, too, are not to be found on their ideal lattice sites.

Many properties of materials with a technological interest are the outcome of the presence and mutual interaction of these kinds of imperfection. Among these we will find the colour of gems, the electrical properties of semiconductors and the mechanical and thermal properties of metals.

Chemical or compositional disorder, which we have already referred to, can be seen in the XY binary alloys which undergo order-disorder transitions.

For materials whose atoms have intrinsic magnetic moment, such as ferromagnetic and anti-ferromagnetic materials, disorder is associated with alterations in atomic spin orientation above the critical point of magnetic ordering.

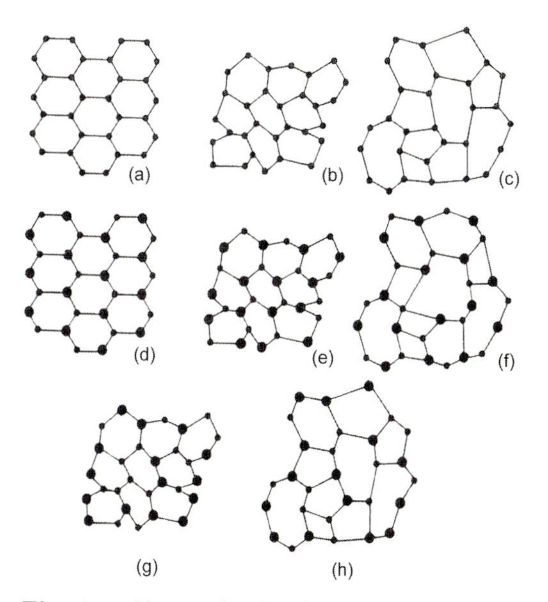

Fig. 2.5. Various kinds of disorder that can be found in a perfect crystal (**a**) elemental and (**d**) made of an equiatomic binary alloy; (**b**) the bond lengths vary generating bond disorder; (**c**) bond coordination also varies giving rise to dangling bonds and topological disorder; (**e**) (**f**) (**g**) and (**h**) present various combinations of the above kinds of disorder with varying weights

Disorder may occur in molecular arrangement as well. This is observed, for example, in the semiconductor compound CsPb, of which it has recently been noticed that melting takes place in two distinct phases. At higher temperature the system is a normal liquid, showing orientational and translational disorder; at lower temperatures, structural units Cs_4Pb_4, each characterised by a tetrahedron (Pb_4^{4-}) inside a tetrahedron (Cs_4^{4+}) oriented in the opposite direction have been identified. Such relatively large units, possessing rotational symmetry, have long range translational order though they are subject to fast re-arrangement along four non-equivalent directions in the lattice. Such an intermediate phase, called a plastic crystal, is common of organic molecular solids and has also been observed in the NaSn semiconductor.

At a temperature $T = 0$ K, in those systems with chemical or orientational disorder, both molecular and involving atomic spins, perfect order sets whereas disorder is prevalent at temperatures above the pertinent critical temperature, T_c. To understand and to describe the disappearance of order at $T > 0$ K and the growth of disorder as the system approaches T_c are both problems that lead to the study of the dependence of the order parameter η on temperature.

The four kinds of disorder we have just examined make up what is called *cellular* disorder. This kind of disorder can be described with reference to a particle placed on an ideal lattice site of a solid. The properties involved are intrinsic, as in the case of spin direction and chemical composition, or pertain to the presence or absence of defects at low or moderate concentration or, as in the case of thermal motion, regard vibrational displacements.

Unlike a crystal, with some degree of disorder, a microscopic sample of an amorphous liquid or a solid, with no defined crystal lattice, is characterised by *topological* disorder. This is the outcome of a relevant property in a liquid, such as the absence of translational atomic motion. In these kinds of system we must consider the *distribution* of the relative positions of the molecules.

Furthermore, if the material is a mixture or an alloy of X and Y then it will also have chemical disorder. The preferential attraction of an X atom for atoms (Y) of the other kind leads to the formation of structural units characterised by well defined geometrical shapes, sizes and structures (often based on tetrahedral packing) with partial Chemical Short Range Order.

In the very same way, topologically disordered structures may display complete *order* regarding the spin orientation, as can be seen in saturated amorphous ferromagnetic materials.

2.5 Structurally Disordered Materials

The most common class of disordered materials are *liquids*. The greatest problem with liquids is understanding their structural properties and macroscopic behaviour based on atomic and molecular interaction. We shall examine various classes of materials based on their specific microscopic interaction:

1) "*simple*" liquids characterised by Van der Waals molecular forces (Ar)
2) *fused salts* where the electrostatic ionic forces prevail (NaCl)
3) liquid *semiconductors* (Se)
4) liquid *metals* whose properties are largely determined by a Fermi gas of conduction electrons

The first investigation, both experimental and theoretical, focused on the structural properties and the equation for the state of the simple liquids. Over the last twenty years research has been widened to include semiconductors and metals. However, a clear understanding of the structure of materials in the liquid state is still some way off.

While liquids are usually thermodynamically in a state of equilibrium the amorphous solids, which comprise the second family of macroscopically structurally disordered systems, are not. In fact, in theory any material can be vitrified, provided

$$\Delta G_{\mathrm{proc}} \geqslant \Delta G_{\mathrm{a\text{-}c}} \tag{2.4}$$

namely that the difference in free energy ΔG_{proc} between the undisturbed crystal and the state of the crystal caused by any external processes (high pressure compression, deforming and mechanical alloying, irradiation, diffusion at the interface, hydrogen charging up to high concentrations) is greater than the free energy difference between the glassy phase and the crystal phase, $\Delta G_{\text{a-c}}$. When condition (2.4) occurs the crystal atoms will be disordered. We have to inhibit them from returning to the ordered crystalline arrangement associated with the minimum value of free energy in the system. The preparation techniques must thus be able to "arrest" the atomic movement and freeze the atoms in their *metastable* non-equilibrium arrangement. The techniques employed often make use of ultra-rapid cooling; among these techniques we can outline vapour condensation onto a layer kept at cryogenic temperature ($10^{11}\,\mathrm{Ks}^{-1}$), ultra-rapid quenching from the liquid ($10^{6}\,\mathrm{Ks}^{-1}$) quenching a thin surface layer, liquefied by a laser pulse ($10^{8}\,\mathrm{Ks}^{-1}$) and ion implantation ($10^{14}\,\mathrm{Ks}^{-1}$).

When a material can be rendered amorphous by a technique with a given cooling rate it can also be vitrified by any technique characterised by a greater cooling rate. What actually changes is the compositional range through which the amorphisation process occurs; this range increases with the cooling rate. The problem to compare the structure of specimens of the same material, amorphised by different techniques, which could lead to freezing of different degrees of disorder, is still relatively open to discussion. Though not much research has been done on this aspect the amorphous alloys $Ni_{63.7}Zr_{36.3}$, obtained by melt spinning, and $Ni_{65}Zr_{35}$, obtained by mechanical alloying, present the very same partial structure factors. Notice that the last is an amorphisation process in the solid state, where the kinetics are very slow compared to the quenching process. For the $Zr_{(1-x)}Ni_x$ alloys, where $x = 28; 33; 40$ at.%, amorphised by the same two very different techniques, the interatomic distances calculated from the deconvolution of the first peak of the radial distribution function are equal. Samples of the metal glass $Zr_{66}Ni_{34}$, obtained by melt spinning and sputtering, namely condensation from the gas phase, are structurally the same. In the amorphous $Ti_{(1-x)}Ni_x$ systems, where $x = 24; 35$ at.%, as obtained by solid state amorphization reactions, or by ultra-fast quenching from the liquid state, though the total structure factors differ with composition they are the same as regards differently prepared specimens with the same stoichiometry. Lastly, even when the CuZr alloy is vitrified using quenching from the liquid state or by low temperature proton bombardment or even by mechanical alloying, no evident structural difference is observed, as shown in Fig. 2.6.

While local structural order in a vitrified metallic system depends on its nature alone, it appears that a number of distinct amorphous phases may form, especially in materials with tetrahedral coordination, such as H_2O and SiO_2. The capability to separate the various phases in the liquid in equilibrium is responsible for amorphous polymorphism.

There are only a few pure elements among the metallic amorphous systems, such as tin, germanium and bismuth. The last two are semi-metals

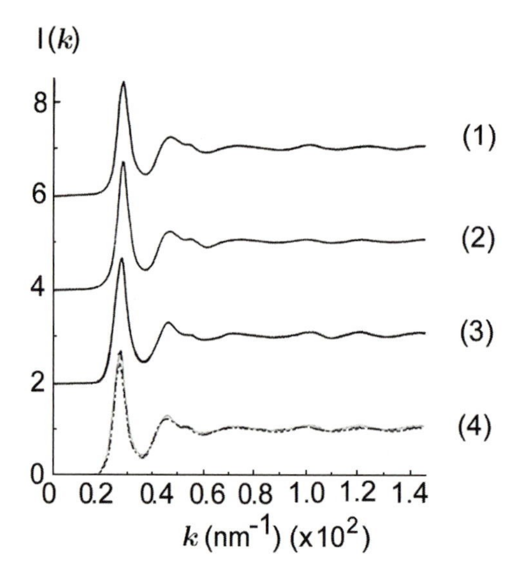

Fig. 2.6. Trend of the scattering intensity $I(\boldsymbol{k})$ versus wave vector $\boldsymbol{k}$, in $Cu_{50}Zr_{50}$ samples amorphised by various techniques; (1) energetic particle irradiation; (2) fast quenching from the liquid; (3) mechanical alloying; (4) superposition of curves (1) dots, (2) dashes and (3) continuous line (adapted from [2.3])

whereas the first has various crystal allotropic forms with free energy values close to the amorphous phase, which is thus competitive with them. Even though we often read that amorphous elemental metals have been obtained by way of rapid quenching techniques, we have to treat these results with some degree of caution since if there is even a minimum amount of impurity, mainly gaseous, in the specimens this impurity plays an important role in the heterogeneous nucleation process of amorphisation.

It has been ascertained, though, that the chemically pure specimens of some transition metals, such as cobalt, titanium and zirconium, are vitrified when bombarded with low doses, at low current, of accelerated heavy ions with energy in the order of GeV. In these experiments, however, the mechanism that destabilises the crystal structure is *not* lattice disordering, rather it is electronic. The interaction between the projectile and the electrons of the target atoms situated along its trajectory is the relevant process, whereas the atoms themselves will not be displaced from their lattice positions owing to the direct effect of the interaction with the ion.

The physical-chemical structure and properties of many amorphous binary alloys have been studied in depth. Among the considered systems, the simple metal and polyvalent metal alloys (CaZn), the inter-transition metal alloys (NiZr), the noble metal and semiconductor alloys (AuSi) and, lastly, the transition metal and metalloid alloys (FeB). Amorphous alloys with many elements (up to eight) are more easily obtained and have high (meta)stability; such materials have a number of technological applications, but they are not present in basic study owing to their complexity.

Hereafter such terms as *non-crystalline* solids, *amorphous* solids, *vitrified* solids, *glassy* solids and *glass* will be taken as synonyms.

Traditionally, glass is taken to mean an amorphous solid obtained by *supercooling* a liquid until it solidifies. Since the viscosity of a solid is different to that of a liquid, then by way of the macroscopic definition of glass we assume that a value of viscosity of 10^{12} Pa s is so high that we can consider any form of fluidity will be inhibited, at least on an infinite time scale as compared to what is experimentally accessible.

From a macroscopic viewpoint, and with reference to the structure, glass is often defined as an amorphous solid. Thus it shows, both in the *length* and in the *angles* that characterise the chemical bonds between first neighbours, a degree of variability. Such a variability, although meeting the constraints of the chemical bond present, which lead to a specific sort of short range order, is, however, sufficient for the constraints are not able to cause long range order.

The silicate glasses such as SiO_2, B_2O_3, and the chalcogenide glasses, including As_2S_3, As_3Se, $GeSe_2$, are part of the traditional glass family. On principle, the metallic glasses are a sub-group of the amorphous metallic solids. These solids can be produced using a wide range of methods, which do not necessarily imply the transformation from a liquid to a supercooled liquid state. To this aim, the term metallic glass will hereafter bear the same meaning as the expression non-crystalline metallic solid. This choice is based on the fact that the re-crystallisation of an amorphous metallic material, caused by heating, takes place through very distinct phases of nucleation and growth, exactly like the crystallisation of a liquid and the re-crystallisation of silicate glasses.

Until the mid 1980s liquids and amorphous materials were the only classes of structurally disordered materials known. In actual fact structural disorder was recognised in *atomic clusters*, and a great deal of research has been done on them in the last twenty years. Having produced them, the aim is to characterise the structure, stability and properties, in particular electronic, of free atomic clusters that contain from a few atoms to several thousands of atoms. The changes in properties in these clusters, based on the number of atoms and possibly on their stoichiometry, allow us to experimentally investigate a topic on which, until a few years ago, it was only possible to draw up models and theories. This is the transition from a single atom to an extended solid through the progressively more complex phases of molecule, atomic cluster, surface.

Not too surprisingly we will see that in rare gas atomic clusters the structure of small neutral clusters, during their spontaneous evolution, is non-crystalline; the structure is dominated by tetrahedral and polytetrahedral packing, largely the same as in amorphous materials. Only when the clusters are larger than a certain size will the structures, which are progressively changing with the size of the clusters, take on the fcc structures present in the solidified crystals of the rare gases.

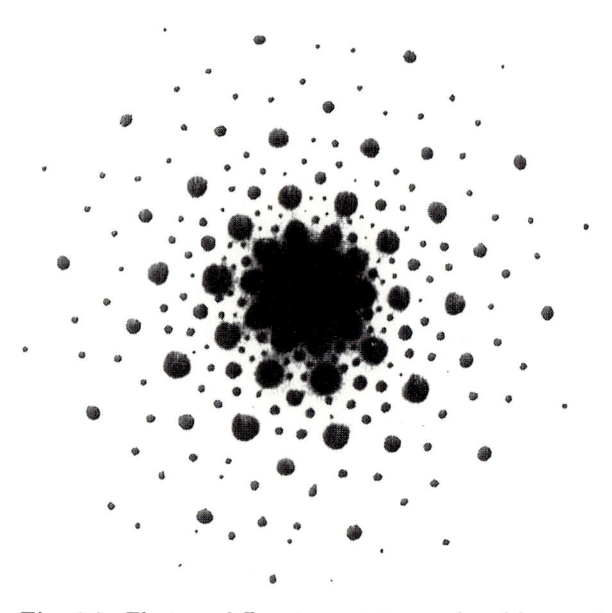

Fig. 2.7. Electron diffraction pattern produced by quasicrystalline $Al_{84}Mn_{16}$

The dependence of the structure on the electronic properties of the constituent atoms, as seen in the small clusters of alkali metals, is also particular. The shell models developed to analyse these clusters are the same successfully used in the study of nuclei and their stability. Basically, the structural stability of the clusters depends on the degree of electronic shell filling, namely on the number of atoms contained in the shell. For those cluster mass numbers, which correspond to shell closing, we will notice peaks in the mass spectra.

In 1984 samples of ribbons of an alloy with $Al_{84}Mn_{16}$ composition were produced using a fast cooling technique. Electron diffraction (Fig. 2.7) showed the presence of rotational three-dimensional fivefold symmetry, that is one of the "forbidden" symmetries in classical crystallography.

From peak width analysis it was clarified that the fivefold symmetry was not an effect due to the presence of multiple twinning, in which case the structure of the materials should have been fcc. It was possible to conclude that in Nature *quasicrystals* with *orientational* long range order and rotational icosahedral fivefold symmetry exist; such a symmetry is responsible for the absence of translational periodicity.

This combination of properties was thought to be present only in geometric model structures, both two-dimensional and three-dimensional, obtained artificially. Among these, in particular, we can mention the Penrose tilings, which uniformly cover the plane provided exact matching rules are used to combine together a small finite number of aperiodic "tiles", thus generating non regular coverings (Fig. 2.3).

Over the last ten years a few hundred compounds have been discovered having quasicrystalline structures, for specific composition and preparation techniques, thus demonstrating that quasicrystal formation is much less pathological than one might think. We are aware of materials with fivefold, eightfold, tenfold and twelvefold rotational symmetry, all of which are "forbidden" by classical crystallography. Much work has been carried out to understand the structural stability of the quasicrystals in relation to the various crystal phases and the glassy state, the microscopic causes at the root of quasiperiodicity and the peculiar physical properties, in particular the electronic ones, of the quasicrystalline materials.

2.6 Description of Disorder Through Entropy

Entropy S is often described as a quantitative index of the disorder present in a system and is frequently used to characterise the glass transition and amorphous materials. From an experimental viewpoint, the trend of the specific heat at constant pressure, c_p, measured against the absolute temperature when, for instance, liquid–to–solid crystal transition occurs, or when an amorphous solid transforms into a crystal, provides an example of how the reduction in entropy is tied to the reduction in disorder in a given material.

Entropy is connected to the fact that in most thermodynamic states a system is not found to be in a unique, well defined quantum state; it is distributed over a large number of microscopic configurations that are compatible with the considered macrostate, according to a certain probability distribution.

The problem lies in giving the right definition to the quantum multiplicity M, namely the true number of quantum states over which the probability is distributed at any time.

Entropy S is defined by

$$S' = k \ \ln M. \tag{2.5}$$

In the simplest case the probability is distributed uniformly over M states, thus the probability that a particular state is occupied, is $p' = 1/M$, hence we obtain

$$S' = -k \ \ln p'. \tag{2.6}$$

Using (2.5) and (2.6) entropy supplies us with a measurement of the order of a system, or rather it indicates the lack of detail in the knowledge of the microscopic state the system is in. Even though the system may be in a *particular* configuration at any time we ignore which state it is in. This is the same as acknowledging that the system is considerably disordered.

If the system is not in a state of equilibrium, or if it can interact with a heat reservoir, then (2.6) must be expressed in general terms with due consideration to the fact that probability p'_i is not the same value for all the states under examination. The definition of S' thus becomes, according to Boltzmann–Gibbs

$$S' = -k_{\mathrm{B}} \sum_i p'_i \ln p'_i \tag{2.7}$$

where $k_{\mathrm{B}} = R/N_{\mathrm{Av}}$, R is the perfect gas constant and N_{Av} Avogadro's number. This function is the *average* of (2.6) carried out on the states based on their probability.

In a system at a temperature T and where p_i is the equilibrium probability according to Boltzmann,

$$p_i = \frac{\exp[-E_i/k_{\mathrm{B}}T]}{\sum_i \exp[-E_i/k_{\mathrm{B}}T]},$$

the entropy defined in (2.7) becomes the equilibrium entropy

$$S = -k_{\mathrm{B}} \sum_i p_i \ln p_i. \tag{2.8}$$

The states of a macroscopic system usually make up a continuum energy distribution and it is not possible to identify any one countable group of states with the same energy.

The opposite case is more interesting where a macroscopic system has a defined countable number, W, of isoenergetic configurations.

As such we obtain a configurational contribution to entropy,

$$S_{\mathrm{conf}} = k_{\mathrm{B}} \ln W \tag{2.9}$$

where W is the number of quantum states of the macroscopic system that are compatible with the set of values for the thermodynamic variables that define the state under examination. W is thus the number of microstates that are compatible with the macrostate. If we work within the microcanonical ensemble then W is the number of microstates that are compatible with the volume V, the number of particles N and the total energy E in the system.

For the canonical ensemble, W is the number of quantum states that contribute significantly to the average energy in the system, where V, N and the absolute temperature T are fixed.

In a macroscopic system, N is extremely large, somewhere in the order of N_{Av}. As such all the quantum states in the ensemble W, namely those that are compatible with the macroscopic variables (thus with "reasonable" probability they can be realised) define a very narrow band of energy values, centred on the average energy. This means that in practice the energy differences are negligible compared to the average energy.

One example of configurational entropy (i.e. due to the various locations of the component particles in space) of a condensed system, liquid or solid, is given by mixing entropy in an ideal XY binary system where the mixing entropy ΔH_{mix} is zero and the variation in free energy during mixing is *only* due to entropic change

$$\Delta G_{\text{mix}} = -T\,\Delta S_{\text{mix}}. \tag{2.10}$$

This representation of an ideal solution, which is easy for us to imagine in a diluted gas with no intermolecular forces, is only acceptable for a solid if the atoms (or molecules) of both species are so similar that we can substitute the atoms of one species with the atoms of the other without causing any appreciable change in either the space structure or in the energy associated with the interatomic interactions in the solution. This implies that the shape and size of the X and Y atoms are practically the same and that the interaction energy between pairs of atoms $X - X$, $Y - Y$, and $X - Y$ coincide.

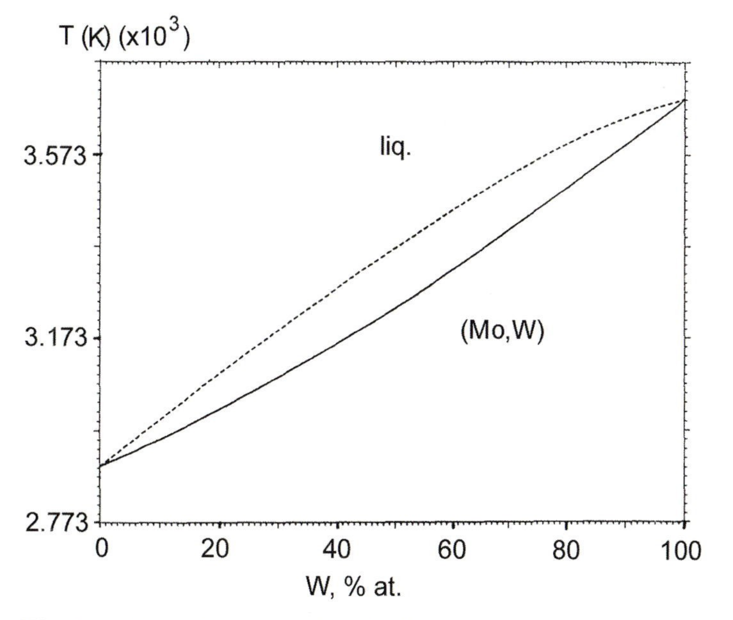

Fig. 2.8. Phase diagram of the Mo–W system: an example of a "mixed crystal"

A typical example of such a "mixed crystal" is given by the binary system Mo – W, which shows a continuous series of solid solutions with bcc structure throughout, as can be seen in the phase diagram in Fig. 2.8. The molybdenum atoms, like the tungsten atoms, can occupy any lattice sites; this gives us a substitutional solid solution, and thus we obtain

$$W = (N_{\text{Mo}} + N_{\text{W}})!/N_{\text{Mo}}!N_{\text{W}}! \tag{2.11}$$

where N_{Mo} and N_{W} are the number of atoms of Mo and W respectively.

If we refer to a mole of solution, and if X_{Mo} and X_{W} are the molar fractions of the two elements, then

$$N_{\text{Mo}} = X_{\text{Mo}} N_{\text{Av}}; \qquad N_{\text{W}} = X_{\text{W}} N_{\text{Av}}.$$

When we substitute this in eqs. (2.9) and (2.11), and we make use of the Stirling approximation ($\ln N! \approx N \ln N - N$),

$$\Delta S_{\text{mix}} = -R \times (X_{\text{Mo}} \ln X_{\text{Mo}} + X_{\text{W}} \ln X_{\text{W}}). \tag{2.12}$$

Since X_{Mo} and X_{W} are fractions of the unit, ΔS_{mix} is positive, namely an increase in entropy is associated with the mixing of the two subsystems of the atoms Mo and W.

If there is neither interatomic interaction of a chemical nature nor change in volume owing to the mixing process the increase in the system disorder is due solely to the increased uncertainty regarding the position occupied by each atom in the crystal lattice. There are many more possible sites for the atoms to settle on than there are separately available in each of the uniform subsystems Mo and W. From a thermodynamic viewpoint, the above situation is similar to mixing two ideal gases except for the presence of the crystal lattice. Disorder in mixed crystals is much less than topological disorder in an amorphous solid due to the existence of the crystal lattice whose translational periodicity is maintained.

Should the chemical interactions lead to the preferential formation of clusters of atoms, either homogeneous (clustering), or according to some stoichiometric rates (formation of compounds), then we would obtain partial chemical ordering which could lead to a negative change in the system volume. Both these phenomena play an opposite role to the increase in volume associated with mixing and can even lead to negative changes in entropy.

The analysis of mixing ideal *fluids* X and Y is very similar to that of solid solutions. In the case of mixing fluids X and Y, the uncertainty regarding the position of a given molecule is due to the possibility that the molecule might undergo diffusive translational motions within a certain volume. Once mixing has occurred, there is even greater uncertainty regarding the position of the molecule due to the fact that the volume available for diffusive motion increases. The increase in entropy associated with mixing is calculated by considering the product of each pure macrostate (namely one that has not yet been mixed) of X for the total number of Y macrostates necessary to obtain W_{mix}.

Since the latter is greater than the sum of W_x and W_y, the change in entropy associated with mixing is positive.

Even when we introduce local disorder into a crystal, when we form lattice vacancies while heating the crystal itself, configurational entropy will increase. For a fixed volume V and number of atoms N, as the number of vacancies increases so does the number of equally probable alternative positions for the vacancies on the lattice sites, obtained through an atomic jump mechanism. If we assume that the vacancies are equivalent to an atomic species the resulting system is obtained by mixing vacancies and atoms in the crystal. The total change in the system entropy is

$$\Delta S = -R\left[X_{\mathrm{v}} \ln X_{\mathrm{v}} + (1 - X_{\mathrm{v}}) \ln(1 - X_{\mathrm{v}})\right] + X_{\mathrm{v}} \Delta S_{\mathrm{v}} . \tag{2.13}$$

Here X_{v} represents the molar vacancy fraction we have introduced and (2.13) coincides with (2.12) for mixing entropy in an ideal system, except for the second term that represents the (small) contribution to thermal entropy from a mole of vacancies ΔS_{v}. This term is associated with the slight change in the vibrational frequency of the atoms surrounding a vacancy in the lattice, which is due to the substitution of a vacancy for an atom.

The aim of this discussion, using the examples, is to clarify that, though strictly connected, disorder and entropy are qualitatively different from each other. Entropy is a variable that is intrinsically statistical. The occupational probability associated with a particular state under consideration is an integral part of its definition. As such there is no reason to discuss the entropy of a single configuration whereas determining the amount of disorder associated with that very same configuration does constitute a significant problem.

If the specific configurations of the system exhibit disorder then the examination of the statistical ensemble of the systems will allow us to calculate a *finite* entropy which, in turn, will give us the number of configurations that are compatible with the degree of disorder.

Since a fluid is normally in a state of thermodynamic equilibrium it is possible to calculate its thermodynamic properties using methods that, though complicated, belong to equilibrium statistical mechanics.

The remaining non-crystalline matter (ordinary glasses and metals, systems with frozen spins, quasicrystals ...) *cannot* be found, normally, in thermodynamic equilibrium since it is obtained through fast quenching techniques. The atomic motion is frozen during the fast cooling of the system so that the atoms do not have enough time to reach configurations of equilibrium at the final temperature. Fast cooling, in fact, does not inhibit thermal motions: the atoms vibrate around their positions just as, in a magnetic system, the spins can be realigned. However, the thermal excitation is not sufficient to cause the redistribution of atom positions. The system is thus metastable, in that the atom positions coincide with *relative* minima of the free energy, unlike the lattice positions, which correspond to absolute minima. Thus, an amorphous system is subject to atomic relaxation motions towards equilibrium configurations, with very low atomic re-arrangement rates.

It is not immediately clear how to apply thermodynamics or statistical mechanics to these classes of supercooled systems. One approach consists in dividing the variables, used to describe the microscopic state, into two groups, $\{C\}$ and $\{T\}$. The first group comprises the instantaneously frozen variables, namely those that specify the atomic sites, but do not contribute to thermal motion, whereas the $\{T\}$ variables are necessary to describe the thermal motions, such as vibrations and alignment.

$p\{C\}$ is the probability factor that the variables in the set $\{C\}$ will take on certain values and it is imposed a priori. It describes the system and depends on the method chosen to prepare it. For each $\{C\}$ we can calculate any thermodynamic quantity using the $\{T\}$ variables. The quantity under examination is then turned into an average over the distribution $p\{C\}$ of the frozen variables.

3. The Glass Transition

3.1 The Phenomenology of Glass Transition

We shall now go on to examine some of the fundamental aspects of liquid-glass transition and the glassy state of matter. We shall pay particular attention to the physical ideas in order to point out those conceptual steps any glass transition theory has to follow.

It is commonly observed that when a liquid is cooled to below the melting point, T_{m}, at a constant pressure it crystallises through a process of nucleation and growth. If the cooling rate $-(\mathrm{d}T/\mathrm{d}t) = -\dot{r}$ is fast enough the liquid will supercool within a temperature range where the processes of atomic redistribution that characterise crystallisation will take place ever more slowly. In the end, below the glass transition temperature T_{g}, the macroscopic effects of atomic redistributions can no longer be seen and the liquid looks frozen, just like a disordered solid off thermodynamic equilibrium, namely a glass. Conversely, if the transformation path is reversed, the glass is progressively heated until, at T_{g}, it melts, just as happens with a crystal when it is brought up to the T_{m} temperature.

Many physical properties remain almost constant, or show slight changes when a crystal liquefies. For example, the distance between first neighbour atoms hardly changes, the thermal expansion coefficient increases, but still has around the same value as that of the crystal, the heat capacity changes slightly and the volume increases, typically between 2% and 5% for most materials. What immediately makes a solid stand out from a liquid is, from a macroscopic viewpoint, its ability or not to resist shear stresses, namely the *fluidity*. Thus we can easily find T_{g} within the narrow temperature range where the measured shear viscosity η reaches the typical value of 10^{12} Pa s. This viscosity rises rapidly as the system approaches the transition from the high temperature side. The changes in η as the temperature varies are schematically shown in Fig. 3.1 for a metallic system or an ionic compound. The parallel branches (1) and (2) correspond to different cooling rates, where $|\dot{r}_2|$ is less than $|\dot{r}_1|$, and clearly demonstrate that hysteresis occurs. Typically, $\Delta T_{\mathrm{g}}/T_{\mathrm{g}}$ can vary by about 10% of T_{g}, given as an absolute temperature, when we vary the cooling rate by several orders of magnitude. Conventionally, T_{g} is defined as the temperature where, during heating tests carried out at a rate

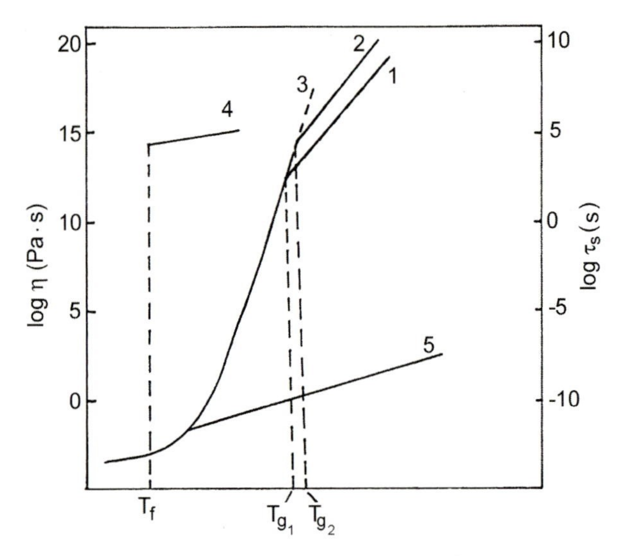

Fig. 3.1. Viscosity trend as a function of the temperature in a metallic system in the solid-liquid transition region. Curves 1, 2: glassy system, obtained with various cooling rates ($|\dot{r}_2| < |\dot{r}_1|$); curve 3: ideal glass; curve 4: liquid-crystal transition; curve 5: computer simulated quenching process from the liquid

of 0.167 Ks^{-1}, we can see an increase in the slope of the specific heat curve at constant pressure, c_p, as a function of the temperature.

The fall in T_g as $|\dot{r}|$ falls confirms that a structural relaxation time, τ, which is representative of microscopic relaxation, depends on temperature. The inverse of τ indicates the rate at which the atomic structure of the system reacts and adapts to the externally set change in temperature. We should note that for a liquid-glass transition, within a typical interval of 200 K, τ varies with continuity from 10^{-13} s to 10^{17} s with values of around 10^3 s at T_g, which correspond to about one atomic jump an hour.

The importance of kinetic factors in the glass transition emerges from the fact that branches (1) and (2) tend to converge on the ideal glass curve, as schematically shown in (3), provided we wait long enough, and even if this trend is never completed.

On the other hand, the dependence of T_g on $|\dot{r}|$, though still not very large, is in any case too marked to indicate that, in the glass transition, kinetic factors alone are the only cause of the change in T_g as the cooling rate changes.

When the temperature is below T_g the configurations of the system do not change appreciably; as such the two states (1) and (2) are defined as isoconfigurational. Curves (4) and (5) respectively schematically show crystallisation and a typical computer simulation of a rapid quenching process.

The viscosity trend shown in the figure is observed in glassy systems where the bonds are *non-directional* (metals, ionic solids or van der Waals solids) and cannot maintain a given medium range order as the temperature varies. Small thermal excitation can produce various structures that compete energetically and fluctuate among many states of particle coordination and orientation. The viscosity of these "fragile" systems is well expressed by the phenomenological relation

$$\eta = \eta_0 \exp\left[\frac{B}{T - T_0}\right]; \tag{3.1}$$

this is called the Vogel–Fulcher–Tammann (VFT) equation. A semi-logarithmic plot of (3.1) as a function of the reduced temperature T_g/T has a hyperbolic trend (Fig. 3.2), with η_0 and B constants. For simple liquids, such as metals, B is relatively small, between $T_m/3$ and $2T_m/3$. Temperature T_0, where η diverges, is called the ideal glass transition temperature. If T_0 is zero, then (3.1) becomes an Arrhenius equation. In this case the constant B is equal to E/k_B and is the activation barrier. If T_0 is positive then the viscosity dependence on temperature is non-exponential, and where $T = T_0$ the relaxation time τ diverges. At any given temperature non-Arrhenius relaxation processes are characterised by an apparent activation energy

$$E = k_B \frac{d(\ln \tau)}{d(T^{-1})}. \tag{3.2}$$

The values of the apparent activation energy for fragile liquids around T_g can reach 500 kJ mol^{-1}. Under these conditions a small temperature change is expected to provoke changes in the system dynamics of 1 decade. A typical temperature change is not greater than 5 K.

It should be noted that viscosity is continuous at T_g (though discontinuous in the case of crystallisation) whereas its singularity at T_0 cannot actually be reached under experimental conditions.

For those "strong" systems (silicate and chalcogenide glasses, elemental semiconductors) that have a network of strongly directional covalent bonds, thus with relevant medium range order (see Sect. 4.8) which survives within the glass transition region, the trend in η is exponential, $\eta = \eta_0 \exp[C/T]$, as seen in Fig. 3.2.

The influence of medium range order on the glass transition has recently been clarified in the exemplary case of the fragile liquids of the family of lithium metaborate, $LiBO_2$, whose structure was examined using infrared and Raman spectroscopies. When the liquid is solidified at 723 K, well above T_g (693 K), $\alpha-LiBO_2$ is obtained and the boron atoms are threefold coordinated. At 683 K, within the glass transition, $\beta-LiBO_2$ nucleates with boron atoms both threefold and fourfold coordinated. Lastly, if the glass is slowly re-crystallised below 673 K the product becomes $\gamma-LiBO_2$ where all the boron atoms are fourfold coordinated. The changes in the coordination number between the crystalline phases reflect the changes that occur in the supercooled liquid during the glass transition. Why the borate systems are so

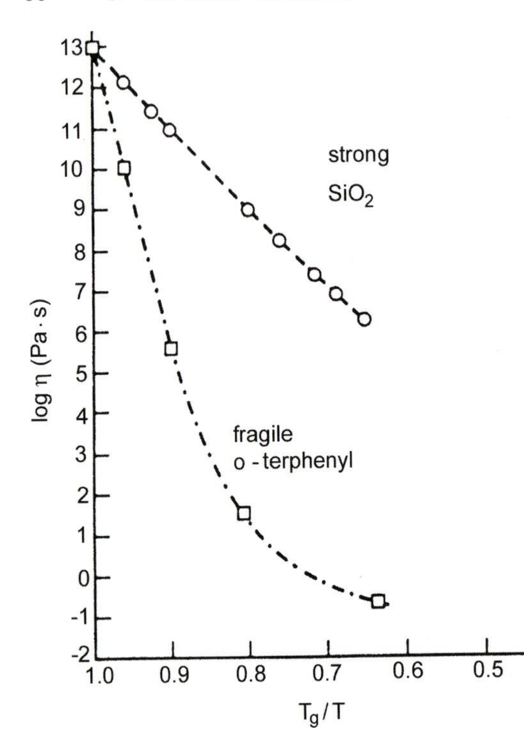

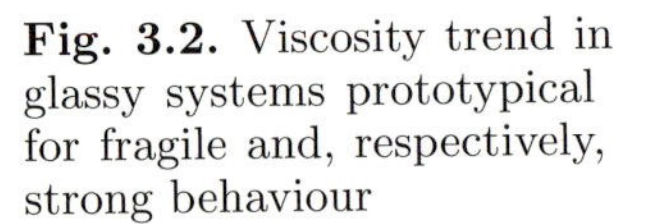
Fig. 3.2. Viscosity trend in glassy systems prototypical for fragile and, respectively, strong behaviour

extremely fragile can be explained if we consider that boron can hardly form π bonds with fourfold coordination, so the medium range order networks can only form at temperatures well below the normal crystallisation temperature.

The special dependence on temperature, as given by the VFT law, whose physical origin is somewhat obscure, is typical also of other transport phenomena in the liquid state, including ionic conduction and atomic diffusion rates. This dependence cannot be obtained by simply starting from an Arrhenius type relation, neither by using suitable energy averages nor from a specific dependence of energy on temperature.

The cooling rate required for a material to transform into a glass is a critical process parameter. This rate spans at least eleven orders of magnitude and depends on the nature of the atomic bond which, in turn, determines the microscopic structural rearrangement rate needed to maintain equilibrium during cooling. In the silicate glasses and the organic polymers the strongly directional bonds allow the system to transform into a glass easily, with $|\dot{r}|$ values around $10^{-2}\,\mathrm{Ks^{-1}}$.

At the other end of the spectrum the liquid metals, with non-directional bonds, may present lower $|\dot{r}|$ values of around $10^3\,\mathrm{Ks^{-1}}$, though they frequently require cooling rates above $10^9\,\mathrm{Ks^{-1}}$. As a consequence, the thickness of the metallic glasses varies from $10^2\,\mu\mathrm{m}$ to less than $0.1\,\mu\mathrm{m}$. The research for metallic systems that can easily transform into the glassy state has recently shifted the lower cooling rate threshold required to inhibit the crystallisation of a liquid alloy to around 10 Ks^{-1}, a value close to those observed in silicate glasses. Not only has this research opened up the way to the preparation of massive amorphous alloys, it has also clarified some of the mechanisms of glass formation and the concept of easy glass formation, too.

The clearest example of the role played by the cooling rate in glass formation is given by the difficulty to transform *pure* metals into glass. Simulation techniques have shown that quenching rates are required in the range of $10^{11}\,\mathrm{Ks}^{-1}$ for metals such as Y, Zr or Rb. By evaporating thin films on cold substrates, where $|\dot{r}| \simeq 10^{13}\,\mathrm{Ks}^{-1}$, we can obtain structurally disordered samples of many transition metals, such as Ni and Mo. However, it is extremely difficult to obtain pure films, particularly as traces of gaseous contaminants are present and these act as stabilisers of the glassy phase.

If we consider that glass is formed by solidification without crystallisation we can then wonder about the nature of the process and the product obtained from the transformation. First of all, for a liquid that has been supercooled just below T_{m}, the structural re-arrangement processes are not significantly different from those just above T_{m}. If the crystalline phase nucleates then the system will be frozen as a crystal below T_{m}. However, while cooling the liquid, this nucleation can be inhibited, for example by kinetically reducing the probability of creating and maintaining the crystalline atomic arrangement. A point will be reached where the experimental times are too short, compared to those required by the supercooled liquid to explore those regions of the configuration space that correspond to structural changes. It is clear that since there are so many possible configurations and rearrangements some of them will require such little energy that we cannot completely disregard them. However, for $T < T_{\mathrm{g}}$, extraction of thermal energy from the system leads to a reduction in vibrational atomic motions *without* any structural changes. This is like saying that for $T < T_{\mathrm{g}}$ the time scale for vibrational motion and the time scale that leads to structural changes depart from each other.

This is the microscopic representation that corresponds to the drastic change in the fluidity of the system at T_{g}. Glass only explores a narrow region of the phase space that corresponds to thermal vibrations, given that atomic rearrangement is impossible or extremely improbable. The accessibility hypothesis (see Sect. 2.3) is no longer valid and the time average of a property, observed on a reasonable experimental time scale, *does not* coincide with the corresponding ensemble average. The system transforms from being ergodic, in the liquid state, to being non-ergodic, in the glassy state.

We can estimate that at T_{g} the time constant τ for structural rearrangement is around the value for the experimental time constant, τ_{exp}, and that this value is connected to the cooling rate $|\dot{r}|$ as

$$\tau_{\mathrm{exp}}\,|\dot{r}| = k_{\mathrm{B}} T_{\mathrm{g}}^2 / E_{\mathrm{a}} \qquad (3.3)$$

where E_a is the apparent activation energy required to structurally relax the supercooled liquid. For most metallic glasses $T_g \simeq 700$ K and $E_a \simeq 4$ eV, so that $(k_B T_g^2/E_a) \simeq 10$ K.

Figure 3.3 schematically shows the trends of free energy, G, specific heat at a constant pressure, c_p, and volume, V commonly observed in the glass transition region. We should notice that the changes in G and V are continuous whereas c_p has an extra positive contribution Δc_p between the values for the glass and the values for the liquid (it is this that allows us to define T_g). Moreover, we can observe a positive discontinuity in the isothermal compressibility K_T, whereas the discontinuity $\Delta\alpha$ of the isobaric thermal expansion coefficient may be either positive or negative.

Bearing in mind that $-S = (\partial G/\partial T)_p$, then the trends in the free energy G for a crystal and a liquid (see Fig. 3.3) are clear since both systems are thermodynamically in equilibrium. The curves for the amorphous solid are dashed (the one with lowest G values pertains to the ideal glass) since the amorphous system is not in equilibrium, as proven by the existence of structural relaxation processes. The glass can at the most transform to a supercooled liquid, which is in internal equilibrium.

The two trends given for c_p at the glass transition are typical of the way the two classes of glass behave. In "strong" (curve a) glasses c_p may be mea-

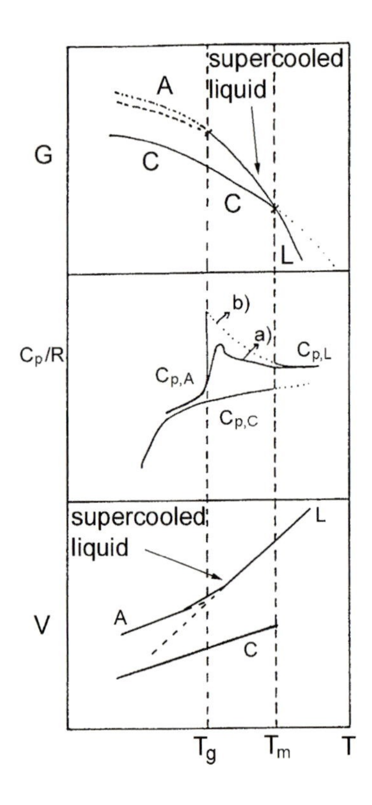

Fig. 3.3. Trend in free energy G, specific heat at constant pressure, c_p and volume, V for crystal (C), amorphous solid (A) and liquid (L) as functions of the temperature, with particular regard to the region between the glass transition temperature, T_g, and the melting temperature, T_m

sured with continuity throughout the critical field. In "fragile" glasses (curve b) part of the curve for the liquid phase is dashed and is an extrapolation of values between just above T_g and just below T_m. Indeed, a little after having exceeded these limits, the system crystallises owing to the low rate of temperature change required to correctly measure c_p.

In both cases the value of the specific heat, which is due to vibrational atomic motions, is similar in the crystal and in its corresponding amorphous solid. The fast decrease from c_p values typical of a liquid to the values for a crystal, generally between 50% and 100% of the vibrational contribution to c_p, indicates that the degrees of freedom typical of liquid have become kinetically inaccessible. It is important to note that when a crystal melts at T_m there is latent heat (c_p diverges), whereas when a glass melts, whether strong or fragile glass, there is no latent heat.

Starting from the trends of the free energy in Fig. 3.3, and supposing that the glass transition is II order, then by the Ehrenfest conditions, at T_g, the entropy values of the liquid, S_l, and of the glass, S_g, coincide with each other thus

$$\left(\frac{\partial S_l}{\partial T}\right)_p \mathrm{d}T + \left(\frac{\partial S_l}{\partial p}\right)_T \mathrm{d}p = \left(\frac{\partial S_V}{\partial T}\right)_p \mathrm{d}T + \left(\frac{\partial S_V}{\partial p}\right)_T \mathrm{d}p.$$

If we bear in mind that the definition of c_p is

$$c_p = T\left(\frac{\partial S}{\partial T}\right)_p = -T\left(\frac{\partial^2 G}{\partial T^2}\right)_p$$

and that of α is

$$\alpha = \frac{1}{V}\left(\frac{\partial V}{\partial T}\right)_p$$

and that Maxwell's thermodynamic relation

$$\left(\frac{\partial S}{\partial p}\right)_T = -\left(\frac{\partial V}{\partial T}\right)_p$$

holds, then, for any variation in T_g, as a function of pressure p, we obtain

$$\frac{\mathrm{d}T_g}{\mathrm{d}p} = \frac{TV(\alpha_2 - \alpha_1)}{(c_{p_2} - c_{p_1})} = TV\frac{\Delta\alpha}{\Delta c_p}. \tag{3.4}$$

Likewise, the continuity condition for volume V at the transition, taking into consideration that the isothermal compressibility K_T is

$$K_T = -\frac{1}{V}\left(\frac{\partial V}{\partial p}\right)_T ,$$

gives

$$\frac{\mathrm{d}T_\mathrm{g}}{\mathrm{d}p} = \frac{\Delta K_\mathrm{T}}{\Delta\alpha}. \tag{3.5}$$

Equation (3.4) has been experimentally demonstrated whereas (3.5) is only valid for computer simulations of the glass transition. For real glasses, on the other hand, the measured value for ΔK_T is always too large.

A comparison between the two relations implies that the glass transition is *not* a simple II order phase transition. On the other hand, in any thermodynamic transition that is controlled by a single order parameter, apart from temperature and pressure, Prigogine and Defay demonstrated that the ratio

$$R_\mathrm{PD} = \frac{\Delta K_\mathrm{T}\,\Delta c_p}{TV\,(\Delta\alpha)^2} \tag{3.6}$$

has value unity.

In any system that undergoes a glass transition R_PD is greater than one, usually between 2 and 5; at least two order parameters are required to study this problem, which, thus, is thermodynamically non-equilibrium. A description is possible with a single order parameter only if the system is considered non-homogenous with respect to both the density and the composition.

We note in Fig. 3.3 that the specific volume of glass does not depend on the temperature alone, since it is continuous at the transition, but also on the formation pressure. In turn, entropy S is continuous; this means that at T_g the free energy surfaces of the glassy (g) and liquid (l) phases are tangent. This, together with the fact that free energy for the liquid, G_l, is lower than for the glass, G_g, means that the intensive variables thermal expansion coefficient $\alpha = (\partial \ln V/\partial T)$, isothermal compressibility $K_\mathrm{T} = -\left(\partial \ln V/\partial p\right)_T$ and specific heat at constant pressure $c_p = -T\left(\partial^2 G/\partial T^2\right)_p$ all show discontinuity at T_g, since they are second order derivatives of Gibbs free energy G.

However, we will also notice that whereas this behaviour seems very similar to what is observed in II order thermodynamic phase transitions, in this particular case the discontinuities are not sharp, but appear to be *distributed* over a small temperature interval. Hence, for example, the round part of the curve $V(T)$ that marks the interval where the transition takes place (the single value of T_g is a graphic extrapolation obtained using the tangent method) corresponds to a narrow step rather than to a vertical discontinuity.

Configurational entropy S_conf, both finite and positive, and which persists until $T = 0$ K, is associated with the amorphous state. This is because in thermodynamic treatment of the system we have to consider the set of "liquid" configurations that are instantaneously frozen at T_g. It is the very existence of S_conf that does not allow us to represent a glass as a specific instantaneously frozen microscopic configuration. On the other hand, this "ill defined" state of matter may be considered as a statistical ensemble of different atomic configurations; it is the statistical weight of these configurations that freezes at T_g. This is the representation of a *liquid* instantaneously frozen in a rigid, isoconfigurational state.

As a consequence, a thermodynamic glass phase is not defined since the supercooled liquid deviates from the (meta)stable equilibrium line at the glass transition. This is a non-equilibrium transition; it connects two different metastable conditions together, i.e. the supercooled liquid and the glass. The crystalline phase is the equilibrium state underlying both liquid and glass, and it is kinetically impeded.

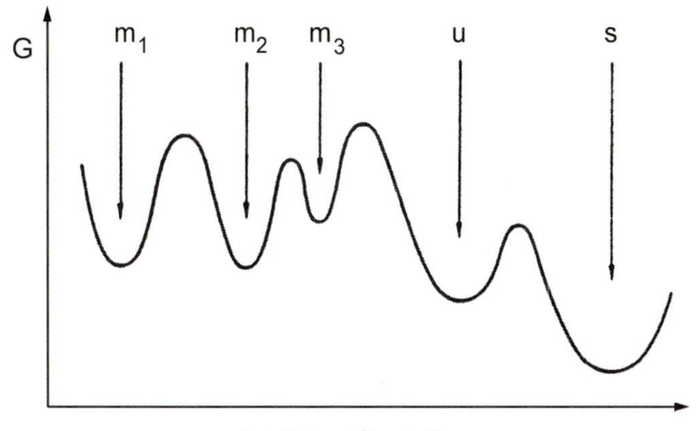

Fig. 3.4. Schematic free energy trend for a system, for metastable (m), unstable (u) and stable (s) atomic configurations

Figure 3.4 compares the free energies for a frozen state, an unstable state and a stable state. The difference between unstable and frozen states is essentially given by the height of the energy barrier in relation to the available thermal energy $k_B T$. Considering that small changes in temperature, or in pressure, induce reversible changes in the properties of the glass, the glassy state may, in practice, be considered stable, namely its properties do not change over time. The existence of amorphous solids from the Moon, frozen and undeformed over billions of years, allows us to use the concept of *false equilibrium*, which only exists because of the extraordinary slow rate of structural relaxation processes of glass. The trend in shear viscosity in Fig. 3.1 backs up this interpretation: the glass transition coincides with the point where the viscous flow of the liquid is inhibited and the value of η diverges.

If we consider liquid cooling at a microscopic level, the increase in viscosity implies an abrupt slowing down of the diffusive atomic motions. When the position of the atoms changes there is a change in the microscopic density. The microscopic shear viscosity η and the microscopic self diffusion coefficient, D, are connected by the Stokes–Einstein relation

$$\eta = \frac{k_B T}{3\pi D \langle x \rangle} \tag{3.7}$$

where $\langle x \rangle$ is the average interatomic distance. The metallic glasses, above T_g, follow (3.7) very well, given the heavy extrapolation required in order to move from the macroscopic scale to the atomic scale.

A realistic representation of the condition of a liquid during cooling is that the continual structural rearrangement required to maintain equilibrium as the temperature decreases favours low energy configurations; these are characterised by increasing short range order. The kind of local order that develops depends critically on both the geometrical hindrances, such as shape and dimensions of the involved structural units, and the nature of interatomic forces. Thus, whereas it is generally difficult to predict the characteristic properties of short range order we can presume that, for a given class of material, e.g. metallic glasses, the short range order is of the same kind.

The relaxation time τ_s for the structural relaxation modes gives us the kinetic evolution for structural changes. The divergence in τ_s (see Fig. 3.1) is connected to the divergence in η, which measures the reaction of the liquid to a suddenly imposed shear stress, through Maxwell's relation

$$\tau_s = \frac{\eta}{G_\infty} , \tag{3.8}$$

where G_∞ is the high frequency shear module. For easy glass forming materials, $G_\infty \simeq 10^{10}\,\mathrm{J\,m^{-3}}$ and, approaching T_g, $\eta \simeq 10^{12}\,\mathrm{Pa\,s}$, so $\tau \simeq 10^2\,\mathrm{s}$. As

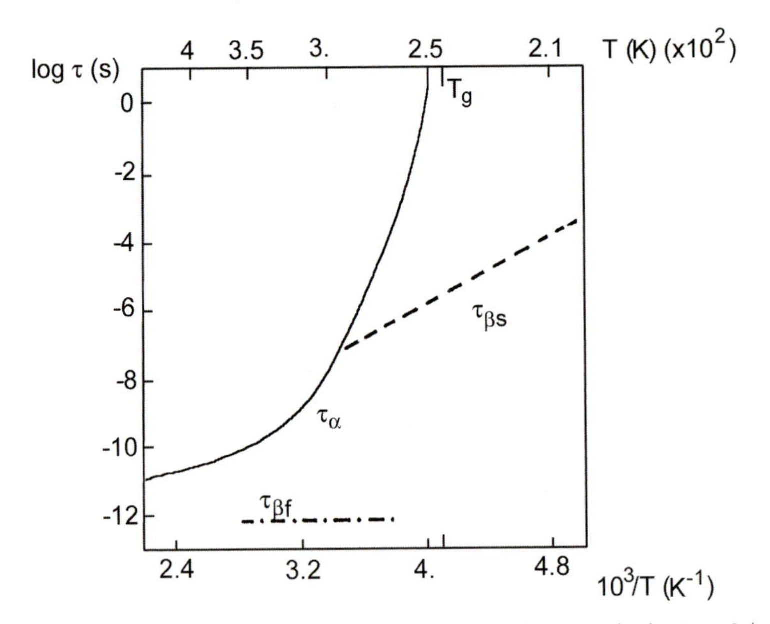

Fig. 3.5. Schematic trend in relaxation times due to α (τ_α), slow β ($\tau_{\beta s}$) and fast β ($\tau_{\beta f}$) processes in *o*-terphenyl, as functions of temperature

we get closer to T_g the behaviour of τ is dictated by (3.8) and explains the kinetic dependence of the glass transition on the cooling rate.

The critical slowing down of the structural relaxation modes is responsible for the separation between the temporal evolution of the fast vibrational modes, which decay well above T_g at the normal value of 10^{-13} s and that of slow configurational modes.

One feature of supercooled liquids with high viscosity is that the temporal dependence of structural relaxation is very different from the exponential trend typical of visco-elastic media. We observe a *distribution* of relaxation times, which broadens as the temperature falls. This feature is shown in Fig. 3.5 for the prototype fragile liquid *o*-terphenyl: the slowest relaxation process, called α, is related to molecular rotation. Secondary relaxation processes occur over shorter time scales; the slow β processes, due to partial reorientation of the molecules, differ from various fast β processes and it is still unclear what mechanisms they should be attributed to. The slowest process gives the dominant contribution to the average relaxation time, $\langle\tau\rangle$. The glass transition is located where $\langle\tau\rangle$ is equal to or greater than reasonable experimental times (for relaxation experiments this is a few minutes at the most).

A fingerprint that the structural fluctuations in the glass have become *partly* static is given by the appearance of an elastic peak in the dynamic

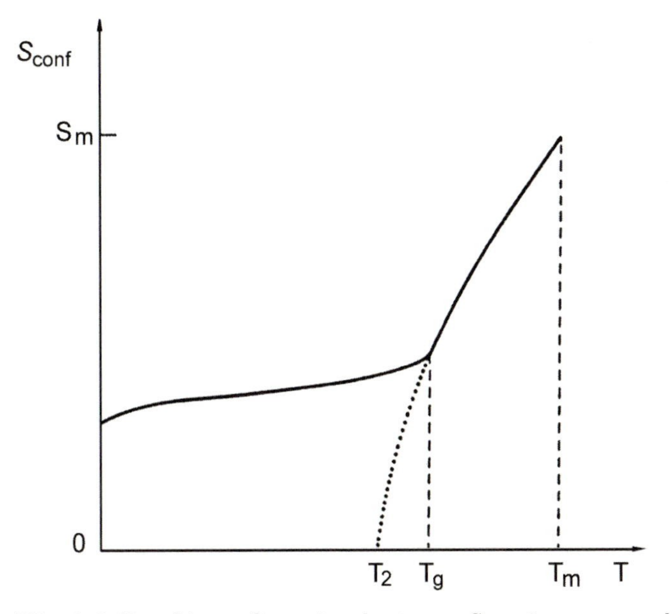

Fig. 3.6. Trend in configurational entropy, S_{conf}, in a supercooled liquid and in the corresponding glass. Notice the value of residual entropy at $T = 0$ K in the glass; extrapolation of the curve for the liquid leads to identify the ideal glass transition temperature T_2

structure factor measured by neutron scattering. Nevertheless, below T_g the slowest structural relaxation modes survive and are gradually frozen. As a consequence we observe relaxation in glasses, too.

So far we have dwelled on the presence of kinetic aspects in the glass transition. If, though, we examine the thermodynamic quantities, the development of structural relaxation in a cooling liquid has an obvious counterpart in entropy behaviour. The excess entropy trend in a liquid, as compared to a crystal (Fig. 3.6), demonstrates that there is a residual entropy associated with the glassy state whose value remains nearly constant from T_g to $T = 0\,\mathrm{K}$. This result does not in any way violate the third law of thermodynamics since the system is not in internal equilibrium. This implies that most configurational disorder in the liquid, which distinguishes it from a crystalline solid, does remain frozen in the glass. If the entropy data for the supercooled *liquid* are extrapolated below T_g, within the region of ever increasing supercooling, and ever lower cooling rates $|\dot{r}|$, then we obtain a transition characterised by an entropy crisis. This region, between T_2 and T_g in Fig. 3.6, is experimentally inaccessible. Temperature T_2, which is lower than T_g, though close to it (for polymer glasses the difference between T_2 and T_g is around 50 K, and, in any case, well above absolute zero, is a point where there is coincidence between the entropies S_l of the liquid and S_c of the crystal.

Apart from vibrational and mixing contributions, liquids have a configurational term S_{conf}; this takes into account any possible configurational changes that represent the distinctive property of the liquid:

$$S_{conf} = S_l - S_c \ . \tag{3.9}$$

At T_2, S_{conf} disappears much the same as what occurs in a second order phase transition. At temperatures below T_2, S_{conf} is negative. This result is paradoxical since the liquid has to be found at least in a configuration; this analysis, developed by Kauzmann and which really is a paradox since it would violate the third law of thermodynamics, is called the Kauzmann paradox.

It has been observed that, within the limits of available experimental data, the ideal glass transition temperature T_0, where we can observe divergence in the transport properties (see (3.1)), corresponds to T_2. This correspondence should constitute an indication of thermodynamic glass transition, which however should be partially concealed by kinetic factors. Indeed the dependence of relaxation times associated with the various transport properties (as described by (3.1)) on temperature implies that all the processes above T_0 would freeze. The singularity is inaccessible. However, in some bulk metallic glasses we observe a discrepancy between values of T_0 and the corresponding values of T_2. Furthermore, one of the most reputed glass transition theories, namely the mode coupling theory (see Sect. 3.2) gives a purely kinetic representation of vitrification of a system and does not allow for any thermodynamic transition phase.

It is important to note that, apart from the theoretical interest in the Kauzmann paradox, the cooling rates required to shift the glass transition from T_g to T_2 (a typical difference is a few tens of degrees) can be evaluated starting from the activation energy for atomic motions (highly limited by viscosity) around T_g. The required cooling time would be in the order of thousands of years. It is doubtful that the system would not crystallise if it were kept in the transition region for such a long time. The disappearance of S_{conf} at T_2 implies that, even if we had infinite time and we could work with a material with an infinitely high crystallisation barrier, it would not be possible to reduce T_g indefinitely; to avoid the absurd condition of a liquid phase with entropy below that of the stable crystalline phase, at T_2 the liquid phase cannot exist and glass transition must occur.

Temperature T_2 is the lowest threshold for the T_g value. This means that thermodynamic factors set a limit to any process kinetics. On the other hand, since the residual entropy term in glass is small and in any case not much greater than crystal entropy, the liquid-glass transition must occur quickly enough for the entropy curve to take on the shape given in the graph in Fig. 3.6. This corresponds to a sharp step, as shown in Fig. 3.3 for the c_p trend at the transition.

These thermodynamic conditions merge on the idea that T_g does not signal glass transition but its final point. The transition from liquid to amorphous solid occurs over a wide temperature interval by way of a progressive loss in internal degrees of freedom in the system. T_g coincides with the prevailing slow kinetics which brings the solidification process to an end. Further energy release from the system would only lead to cooling, namely a change in the vibrational atomic dynamics, without causing any structural effects.

3.2 Theories of the Glass Transition

Analysis of the above examined phenomena shows that kinetic and thermodynamic features are present in the glass transition and that they are interwoven. For this reason it is extremely difficult to put forward a theoretical representation of the transition. This is one of the most complicated problem in solid state physics. As such, this problem has not yet been satisfactorily resolved. Those models, often phenomenological, that have been developed have put much light on the trend of (kinetic) relaxation phenomena or, respectively, on thermodynamic quantities; only recently have theories been put forward that, though complicated, merge relaxation and thermodynamic aspects in a consistent way.

We are now going to examine the phenomenological theory of free volume, both because of its historical importance and because of its pictorial simplicity. We will then move to a thermodynamic theory developed in the framework of scaling, and finally we will give an introduction to the theory of mode coupling.

The *free volume theory* was originally developed for a fluid of hard spheres which simulate the molecules in a liquid. The molecules vibrate due to thermodynamic effects in cages that correspond to Voronoi polyhedra (see Chap. 4). The total volume V_l of the liquid is divided into one part that is occupied by the spheres, V_0, and one part of free volume, V_L, where the spheres can carry out diffusive motion. It has been postulated that transport takes place only if the sum of all voids, V_L, is greater than a critical value, $V_{L,c}$. Furthermore, since the free volume consists of variable sized voids, whose distribution is the result of random molecular motion, there are no local concentration variation terms, which would otherwise lead to space rearrangement of the voids. Space concentration of free volume is thus uniform.

As the temperature falls, both the occupied volume and the free volume contract.

The liquid is different from a glass because in the glass V_L is independent of temperature and is not redistributed, being frozen in the set of positions it occupies at glass formation. Glass transition occurs if $V_L < V_{L,c}$.

It is interesting to estimate the fraction of free volume at T_g. This can be achieved if we remember that both the free volume V_L and the occupied volume V_0 contribute to the heat expansion coefficient of the liquid α_l whereas the heat expansion coefficient of glass, α_g, only depends on V_0. As such, $\Delta\alpha = \alpha_l - \alpha_g$ is the heat expansion coefficient for the free volume V_L. If we take the total volume V_g of the glass at T_g as our reference, the free volume V_L at a temperature T above T_g is

$$V_L = V_{L,g} + V_g \, \Delta\alpha(T - T_g) \; .$$

.

Now, if we equalise V_l, extrapolated at $T = 0$ K, with V_0, again at $T = 0$ K, we obtain

$$V_0(T = 0\,\mathrm{K}) = V_g(1 - \alpha_l T_g).$$

The volume of the glass at $T = 0$ K is $V_g(1 - \alpha_g T_g)$, so we can assume that, for the free volume of the glass,

$$V_{L,g} = V_g(1 - \alpha_g T_g) - V_g(1 - \alpha_l T_g) = V_g T_g \Delta\alpha \tag{3.10}$$

and so we obtain

$$\frac{V_{L,g}}{V_g} = T_g \Delta\alpha. \tag{3.11}$$

The product $T_g\Delta\alpha$, measured for polymer and chalcogenide glasses, varies between 0.08 and 0.13. Likewise, the estimated fraction of free volume at T_g is around 10% of the total volume. The fluidity depends on the probability, P, that volume fluctuations occur locally and will cause a change in the position of an atom. The probability, P, that a volume fluctuation is greater than ΔV is

$$P(\Delta V) = \frac{V}{V_{\mathrm{f}}} \exp\left[-\gamma \frac{\Delta V}{V_{\mathrm{f}}}\right] \tag{3.12}$$

where γ is a geometrical factor and V_{f} the specific free volume, so $V_{\mathrm{f}} = V_{\mathrm{L}}/N$, where N is the number of atoms.

If we postulate that the fluidity is proportional to the number of fluctuations with amplitude greater than a critical value V^*, then for diffusivity D we obtain the following

$$D = g\, a^* \left(\frac{3k_{\mathrm{B}}T}{m}\right)^{1/2} \exp\left[-\frac{\gamma V^*}{V_{\mathrm{f}}}\right] \tag{3.13}$$

where g is a geometric constant, a^* is in the order of an atomic diameter and m is the atomic mass.

When we match (3.12) with the Stokes–Einstein equation

$$D = \frac{k_{\mathrm{B}}T}{3\pi a \eta}$$

where a is the diameter of a diffusing atom, the viscosity is given by

$$\eta = \frac{1}{3\pi g a^{*2}} \left(\frac{m k_{\mathrm{B}}T}{3}\right)^{1/2} \exp\left[\frac{\gamma V^*}{V_{\mathrm{f}}}\right] \tag{3.14}$$

In order to agree with the experimental data the product $\gamma V^* \simeq V$ in (3.14) must be around the atomic volume.

From (3.14), keeping the volume constant, the viscosity η varies with $T^{1/2}$. The model predicts a *continuous* increase in η with the temperature.

If we keep the pressure constant instead of the volume, then thermal expansion must be considered, and the dominant temperature dependence is given by V_{f}; in practice (3.14) becomes the Vogel–Fulcher–Tammann equation.

The theory considers the high levels of entropy associated with a liquid by introducing the cumulative entropy, starting from a crystal where each atom moves in a potential finite square well such as the Kronig–Penney one. The entropy S for each atom is

$$S = k_{\mathrm{B}} \ln V_{\mathrm{s}} \tag{3.15}$$

where V_{s} is the ratio between the volume of the cell the atom can move in and the volume V_{a} of the atom itself. For a mole of atoms ($N = N_{\mathrm{Av}}$)

$$S^{\mathrm{s}} = R \ln V_{\mathrm{s}} \tag{3.16}$$

where R is the gas constant. In the simplest model a liquid is characterised by the possibility for each atom to move not only in its own cell, but within the whole volume. The associated entropy thus becomes

$$S^{\mathrm{l}} = k_{\mathrm{B}} \ln \left[\frac{(NV_{\mathrm{s}})^N}{N!} \right] \tag{3.17}$$

where the $N!$ term is due to the fact that when two identical atoms exchange their positions the entropy does not change. By applying the Stirling formula, the difference

$$S^{\mathrm{l}} - S^{\mathrm{s}} = k_{\mathrm{B}} \ln \left[\frac{N^N}{N!} \right] \tag{3.18}$$

is R.

This term is the *cumulative entropy* and is present in a liquid due to the complete atom delocalisation throughout the available volume.

The model, in its simplicity, supplies us with an interesting picture of the liquid, though it has some severe limitations.

From the experimental point of view it is known that when $T \simeq T_{\mathrm{g}}$ a supercooled liquid has lost almost all its excess entropy as compared to a crystal; the model does not take this into consideration since the cumulative entropy is independent of temperature. Furthermore, it is difficult to introduce a mechanism for the gradual reduction in cumulative entropy; a simple reduction in atomic mobility is indeed useless since changing the mobility is just like changing the time scale, thus leaving the thermodynamic properties unchanged.

A revised version of this theory considers the atoms in motion within cells formed by the surrounding atoms, but introduces liquid cells larger than some critical size. The potential profile within such cells is flatter and allows wider atomic fluctuations. A free volume is only associated with the liquid cells. The essential feature of this version of the model is that the liquid cells can form agglomerates.

If we use the percolation theory, then above a critical fraction of agglomerates in the volume of the system some agglomerates, which were initially separate, coalesce to form a single infinite agglomerate, i.e. large enough to span the entire system.

The solid-liquid transition coincides with the formation of the infinite agglomerate. Since all the atoms in the same agglomerate have the same free volume then the model is able to provide for a gradual increase in cumulative entropy.

Since the agglomerates have greater entropy, due to the cumulative term contribution, they are favoured at high temperature; on the other hand, since their energy content is greater they are less advantaged at low temperatures. The model provides for the amorphous solid melting through forming drops of liquid agglomerates whose sizes will grow as the temperature increases.

The drawback to the free volume theory is that the liquid-glass phase transition is first order. However, this kind of transition can always be inhibited by slowing down the kinetics; thus it should be possible to keep the liquid supercooled at temperatures below T_g provided the cooling rates are high enough, which is contrary to any experimental results.

In the framework of the standard theory of *scaling* at a second order transition we may consider the glass transition as being a classical continuous transition, with dynamic (divergence) and static (anomalies in the specific heat c_p, Kauzmann paradox) effects. Both these effects reflect divergence of a correlation length ξ.

By adopting the following relation for ξ:

$$\frac{\xi}{\xi_0} = \left(\frac{T - T_c}{T}\right)^{-v} \tag{3.19}$$

where the critical temperature T_c coincides with the glass transition temperature T_g, and by assuming that the ξ divergence follows a power law, we obtain an equation for the critical slowing down which includes the dynamics of a second order phase transition

$$\left(\frac{\xi}{\xi_0}\right)^p = \left(\frac{T - T_c}{T}\right)^{-vp} = \frac{\tau}{\tau_0}. \tag{3.20}$$

A Taylor expansion of (3.20) gives

$$\ln\left(\frac{\tau}{\tau_0}\right) \simeq vp\frac{T_c}{T}\left(1 + \frac{T_c}{2T} + \frac{T_c^2}{3T^2} + ...\right) \tag{3.21}$$

At high temperature, well above the critical temperature T_c, the trend of $\ln(\tau/\tau_0)$ in (3.21) is comparable to the trend obtained for $\ln(\tau/\tau_0)$ using the Vogel–Fulcher–Tammann law

$$\ln\left(\frac{\tau}{\tau_0}\right) \simeq \frac{vpT_c}{T - T_c(1/2 + T_c/3T + ...)} \tag{3.22}$$

provided we require $B = pT_c$. In the high temperature limit the two curves are thus asymptotic. Since the ratio B/T_0 may be considered constant for a given family of systems then we can use the concept of universality class: all systems in a single family are characterised by the same dynamic exponent and are part of a specific universality class.

The theory takes into account both the specific heat singularity and the Kauzmann paradox starting from the assumption that the proximity to T_g of the lower limit for the glass transition temperature T_2 is by no means accidental. Furthermore, the divergence in the relaxation times around T_g leads to underestimate T_2 since both c_p and S are underestimated. If we consider that the leading contribution to the entropy at the glass transition comes from the high temperature region ($T > T_c$), the correlation length allows us to define n renormalised pseudo-particles in d dimensions, according to the relation $n = V/\xi^d$, where ξ^d is the volume of the particle. For a gas of such particles, if we differentiate with respect to the temperature the free energy G, whose expression is

$$G \propto T\frac{V}{\xi^d} = TV\left(\frac{T - T_c}{T}\right)^{dv} \tag{3.23}$$

we obtain the entropy S and the specific heat c_p as

$$S = -A\left(\frac{T - T_c}{T}\right)^{1-\alpha}\left[1 + (1-\alpha)\frac{T_c}{T}\right] \tag{3.24}$$

where $\alpha = (2 - dv)$ and

$$c_p = \frac{A}{T^2}\left[(dvT_c)^2 - dvT_c^2\right]\left(\frac{T - T_c}{T}\right)^{\alpha} \tag{3.25}$$

with the conditions $A \geq 0$ and $T_c \leq (dvT_c)$.

In a second order transition entropy contributions are found both from the low ($T \leq T_c$) and from the high ($T > T_c$) temperature sides. Thus, there is an anomaly on both sides of T_c.

In (3.25) parameter A is the difference between the entropy at infinite temperature and the entropy at the critical temperature and gives us the entropy change in the high temperature field ($T > T_c$). When $A = 0$, there is no entropy change in the high temperature field; this is like considering a transition where any entropy change is concentrated below T_c, in analogy with the mean field approximation for ferromagnetic materials. If $A = S_{\text{conf}}$, i.e. it consists of all the excess configurational entropy in the supercooled liquid, as compared to the crystal (see (3.18)), we assume that the entire entropy change takes place in the high temperature region and that it becomes zero at T_c. This picture of the Kauzmann paradox corresponds to the freezing of *all* the degrees of freedom at a finite temperature.

A different approach to the glass transition is given by the *mode coupling theory* (MC). This theory applies to fragile materials that can transform into glass easily and can thus be cooled slowly in the transition region, so that it is possible to measure both the thermodynamic and structural quantities, the latter using neutron scattering. What makes the MC theory different to the other theories is that it is dynamic. It was developed starting from the assumption that, apart from (3.16) there is another independent relation between the coefficient of self diffusion, D, and the shear viscosity η, since the viscous flow requires diffusive atomic motions.

The MC theory has its roots in the theory that was developed to describe the dynamics of simple liquids that only need a few correlation functions, and where density fluctuations and both longitudinal and transverse current fluctuations are considered.

In the case of glass transition the density-density correlation function plays the fundamental role since the trend of η at T_g implies freezing the density fluctuations. The theory supplies us with a new relation between D and η.

We consider that the damping of the density fluctuation modes is the result of their decaying into other modes of the same kind. This is an extension of the Mori–Zwanzig formalism that describes the correlation functions for equilibrium fluctuations in simple liquids. It is necessary to determine the bare interaction potential and the static structure factor $\mathfrak{S}(\boldsymbol{k})$ for the liquid at equilibrium separately.

The density correlation function, which depends on the time and on the wavevector, is introduced for a dense, single component liquid

$$F(\boldsymbol{k},t) = \langle \partial n(\boldsymbol{k},t)\, \partial n(-\boldsymbol{k},0)\rangle \tag{3.26}$$

where $\partial n(\boldsymbol{k},t)$ is the microscopic density fluctuation with respect to the constant equilibrium value n.

The Laplace transform for $F(\boldsymbol{k},t)$ is

$$F(\boldsymbol{k},\omega) = \int_0^\infty \exp\left[\mathrm{i}\omega t\right] F(\boldsymbol{k},t)\mathrm{d}t \tag{3.27}$$

and this can be rewritten as

$$F(\boldsymbol{k},\omega) = -\mathfrak{S}(\boldsymbol{k})\frac{\omega + \boldsymbol{k}^2 m(\boldsymbol{k},\omega)}{\omega^2\left(\boldsymbol{k}^2 k_\mathrm{B}T/m\mathfrak{S}(\boldsymbol{k})\right) + \omega\boldsymbol{k}^2 m(\boldsymbol{k},\omega)}\ , \tag{3.28}$$

where $m(\boldsymbol{k},\omega)$ is the generalised viscosity that at T_c shows a singularity with the ω^{-1} dependence.

The main assumption behind the MC theory is that after a microscopic time interval, lasting about the same time as the inverse of the Debye frequency, $m(\boldsymbol{k},t)$ may be expressed as a function of $F(\boldsymbol{k}',t)$ only.

It is convenient for us to represent relation (3.26) as the equation of motion for a harmonic oscillator, including a term with a memory function $M(\boldsymbol{k},t)$ which is considered a damping function

$$\left[\frac{\partial^2}{\partial t^2} + \frac{\partial}{\partial t}\gamma(\boldsymbol{k}) + \omega_0^2(\boldsymbol{k})\right] F(\boldsymbol{k},t)$$
$$+\omega^2(\boldsymbol{k})\int_0^t M(\boldsymbol{k},t-t')\frac{\partial}{\partial t'}F(\boldsymbol{k},t')\mathrm{d}t' = 0. \tag{3.29}$$

In the elementary version of the theory there is no dependence on the wavevector and the memory function reduces to

$$M(t) = 4\lambda\omega_0^2\left[F(t)\right]^2 \tag{3.30}$$

where the parameter λ expresses the coupling force and is a linearly increasing function of the density. The memory function $M(\boldsymbol{k},t)$, which is generally expressed as

$$M(\boldsymbol{k},t) = \sum_{\boldsymbol{k}'} V_{\boldsymbol{k}\boldsymbol{k}'} F\left(|\boldsymbol{k}+\boldsymbol{k}'|, t\right) F\left(-\boldsymbol{k}', t\right) \tag{3.31}$$

gives rise to a non-linear closed coupling of all the modes. The essential parameters are the static coupling terms $V_{\boldsymbol{k}\boldsymbol{k}'}$, which are postulated to vary slowly with the temperature.

When the non-linear coupling force is increased above certain thresholds, (3.29) leads to partial freezing of the dependence of the intensity of the fluctuations on $M(\boldsymbol{k},t)$. However, we observe that such an intensity is variable in space. Since there are no changes in the structure, which is and remains a liquid, this freezing is considered an index of the glass transition. Thus, apart from the obvious condition that, when the time diverges, $F(\boldsymbol{k},t_\infty) \to 0$ when $D \neq 0$, the solution of (3.29) leads to the critical slowing down of $F(\boldsymbol{k},t)$ at a critical temperature T_c where $D \to 0$ and $\eta \to \infty$, so that

$$F(\boldsymbol{k},t\to\infty) \begin{cases} = 0 & \text{for } T > T_c \quad \Longrightarrow \text{ergodic behaviour} \\ > 0 & \text{for } T \leq T_c \quad \Longrightarrow \text{non-ergodic behaviour.} \end{cases} \tag{3.32}$$

Thus, $F(\boldsymbol{k},t_\infty)$ coincides with $\mathfrak{S}_c(\boldsymbol{k})$, the Debye–Waller factor for the elastic peak of the dynamic structure factor; this is a wavevector dependent order parameter.

It is the partial freezing of the fluctuations that makes the glass a non-ergodic system: this cannot be described macroscopically within the MC sce-

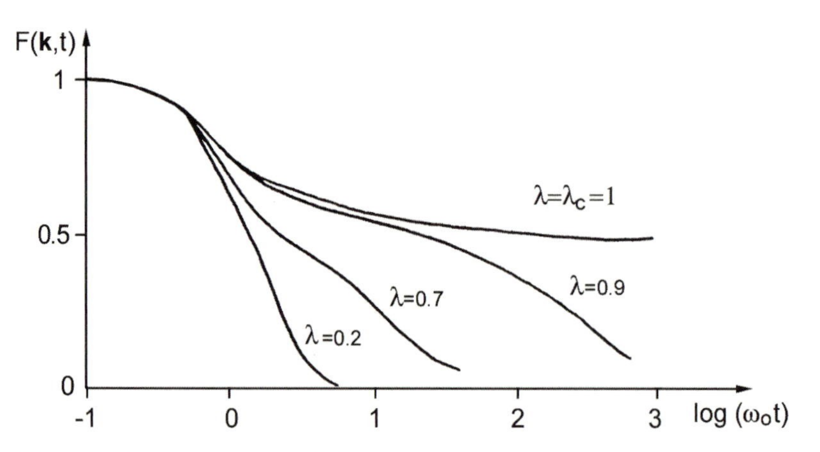

Fig. 3.7. Response function $F(\boldsymbol{k},t)$ as a function of time, for various coupling force values λ, each being constant. Notice the critical slowing down in the decay at the critical value λ_c

nario simply because there is no thermodynamic description of the glass transition.

Even in the simplest form the mode coupling theory provides for correct time decay for the response function $F(\boldsymbol{k}, t)$ as λ approaches $\lambda_c = 1$ from below, namely starting from the characteristic values for a liquid system.

The $F(\boldsymbol{k}, t)$ limit for long times changes quickly from zero, above the transition, to a non-zero value below the transition, as schematically shown in Fig. 3.7; the very fact that the density response function presents a critical decay slowing down at the critical value for the coupling constant means that the density fluctuations have been frozen by purely dynamic mechanisms. At temperatures below T_c the atoms are trapped in cages made of their own local surroundings, which are heterogeneous both in space and dynamically, and the diffusion process is heavily slowed down. The relaxation of the supercooled liquid changes its behaviour from a liquid to a solid; as the temperature falls the system properties characteristically come close to those of a solid.

As we approach the transition from the liquid side, where $\lambda < \lambda_c$ a singularity with power law dependence is found both in relaxation time and in viscosity:

$$\tau \simeq \left| 1 - \frac{\lambda}{\lambda_c} \right|^{-\upsilon}$$

where $\upsilon = 1.765$. This expression does not agree with the expression obtained from the Vogel–Fulcher–Tammann phenomenological law,

$$\tau \simeq \exp\left[(\lambda_c - \lambda)^{-1} \right] .$$

Among the most significant corrections to the idealised MC theory, a correct treatment of the dependence on the wavevector $\boldsymbol{k}$ indicates that only those density fluctuations with wavevector value close to that of the first peak in the static structure factor significantly contribute to amorphisation. Furthermore, coupling has also been introduced among other modes in the liquid; as a result the sharp singularity characteristic of the ideal transition (see 3.32) is removed, leaving a "rounded" transition similar to what has been observed experimentally.

However, to obtain this result we have to push the system to follow trajectories in the entry parameter space that correspond to correlation functions for density fluctuations which, invariably, disappear as time diverges. Consequently, ergodicity is not broken. On the other hand, in real systems the thermally activated atomic jumps smear out the singularities and restore ergodicity by way of very slow relaxation of the underlying disordered structure.

Just as for any theory the crucial point to the MC theory occurs when its predictions are compared to experimental results.

We have to consider that the significant time scale required to evaluate the theoretical predictions, even for fragile systems that can transform into glass easily, is around 10^{-9} s, which is less than the usual experimental time scale. Moreover, the relaxation spectra extend over many decades. Thus, inelastic neutron scattering, using the spin echo technique, allows us to analyse relaxation spectra in the frequency interval between 0.1 GHz and 10 THz. Besides this, only the results of Molecular Dynamics (MD) simulations performed in the region of temperatures approaching T_c from above can be directly compared to experiments.

The kinetic transition observed in MD calculations for a binary alloy, simulated using spherical atoms with two different diameters, d_1 e d_2, which interact by a soft purely repulsive pair potential,

$$V_{1,2}(x) = \varepsilon \left(\frac{d_{11} + d_{22}}{2x} \right)^{12}$$

shows that the cooling rate does not influence the dynamic behaviour around the critical temperature. The shear viscosity η and the self diffusion coefficient D for the two atomic species allow us to follow the Stokes–Einstein relation (see 3.7) in the supercooled liquid within a very wide temperature field. The dependence on the wavenumber of the order parameters for the glass matches the results predicted by the mode coupling theory.

Also, coherent and incoherent neutron scattering results from the supercooled liquid state and the glassy state of the prototype *o*-terphenyl (OTP) system, compare in a satisfactory way with the MC theory. To obtain this, the data are analysed in the time domain, under the assumption that the correlation times τ scale with the shear viscosity η according to $\tau(T) \simeq \eta(T)/T$.

In the temperature dependence of the Debye–Waller factor $F(\boldsymbol{k}, t_\infty)$ a cusp appears; this is interpreted as an indication of the critical temperature T_c, whose existence is an essential point in the MC theory. As predicted, below T_c, $F(\boldsymbol{k}, t_\infty)$ follows a $(T_c - T)^{1/2}$ behaviour, whereas the structural relaxation above T_c is well parameterised by a stretched exponential function. The existence of the critical temperature $T_c > T_g$ ($T_c \simeq 1.15$–$1.2\ T_g$) is one of the strongest arguments in support to the mode coupling theory. The very same scenario observed for fragile liquids characterises the dynamics of strong liquids at high temperature, such as B_2O_3. These systems, too, show a critical temperature below which the fast relaxation processes become temperature dependent. T_c is, though, much higher than T_g ($T_c \simeq 1.15$–$1.6\ T_g$).

Lastly, a recent analysis of the variation with temperature of the structural relaxation time τ has proven that the critical temperature T_c does indeed exist. As the temperature falls towards T_g, τ varies in ten orders of magnitude. The parameter $\varrho = [\mathrm{d}(\log \tau)/\mathrm{d}T]^{-1/2}$, which according to the VFT law should depend linearly on temperature, is shown to clearly deviate from its linear trend around T_c for various kinds of glass forming compounds.

One of the most severe drawbacks to the MC theory lies in the choice to study systems where the interatomic interactions are characterised by spherical symmetry. This ignores the complexity of real easy glass forming systems in which interatomic interactions are covalent and thus non-central. Furthermore, the theory gives a quantitative description of the behaviour of liquids in the low viscosity region (η is less than 10 Pa s) in reasonable agreement with experiments, but it predicts the glass transition at too low η values, around 10 Pa s.

However, apart from highlighting the most significant characteristics of the dynamics of a supercooled liquid in the glass transition region, the MC theory allows us to perform a detailed study of the various relaxation processes around the critical temperature. One of these processes is the so-called slow relaxation regime, β. This regime is characterised by the fact that at low temperature the trend of the correlation functions for the supercooled liquid, once they are represented as a function of the logarithm of time, exhibits a wide shoulder and sometimes even a terrace (Fig. 3.7). The theory predicts that in the β regime the correlation function $F(t)$ can be expressed as

$$F(t) = f - A\left[\frac{t}{\tau(T)}\right]^{b} \tag{3.33}$$

where f is the terrace height, often defined as the non-ergodicity parameter, and A and b are positive constants. In the critical region exponent b should have the same value for all the correlation functions, and the relaxation time $\tau(T)$ should diverge at T_{c}, in agreement with a power law where the value of the exponent γ is linked to b.

MD simulations of a binary liquid made of a mixture that is controlled by a Lennard–Jones potential (Fig. 4.44), have clarified that (3.33) strictly holds for this simple liquid, which is particularly suited to quantitatively test the theory. Furthermore, the value of b is actually the same for both liquid components, thus verifying a second fundamental prediction of the theory.

There is, though, disagreement between the simulation results and the theory predictions when we consider the longer time scales. Indeed the MC theory predicts that the diffusion constant diverges at T_{c} with the same exponent γ as the relaxation times. However, the simulation shows that at T_{c}, γ value in the diffusion relation considerably differs from the value found in the relaxation relation, even though the relaxation time and the diffusion constant are the same.

According to the MC theory, the glass transition is characterised by qualitative changes in the dynamics of the liquid around T_{c}. Above T_{c} the system is a normal liquid where the diffusive motions dominate, whereas at T_{c} a transition towards a solid begins, and this becomes even more pronounced as the temperature falls. The difference between fragile and strong glasses may be put in relation to the width of the temperature interval between T_{c} and T_{g}, namely to the temperature field where the glass transition takes place; this is large for strong systems and small for fragile systems.

3.3 Ease of Glass Formation

The empirical observation that some systems can be brought into the glassy state more easily than others has stimulated research, and technological interest, into criteria and theories on the formation of non-crystalline phases that can both interpret and predict glass formation ability.

Whenever this kind of criteria appears there should already be a firmly established glass transition theory which, however, is still lacking. This is why models have been developed with validity generally limited to certain classes of materials, usually with no more than two components, and for some specific amorphisation processes. These models are given more to post-experimental interpretation than to the prediction of how materials will behave when they possibly undergo vitrification.

Before we proceed to examine the various criteria we should bear in mind that amorphisation can be obtained by way of fast quenching, starting from the liquid phase, or by a direct reaction in the solid state. Whereas kinetic and thermodynamic factors dominate in the first case, the link between the amorphisation reactions in the solid state and the quenching processes in the liquid state is based on the fact that even in the former *heterogeneous* nucleation of the non-crystalline phase is essential. This is associated with the breakdown in the long range chemical order, similarly to what is observed when a crystal melts. In this case the heterogeneous nucleation is an essential mechanism that many models of the phenomenon are based on.

The loss of extended chemical order is an important condition to drive amorphisation. To this aim the relevant factors are the defects, the degree of chemical interaction between system components and its elastic properties.

The difference between the above glass forming processes is actually much less marked than might appear, because first in many systems considerable traces of the short and medium range structural organisations present in the solid crystalline phase still remain (see Chap. 4) in the liquid phase. Second, in many amorphisation reactions in the solid state, it is quite plausible for the system, which is highly energised, not to stay in the solid state but, for very short time intervals and at a local scale, to liquefy or even vaporise. Moreover, some vitrification processes occur over quite long time scales so that thermodynamic factors may come into play.

In general the tendency to readiness of glass formation is most evident in metallic alloys with large, negative enthalpies of mixing, such as Co–Nb, Cu–Zr and Ni–Ti. In such systems where compound formation is favoured, amorphous phases are obtained over wide composition ranges. On the other hand, when systems with positive enthalpies of mixing are processed under non-equilibrium conditions, supersaturated solid solutions are usually obtained. Amorphisation is observed as an exception, when the enthalpies of mixing of the solid solutions and the amorphous phase are considerably different in magnitude, or they differ in sign. In these systems the Gibbs free energy curve of the amorphous phase, calculated by extrapolation of the liquid curve to low temperatures, lies above the curves for the solid solutions over the whole composition range. In fact the amorphous phase is usually less metastable than assumed on the basis of the free energy curve of the liquid extrapolated to the supercooling region. The increased metastability is attributed to the short-range order that evolves in the liquid during supercooling. In systems which are immiscible even in the liquid state the formation of an amorphous phase is at variance to the thermodynamics of "normal" liquids. The glassy phase free energy is much lower than deduced from the phase diagram; this implies that during cooling the enthalpy of mixing of the liquid changes its sign. A prototypical study on this kind of systems was performed on $Nb_{(1-x)}Cu_x$ (32 at. % $\leq x \leq$ 77 at. %), whose enthalpy of mixing in the liquid state is +12 kJ g atom^{-1} at 1800 K. X-ray diffraction on sputter deposited films of the alloy shows that they are amorphous and by differential scanning calorimetry crystallisation enthalpies between 4.5 and 7.6 kJ g atom^{-1} were measured. This result shows that there is a stabilisation of the liquid phase, which implies that the positive enthalpy of mixing at high temperature has changed sign becoming negative at low temperature.

The glass becomes unstable at T_g, compared to the supercooled liquid, and the supercooled liquid exists as such above T_g for a short temperature interval, usually around 20 K, before it crystallises again, either directly or by way of a sequence of metastable phases. Thus the narrower the interval between the liquidus and the glass transition temperatures T_l and T_g, respectively, the easier it is to vitrify the system.

This criterion leads us to define the *reduced glass transition temperature*, $T_{rg} = T_g/T_l$, as the dominant parameter for amorphisation. This parameter is favoured the closer T_{rg} approaches the unit. For systems with relatively weak chemical interactions, T_g is not very sensitive to the composition, so T_{rg} depends mostly on the trend of T_l. For many systems that undergo vitrification we observe that the critical amorphisation cooling rate $|\dot{r}_c|$ is a single monotonic function of T_{rg} that decreases rather rapidly. For those systems with stronger chemical interactions, thus with thermodynamic functions that are more markedly dependent on temperature and concentration, we expect similar behaviour, except for stronger T_g dependence on the composition.

It has commonly been observed that a compound, whether organic or inorganic, with stoichiometry close to a *eutectic* composition, undergoes easier transition from the crystalline to the amorphous phase. The melting point T_m for this composition is depressed so that at T_g the liquid is less supercooled and it is more difficult for crystallisation to take place. If, though, $|\dot{r}|$ is not high enough, the liquid eutectic alloy will crystallise and will form, in the simplest of cases, a mixture of two components. We note that around the eutectic alloy the excess entropy is negative; this means that locally there is a considerable degree of chemical/structural order. This is confirmed by the fact that, at least in the metallic systems, the number of eutectic structures observed is particularly high.

The first interpretation for the ease of amorphisation in certain oxides, while others resist vitrification, was given by W.H. Zachariasen in 1932 [3.1]. This was an initial attempt to consider both topological and structural factors when analysing vitrification. The ideas at the heart are simple: the internal energy of an amorphous material must only be a little higher than it is in its corresponding crystalline phase, otherwise it would immediately re-crystallise. Furthermore, it is reasonable to consider that the same kind of atomic interaction exists in both states, thus the elementary structural units must be the same, and only at an extended packing level must periodicity develop, or not. For example, both in crystalline and amorphous SiO_2 we must find the same tetrahedral SiO_4 units that are vertex-connected through dihedral (torsion) angles, all equal, or, respectively, variable within a certain interval.

From this rule the so-called *Continuous Random Network* (CRN) structural model for amorphous materials was obtained and we can explain why the crystal is absolutely stable for certain compound compositions. For example, in a hypothetical bidimensional oxide with a XO composition (Fig. 3.8), each oxygen atom would be at the vertex of three different XO_3 triangles. In this case, if we introduced disorder, by distorting the bond angle (equivalent in the plane to the three-dimensional dihedral angle) we would have to spend so much energy that the glass structure is definitely unfavoured.

The above model developed for covalent systems with strong chemical interactions has an equivalent proposed in the 70's for metallic systems. This model is particularly suited to transition metal-metalloid alloys. Once the metal atoms have arranged themselves in a dense, random packing, the smaller metalloid atoms will fill the interstices, thus forming the structure. The compositional ratio is correctly predicted; amorphisation is favoured at about 20% metalloid, even though recent structural investigations point at

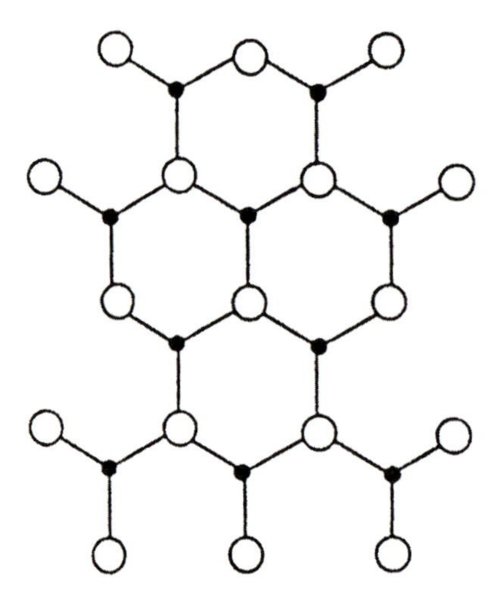

Fig. 3.8. Structure of a hypothetical stoichiometric oxide XO on a bidimensional lattice: each oxygen atom (•) is at the vertex of three different triangles XO_3. In the CRN model, disorder is introduced, in the simplest way, by distorting the bond angle

a higher degree of steric correlation between the metal and the metalloid atoms, and in particular rule out the random interstice occupation by metalloid atoms (see Sect. 4.7).

The above model also incorporates the idea that steric hindrances may play a significant role in whether a system can undergo the amorphisation process or not. Indeed, if the atomic sizes of the compound constituents differ significantly from each other, the cooling down kinetics of a melt is considerably reduced. In fact, it has been observed that a minimum value in the *difference in the atomic radii* of the constituents of the binary alloy, or compound, coincides with glass formation ease. Experiments with model systems (hard spheres and soap bubbles) give a 10% difference in the "atomic" radii as the lower threshold. The atomic radius difference has been associated with other parameters connected to bond strength, such as electronegativity, or vaporisation enthalpy, in order to draw up two parameter maps where intervals for parameter values that correspond to easy amorphisation are identified. In general the size criterion better fits experimental results than the energetic criteria, although a number of exceptions are found in the maps.

Very recently *massive* amorphous alloys with wide compositional intervals have been produced starting from the dimensional difference between constituent elements. These alloys are the ternary La–Al–(Ni, or Cu), Mg–Y–(Ni, or Cu) and Zr–Al–(Ni, or Cu) systems and have exceptional supercooling intervals, higher that 50 K, (ΔT = 77 K, for $Zr_{60}Al_{15}Ni_{25}$), and lower cooling rate thresholds to induce vitrification, $|\dot{r}_c|$, up to around 90 Ks^{-1} ($|\dot{r}_c| = 87$ Ks^{-1}, for $La_{55}Al_{25}Ni_{20}$). The glass forming ability has been correlated in this case, to what occurs in Al–lanthanide–transition metal alloys (e.g. $Al_{87}La_8Ni_5$) where the chemical interaction between aluminium and lanthanide is strong, which is an index of relevant short range order. This has been confirmed by the formation enthalpy values for this family of alloys, negative, with large absolute values.

The alloy $Zr_{41.2}Ti_{13.8}Cu_{12.5}Ni_{10}Be_{22.5}$ has been produced using the same criteria. This is a peculiar alloy in that it transforms into the amorphous phase at $|\dot{r}_c|$ values below 10 Ks^{-1}, the lowest so far obtained for metallic systems, and much less than the cooling rate usually required to produce metallic glasses. The viscosity η of this system, measured in the supercooled liquid region, is between 10^{10} and 10^6 Pa s, a very unusual range for amorphous metals whose minimum viscosity just reaches 10^{10} Pa s in the most favourable of cases. The η dependence on temperature obeys the VFT law (see (3.1)) where T_0 = 352 K, a very low value which is equal to around $T_g/2$. This behaviour is similar to what occurs with silicate glasses. Results on microstructure and structure evolution during isothermal crystallisation of the alloy between the liquidus and T_g temperatures are reported in Fig. 3.9. The alloy structure changes from Zr_2Cu type at high temperature (above 780 K), to $MgZn_2$ type, between 740 and 710 K, to an fcc solid solution at low temperature, between 670 and 610 K. The characteristic length scale

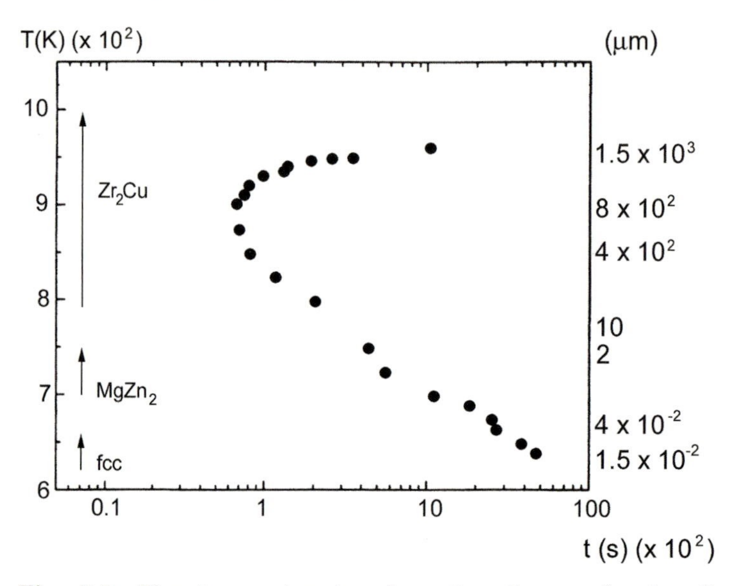

Fig. 3.9. Time-temperature-transformation diagram for the alloy $Zr_{41.2}Ti_{13.8}Cu_{12.5}Ni_{10}Be_{22.5}$. Full circles indicate data points. The figure shows the liquidus temperature T_l, the temperature ranges for the observed structures and the length scales of the microstructures developed at various temperatures after isothermal annealing (adapted from [3.2])

of the microstructure decreases by five orders of magnitude with increasing supercooling, from 958 K to 613 K. The number density of crystallisation nuclei, as estimated from microstructure observations, increases from $10^8\,\mathrm{m}^{-3}$ at 958 K, to $10^{23}\,\mathrm{m}^{-3}$ at 613 K. Such high values for nuclei density mean a high nucleation rate, in contrast to the observed slow crystallisation kinetics shown in Fig. 3.9. A possible explanation for the contradiction is that decomposition occurs in the liquid before it crystallises.

If the difference in the atomic radii of the components is too great, which is typical of small solute atoms, such as carbon atoms, then amorphisation will hardly occur, or the process will be incomplete. Indeed high atomic diffusivity of carbon reduces the effect of the steric hindrances; the Fe–C alloys are prototypes of this mechanism. Likewise, hydrogen is a good stabiliser in the glass phase only if it is highly concentrated.

The steric criterion has been made more accurate by considering the minimum concentration of solute required for the amorphous phase to occur in a compound $X_xY_{(1-x)}$. For a set of about sixty glassy systems, the following relation has been verified and is hypothesised as being generally valid:

$$Y_{\min} \simeq A \left[\left| \frac{V_y - V_x}{V_x} \right| \right]^{-1} \tag{3.34}$$

where V is the atomic volume of the alloy components and $A \simeq 0.1$ is a numerical constant. The reason why solid crystalline solutions are unstable when the solute concentrations are above $Y_{\rm min}$ is explained on the basis of elastic deformation, which is locally introduced into the crystalline lattice.

If we consider a solid (metallic) solution as an elastic continuum, we can then define the elastic modulus $E_{\rm m}$ in terms of shear stress μ and compressibility B

$$E_{\rm m} \propto (2\mu^{3/2} + B^{3/2})^{2/3} \propto \varrho \left(\frac{k_{\rm B}\, \Theta_{\rm D}}{h\, k_{\rm D}} \right)^2$$

The Debye temperature $\Theta_{\rm D}$ has been used here along with the corresponding wavevector $k_{\rm D}$ and the macroscopic density ϱ; $k_{\rm B}$ is the Boltzmann constant and h the Planck constant.

Using the *elastic* criterion we find that, if the difference in the atomic radii for the constituents of the alloy is large, then $E_{\rm m}$ undergoes a dramatic lowering as a function of the solid solution stoichiometry. It is thus possible to find a composition at which, by linearly extrapolating the trend of the value of the elastic modulus, it will soften. The system becomes mechanically unstable and vitrifies. In the case of the Zr–Rh alloy, where the atomic size difference is about 20%, the predicted critical concentration of Rh is 12 at. %, and the alloy undergoes glass transition quite easily when $Zr_{85}Rh_{15}$ is formed. On the other hand, if we consider metallic systems with small differences in the constituent atomic radii, no instability is highlighted in the trend of the elastic modulus.

The previously examined *free volume* theory also lies at the base of an instability criterion for crystals, with respect to amorphous solids, that was developed for binary alloys. Basically, if zero, or negative, volume changes are measured when a crystalline alloy melts then it will transform into glass easily under a fast quenching process. In fact, the greater the free volume of the liquid the less the viscosity is, and the higher the atomic diffusion. Thus, if a crystal is denser than the liquid from which it grows, it will push free volume into the liquid whose viscosity, in turn, will be further reduced, thus favouring the growth of new crystallites.

If a material is found in various *polymorphous modifications* then it will easily be vitrified. This amorphisation criterion is particularly important when examining the competition between the various possibilities of local order. SiO_2 is probably the best example among all the non-metallic glasses for this criterion since it has three crystalline phases.

However, this criterion is also true for, for example, elemental tin and for various binary alloys. In these cases the glass forms because while the system is cooling, even below the melting temperature groups of atoms structurally coordinated in different ways are nucleated. Such agglomerates compete with each other, and none of them can grow at the expense of the others to become a centre of extended crystallisation. Though there is much validity to this criterion its theoretical foundations are disputable since it gives us a microcrystalline picture of the amorphous solid. Our current understanding is that the microscopic structure of a covalent, or metallic, amorphous solid is specific and cannot be reduced to assemblies of microcrystals, even though they may have small dimensions (see Sect. 4.8).

An *electronic* amorphisation criterion has been developed for binary alloys on the basis of the interaction between the electronic structure and the ionic structure. If the composition of an amorphous alloy is such that the Fermi level E_F coincides with a minimum of the electronic density of states, then it is particularly (meta)stable with respect to crystallisation. Quantitatively, if $\boldsymbol{k}_\mathrm{p}$ is the wavevector that corresponds to the strongest peak for the structure factor $\mathfrak{S}(\boldsymbol{k})$ (see Chap. 4), where the relation $\boldsymbol{k}_\mathrm{F} = \boldsymbol{k}_\mathrm{p}/2$ is valid, and where $\boldsymbol{k}_\mathrm{F}$ is the wavevector at the Fermi level, then an energy gap is formed. This mechanism is associated with the stabilisation of the glassy structure.

We assume that the crystal Brillouin Zone corresponds to a pseudo-zone with spherical symmetry in the structurally disordered amorphous system. Furthermore, in metals the first peak of $\mathfrak{S}(\boldsymbol{k})$ is often so narrow that only one Fourier component is dominant, namely the one with wavevector $\boldsymbol{k}_\mathrm{p}$, whose role is similar to that of a reciprocal lattice vector in a crystal.

The key point is the hypothetical relation between $\boldsymbol{k}_\mathrm{p}$ and $2\boldsymbol{k}_\mathrm{F}$. In general $\boldsymbol{k}_\mathrm{p}$ is greater than $2\boldsymbol{k}_\mathrm{F}$ for monovalent metals and less than $2\boldsymbol{k}_\mathrm{F}$ for divalent metals (Fig. 3.10). For an alloy made of a metal X with a valence of $Z_x = 1$ (the transition metals are considered part of this group) and a metal Y whose valence Z_y is greater than one, the effective valence is $Z_e = c_x Z_x + (1 - c_x) Z_y$. This changes with the composition, and the position of $2\boldsymbol{k}_\mathrm{F}$ changes accordingly, until, for a certain composition, it coincides with the $\boldsymbol{k}_\mathrm{p}$ position. The glassy phase of the alloy with such a stoichiometry is particularly (meta)stable with respect to recrystallisation.

If fluctuations occur then in the system long range order may start to develop which would dispel the spherical symmetry of $\mathfrak{S}(\boldsymbol{k})$. The value $\boldsymbol{k}_\mathrm{p}$ takes on would now depend on the direction so that the reduction mechanism

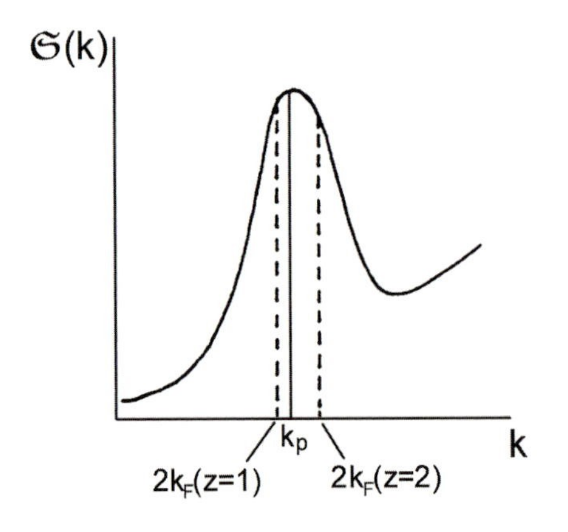

Fig. 3.10. Structure factor $\mathfrak{S}(\boldsymbol{k})$ as a function of the wavevector $\boldsymbol{k}$; the positions of the Fermi wavevectors are shown for divalent and monovalent metals

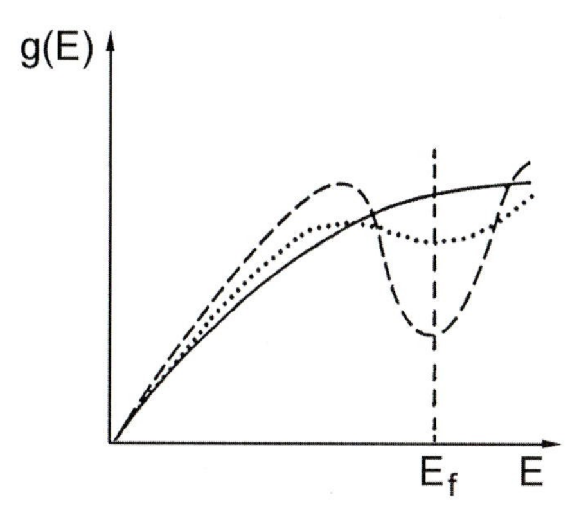

Fig. 3.11. Schematic structure destabilisation mechanism for the solid; the deep minimum (dashed curve) for a system characterised by spherical symmetry is greatly reduced by the structural fluctuations that onset long range order and dispel the spherical symmetry (dotted curve)

for the electronic density of states would be activated in different directions, for different electronic energy values. Since not all the states with $|\boldsymbol{k}| = \boldsymbol{k}_p/2$ would contribute to the stabilisation mechanism at the same time, the deep minimum at E_{F} present when the system is spherically symmetrical would be heavily reduced. The perturbation of $\mathfrak{S}(\boldsymbol{k})$ would imply that system energy would increase when the electrons reached the Fermi level (Fig. 3.11).

This criterion represents an extension of the Hume–Rothery rule for the stability of the various phases in so-called electron compounds, and was subjected to much debate, due to contrasting experimental results. Whereas in some model systems of the noble metal-semiconductor family (AuP; AuSi), the electronic criterion correctly predicts the composition for maximum (meta)stability for the metallic glass, it has been impossible to unquestionably find the pseudo-gap in the density of states of transition metal alloys. However, the approach has been proved valid for amorphous binary noble metal polyvalent metal alloys (AuSn) with varying compositions.

The basic concepts of the model have been successfully transposed into a structural stability criterion in the physical space. In this case the conduction electrons shield the ionic charge and this causes the so-called Friedel oscillations of the pair potential $\Phi(x)$ whose asymptotic trend is cosine: $\Phi(x) \propto \cos(2k_{\mathrm{F}}x)/x^3$. In turn, the pair correlation function $g(x)$ also shows oscillating behaviour since $g(x) \propto 1 + [\sin(k_{\mathrm{p}}x)/x]$, where k_{p} is the wavevector corresponding to the maximum of the strongest peak for the isotropic structure factor $\mathfrak{S}(k)$.

When the maxima of $g(x)$ coincide with the minima of $\Phi(x)$, then the energy term for the band structure

$$E_{\mathrm{b}} \propto \int x^2 \left[g(x) - 1\right] \Phi(x) \mathrm{d}x$$

has a large negative value, thus contributing to the stability of the system. The best coincidence between a cosine curve and a sine curve is obtained when $2\boldsymbol{k}_{\mathrm{F}} = \boldsymbol{k}_{\mathrm{p}}$, i.e. when the wavelength of the two kinds of oscillations are

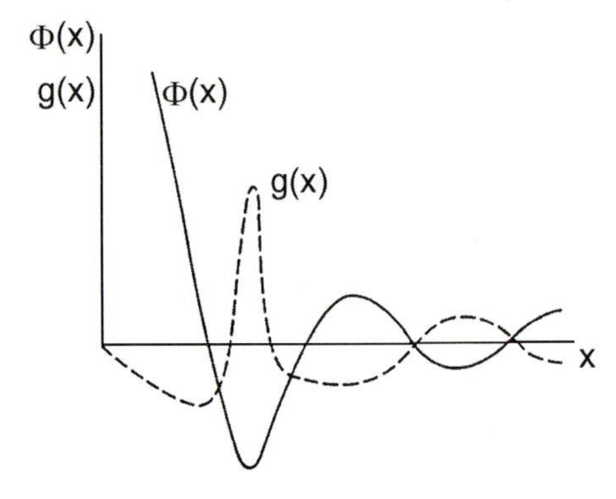

Fig. 3.12. Interatomic pair potential, $\Phi(x)$ and pair correlation function, $g(x)$, in CaZn. When the maxima of a function coincide with the minima of the other, structure stabilisation occurs

equal (Fig. 3.12). This criterion has been proven on simple metal alloys with interatomic potentials basically characterised by spherical symmetry (MgCa, MgZn).

One microscopic theory for the stability of the glassy phases has recently been put forward for binary compounds with both metallic and covalent bonds. The compounds have undergone amorphisation reactions in the solid state, particularly ion bombardment. Presently this is the most highly controlled technique to induce disorder and subsequent vitrification in a system. The theory takes into account the physics of the irradiation process, with special regard to the formation of atomic collision cascades along the trajectory followed by an energetic projectile as it slows down in the target. It is assumed that at the interface between these cascades, where the atoms are highly energised, and the surrounding unperturbed crystalline lattice, atoms of one of the compound components coming from the cascade core preferentially migrate (segregate). As a consequence, both the compositional profile and the electronic charge density at the interface are altered with respect to their equilibrium values. Such profiles can be non-equilibrium over the typical time scale for cascade quenching (10^{-11} s).

Re-equilibration is supposed to occur by way of local charge transfer reactions between pairs of atoms of either compound constituent. The result is that the atom of the segregated component captures an electron from the atom of the other component. In so doing an atomic pair is formed, namely a *dimer* of an effective compound whose electronic and thermodynamic properties are compared with those of the original compound. In particular, for those systems that undergo the glass transition, we observe: a positive energy contribution ΔE_{e^-} associated with the electronic charge transfer and an increase in surface atomic mobility in the effective compound, whose formation enthalpy ΔH_{f} is greater than that of the original compound. The introduction of dimers into the effective compound increases the system energy, and this system becomes unstable. Furthermore, the high surface atomic mobility corresponds to the realisation of many energetically equivalent local configurations, that are off thermodynamic equilibrium, and whose freezing corresponds to the formation of the glassy phase. The opposite is true for

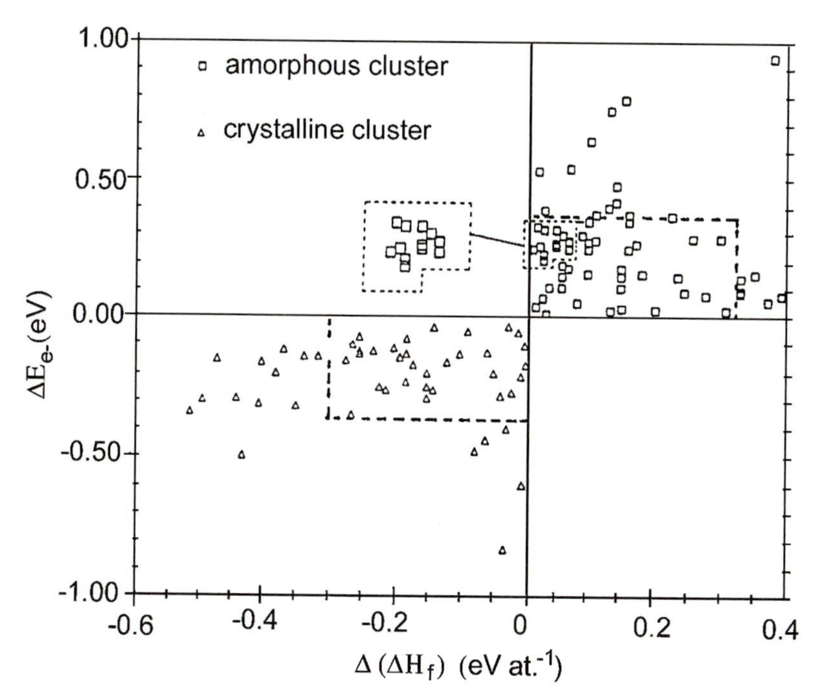

Fig. 3.13. Trend in the electronic energy contribution ΔE_{e^-} associated with the creation of a dimer (cluster) of effective alloy, as a function of the difference in the formation enthalpy ΔH_f between the original alloy and the effective alloy. The (Δ) points for the crystalline systems under irradiation occupy the region of negative parameter values, whereas the (□) points for the amorphised systems lie in the region of positive parameter values. Within both regions dashed lines mark intervals of parameter values most commonly found (adapted from [3.3])

compounds that remain crystalline upon bombardment. By calculating the local deformation associated to the formation of a dimer of effective compound, threshold deformation levels have been associated to vitrification.

The theory has been applied to a large number of metallic systems and nearly no exceptions to the described behaviour have been found. It has also been tested with success on non-metallic compounds, including borides, carbides, nitrides and oxides. Limiting ourselves to metallic alloys, Fig. 3.13 shows a two parameters map: the first is ΔE_{e^-} and the second is the difference between the formation enthalpy for the effective alloy and for the original alloy, $\Delta(\Delta H_f)$. The map clearly shows value intervals for the two parameters that define the regions where amorphous and, respectively, crystalline compounds are most easily formed.

The theory offers us a physical representation of amorphisation starting from the essential processes that lie at the heart of system energisation. In principle this theory may be applied to other experimental typologies characterised by structural disordering followed by glass formation, and thus at least to all the glass-forming reactions in the solid state.

4. The Structure of Disordered Systems

4.1 Why We Study the Structure of Amorphous Systems

Structural research has traditionally had a preferential role in solid state physics since the determination of the arrangements of atoms in space is the foundation stone for subsequent studies in order to determine the different physical, or chemical, properties. As far as crystalline solids are concerned, determining the structure has been made simpler by the possibility to limit ourselves to the relatively few atoms in the unit cell. Due to spatial periodicity, these atoms make up the fundamental unit the whole structure is based on. Conversely, for those solids that lack translational periodicity, the unit cell has *infinite* spatial extension, which makes it of no use to us in analysing the structure of amorphous systems.

If the structure of an amorphous solid were random and chaotic, in the strictest sense of the term, and we were only able to carry out average statistical calculations regarding the position of the atoms, then the structural analysis would be meaningless. However, a structurally disordered solid is an extremely complex system, it certainly is not crystalline, but, at the same time, it is not purely random.

The amorphous solids, particularly the metallic amorphous solids, are almost invariably systems with more than one component (at least two); they are thus made up of a mixture of different atomic species. In general we need to know how many atoms of each species surround the atom under consideration, and to define the structure and the symmetry of the shell, which is made of the atom's first neighbours. Only if a structure can be described in a satisfactory manner in terms of a random collection of atoms, will the problem in determining the environment around each atom be resolved from the knowledge of the stoichiometry of the material and the atomic radii of the constituent elements. In this case, short range order, both chemical and topological, does not exert any significant influence on the local structure.

In reality, the structure of many amorphous solids is *non-random*, at least on certain length scales. The structure has a considerable degree of order even though it lacks global periodicity. For example, the structure of the silicate glasses consists of tetrahedral structural units with SiO_4 stoichiometry, disorderly connected together. The existence of this structural unit, which can be observed using various experimental techniques, implies that the system is not totally random.

Even the metallic glasses made of a transition metal and a metalloid (e.g. NiB) show practically *perfect* local order. The boron atoms are likely to co-ordinate with a first shell of nickel atoms, whereas they are correlated among themselves at a first neighbour level only for high metalloid concentrations, above 25 at.%. Furthermore, it has been experimentally ascertained that there is definite topological organisation of the atom shells of nickel and the first neighbour atoms of boron. Again, referring to medium range order, these amorphous alloys present a high degree of structural organisation, to a scale of around 2 nm.

However, we must remember that these observations do not have general validity. Some amorphous transition metal alloys only exhibit a weak degree of chemical organisation and no topological organisation.

The presence of *chemical* bonding between the atoms of a solid, resulting in highly directional covalent bonds is usually responsible for local order. This is true also in metals and ionic compounds, where chemical bonding adds to predominant non-directional forces. Structural research on the amorphous systems is intended to ascertain the degree of local order, what is responsible for it, and why it shows up, however strongly. If we can answer these questions we will gain detailed knowledge of each amorphous system and be able to identify the relevant factors in a general theory for the non-crystalline phases.

It is important for us to specify the *scale* of length used to study the structure. Bearing in mind that the various reciprocally incompatible experimental techniques used in structural research are efficient for lengths of below and above 10 nm, we assume that lengths of this order are discriminatory for a microscopic structural approach. For example, X-rays are sensitive to structural changes of a few tenths of a nanometre (microscopic region), whereas the optical microscope is efficient in observing lack of structural homogeneity over hundreds of nanometres, or even more (macroscopic region). Thus, X-rays allow us to determine the position of the atoms around an arbitrarily chosen "origin" atom, for example by the EXAFS technique. By contrast, the meaningful lengths required in studying the re-crystallisation of a glassy material are macroscopic and an optical microscope is generally used to count the number of micro-crystals that have formed in the amorphous matrix.

Here we shall concentrate on the microscopic structure of structurally disordered systems, and we shall make an additional distinction between short range order (SRO), which extends over a few tenths of a nm, and medium range order (MRO), between 0.5 and 2 nm.

An investigation into the structure of non-crystalline materials has to answer three basic questions:

1) What *elements* are necessary for us to understand the amorphous structure?

2) What *structural features* do the various physical-chemical properties depend on? It is essential we answer this question to be able to prepare new materials with ad hoc composition for particular applications.
3) Among the specific elements in the structure of an amorphous solid, which are relevant in describing the *very nature* of the glassy state of the material?

In this area of research we encounter certain specific problems. The first problem concerns what mathematical tools can be best used to represent the disordered structure, in particular the topologically disordered structures, in the most rational way.

We have to turn to *statistical* data such as the distribution in the number of first neighbours. The very definition of first neighbours for a given atom requires caution. Moreover, it is difficult to obtain the statistical data for real systems. Diffraction experiments give us the interparticle *distances* but not the *spatial arrangements*. For example, if we consider groups of three atoms, the structure and size of the triangles they form are unknown and have to be calculated approximately by starting from the distribution of the interparticle distances. Furthermore, for fluids, the configuration taken on by the particles is continuously changed by molecular motion. Therefore, experiments and theories have to consider time, or configuration *averages*, for the observed physical quantities. It is not easy to obtain these average values in a correct way; ultimately, it reduces to solve the Schrödinger equation when the potential energy varies irregularly from one point to another. In this case the statistical information, such as the distribution of the interparticle distances, is really useful.

Lastly, there is a practical hurdle to characterising samples of the amorphous systems. As discussed in Chap. 2, even though there are strong indications that a material, prepared using various techniques, takes on non-equilibrium configurations that, essentially, are *indistinguishable*, yet the history of a sample may considerably affect its properties.

The above difficulties slow down any progress towards understanding the structurally disordered systems. Non-crystalline matter is some way off from the conditions where we can apply the simplified procedures allowed by translational periodicity, or, at the other end of the spectrum, by the almost perfect randomness that the kinetic theory of gases is based on.

4.2 The Distribution Functions

In the following discussion, we shall first refer, for simplicity, to an ordered structure, namely an elemental face centre cubic (fcc) crystal where each atom is represented by a hard sphere with a radius r. Each sphere is in contact with 12 first neighbour spheres; as such the space is fully occupied. Hence, $2r$ represents both the spacing between the centres of two adjacent spheres and the spacing between the positions of the first neighbours in the fcc lattice.

shell order	R_i [2r]	Z_i
1	1	12
2	$\sqrt{2}$	6
3	$\sqrt{3}$	24
4	2	12
5	$\sqrt{5}$	24
6	$\sqrt{6}$	8
7	$\sqrt{7}$	48
8	$2\sqrt{2}$	6
9	3	36
10	$\sqrt{10}$	24
11	$\sqrt{11}$	24
12	$2\sqrt{3}$	24
13	$\sqrt{13}$	72
14	–	–
15	$\sqrt{15}$	48
16	4	12

Table 4.1. Features of the atomic coordination shells of increasing order for the fcc lattice.

If we take a lattice site as the origin, 0, we build a sphere, and progressively increase its radius R. As we look for the values of R we observe intersections between the "explorer" sphere and the atoms of the structure being examined. The first intersection, with 12 atoms, is observed for $R = R_1 = 2r$; this coordination shell, for the first neighbours, is characterised by two parameters, $R_1 = 2r$ and $Z_1 = 12$. As R grows, and $R_2 = 2r\sqrt{2}$, we obtain the following coordination shell, for the second neighbours, where $Z_2 = 6$. Continuing, the third coordination shell is given by $R_3 = 2r\sqrt{3}$ and by $Z_3 = 24$. For the fcc structure we build a discrete sequence of coordination shells characterised by the pair $(R_i; Z_i)$, until $i = 13$. Generally speaking the spacing between following, adjacent shells shortens as the order of the coordination shell increases. In the fcc case, there are no neighbours for $2r\sqrt{14}$ and, successively, for $2r\sqrt{30}$. Table 4.1 gives the R_i values and the distribution of the neighbour number of order i, Z_i, for the first 16 coordination shells in a fcc lattice.

Once this construction is generalised, we obtain the so-called Radial Distribution Function (RDF), which is given the symbol $J(x)$. For an ideal crystal, with no defects or thermal noise, each atom is rigidly locked onto its correct lattice site and $J(x)$ is thus a summation of Dirac δ functions, as shown schematically by curve (a) of Fig. 4.1. In a dilute gas, the RDF has an increasing monotonic trend since there are no interatomic distances where particular interparticle correlation occurs. This condition corresponds to the dashed curve (b) in Fig. 4.1. A non-crystalline system, whether liquid or solid, shows a *distribution* of irregularly spaced atomic positions. In this case the qualitative trend for $J(x)$ is the curve (c) in Fig. 4.1.

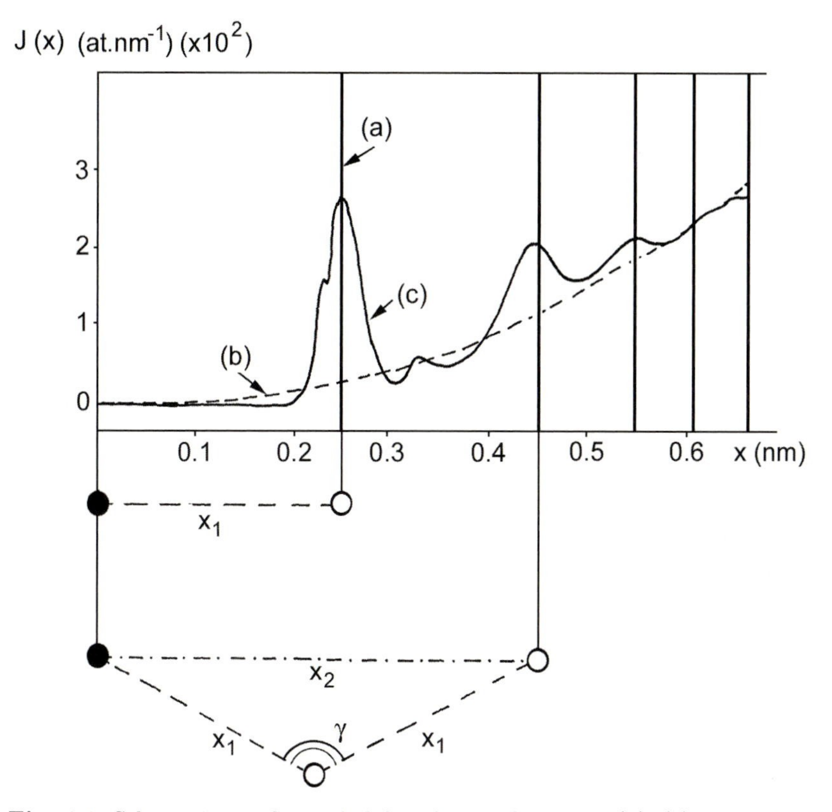

Fig. 4.1. Schematic trend in radial distribution function $J(x)$: (a) an ideal crystal; (b) a dilute gas; (c) a non-crystalline condensed system. An origin atom is represented by the full circle. Two atoms, that belong respectively to the shells of the first and second neighbours, here schematised by open circles, are shown to illustrate the definition of the correlation distance between first neighbours, x_1 and between second neighbours, x_2; γ is the bond angle between first neighbour atoms

In the description of the microscopic structure, particularly in determining the short range order ($0.2 - 0.5$ nm), the distribution functions for the interparticle distances are an essential mathematical tool. We can now define some of the most commonly used functions.

Let us consider N point-like, identical, indistinguishable particles, which lie within a volume V. The position vectors relative to their centres, are $\{\boldsymbol{X}_i\}$. The average number density is

$$N_0 = N/V. \tag{4.1}$$

We require each point in space to be occupied by a *single* particle centre, or *empty*. Hence, only *one* particle centre may occupy *one* point. Using the Dirac δ function the single-particle density function is

$$\zeta^1(\boldsymbol{x}) = \sum_{i=1}^{N} \delta(\boldsymbol{x} - \boldsymbol{X}_i) \tag{4.2}$$

and

$$\int_V \zeta^1(\boldsymbol{x})\mathrm{d}\boldsymbol{x} = N. \tag{4.3}$$

In a single-particle approach, this means that $\zeta^1(\boldsymbol{x})$ expresses the number of ways the particle can be located in the set of positions $\boldsymbol{X}_i$, without regard to the other particles.

The *two*-particle density function is

$$\zeta^2(\boldsymbol{x}_1, \boldsymbol{x}_2) = \sum_{i=1}^{N} \sum_{j\neq 1}^{N-1} \delta(\boldsymbol{x}_1 - \boldsymbol{X}_i)\, \delta(\boldsymbol{x}_2 - \boldsymbol{X}_j) \tag{4.4}$$

Equation (4.4) is null unless two particles are respectively located in $\boldsymbol{X}_i$ and $\boldsymbol{X}_j$. We must obviously disregard any possibility that the positions of any other particles may coincide with each other. If we integrate over $\boldsymbol{x}_2$ we will obtain a contribution whenever the following occurs,

$$\boldsymbol{x}_2 = \boldsymbol{X}_j \neq \boldsymbol{X}_i$$

thus

$$\int_V \zeta^2 \mathrm{d}\boldsymbol{x}_2 = (N-1) \sum_{i}^{N} \delta(\boldsymbol{x}_1 - \boldsymbol{X}_i) = (N-1)\zeta^1(\boldsymbol{x}_1). \tag{4.5}$$

So, having fixed the position of a particle, as given by $\zeta^1(\boldsymbol{x}_1)$, ζ^2 is the number of arrangements such that particle 1 is any of the points, with coordinates $\boldsymbol{X}_i$, and particle 2 is located in any one of the other points in the integration volume V.

If the particles under examination are atoms, then they are not spatially locked into rigid positions, so we have to turn to the ensemble, or time averages. These are given as

$$\langle \zeta^1(\boldsymbol{x}) \rangle = N^1(\boldsymbol{x}) \,.$$

$N^1(\boldsymbol{x})d\boldsymbol{x}$ thus represents the *average number* of particle centres in $d\boldsymbol{x}$ and, as such, coincides with the probability of finding a particle in $d\boldsymbol{x}$.

So,

$$\int_V N^1(\boldsymbol{x})\mathrm{d}\boldsymbol{x} = N \tag{4.6}$$

where $N^1(\boldsymbol{x})$ is referred to as the *number* density function or the *one-particle distribution* function.

If each particle is given its own size in order to conform to the hypothesis that the various particles do not coincide with each other we have to consider that $N^1(\boldsymbol{x})\mathrm{d}\boldsymbol{x}$ must be less than the unity. By taking the average of (4.5) we obtain

$$\mathrm{d}\boldsymbol{x}_1 \int_V N^2(\boldsymbol{x}_1, \boldsymbol{x}_2)\mathrm{d}\boldsymbol{x}_2 = (N-1)N^1(\boldsymbol{x}_1)\mathrm{d}\boldsymbol{x}_1. \tag{4.7}$$

In (4.7), $N^2(\boldsymbol{x}_1, \boldsymbol{x}_2)\mathrm{d}\boldsymbol{x}_1\, \mathrm{d}\boldsymbol{x}_2$ represents the number of configurations with the two particles 1 and 2 respectively in $\mathrm{d}V_1$ and in $\mathrm{d}V_2$ at the very same time. This is clarified when we notice that the result of the double integration for N^2 is $N(N-1)$, namely the number of pairs that can form in an ensemble of N elements (when considering pairs $1-2$ and $2-1$ as distinct pairs).

We define N^2 as the *two*-particle *distribution* function.

In general $N^2(\boldsymbol{x}_1, \boldsymbol{x}_2) \neq N^1(\boldsymbol{x}_1)\, N^1(\boldsymbol{x}_2)$ since the probability that $\boldsymbol{x}_2$ will be occupied may be influenced by the probability that $\boldsymbol{x}_1$ will be occupied. This effect is more significant the more effective the interparticle forces are on the interval $(\boldsymbol{x}_1 - \boldsymbol{x}_2)$.

We may define the *pair correlation* function $g^2(\boldsymbol{x}_1, \boldsymbol{x}_2)$ as

$$N^2(\boldsymbol{x}_1, \boldsymbol{x}_2) \equiv N^1(\boldsymbol{x}_1)N^1(\boldsymbol{x}_2)g^2(\boldsymbol{x}_1, \boldsymbol{x}_2) \tag{4.8}$$

As $(\boldsymbol{x}_1 - \boldsymbol{x}_2)$ tends towards the infinity, we expect g^2 to tend towards the unity, since the reciprocal interparticle influence cancels out.

If the system is *homogeneous*,

$$N^1(\boldsymbol{x}_1) = N^1(\boldsymbol{x}_2) = N_0$$

and as a consequence

$$N^2(\boldsymbol{x}_1, \boldsymbol{x}_2) = (N_0)^2\, g^2(\boldsymbol{x}_1, \boldsymbol{x}_2). \tag{4.9}$$

It is often convenient to put $(\boldsymbol{x}_2 - \boldsymbol{x}_1) = \boldsymbol{x}$ and make the origin coincide with the position of particle 1. In this case, eqs. (4.7) and (4.9), supposing that particle 1 is fixed in its official position, give us

$$N_0 \int_V g^2(\boldsymbol{x})\mathrm{d}\boldsymbol{x} = (N-1)\cdot 1. \tag{4.10}$$

The interpretation of (4.10) is that $N_0 g^2(\boldsymbol{x})\mathrm{d}\boldsymbol{x}$ represents the average number of particles and thus the probability that within the elementary volume $\mathrm{d}\boldsymbol{x}$ one particle will lie at a distance $\boldsymbol{x}$ from another particle located at the origin.

It may now be convenient for us to count also the particle located at the origin; in this case we define

$$Z(\boldsymbol{x}) \equiv N_0 g^2(\boldsymbol{x}) + \delta(\boldsymbol{x}) \tag{4.11}$$

and

$$\int_V Z(\boldsymbol{x})\mathrm{d}\boldsymbol{x} = N. \tag{4.12}$$

Since the disordered materials may be considered *isotropic*, even though they are not usually homogeneous, we may then put forward the following simplification

$$g^2(\boldsymbol{x}) \simeq g^2(|\boldsymbol{x}|) \equiv g(x). \tag{4.13}$$

In the last term of (4.13), apex 2, which gives the order for the considered correlation, has been suppressed since $g(x)$ conventionally indicates the *pair correlation* function.

From a formal point of view we can immediately extend the treatment to *three*, or more, particles. The probability of finding three particles in $\mathrm{d}\boldsymbol{x}_1$ $\mathrm{d}\boldsymbol{x}_2$ $\mathrm{d}\boldsymbol{x}_3$ at the very same time is $N^3(\boldsymbol{x}_1, \boldsymbol{x}_2, \boldsymbol{x}_3)$ $\mathrm{d}\boldsymbol{x}_1$ $\mathrm{d}\boldsymbol{x}_2$ $\mathrm{d}\boldsymbol{x}_3$. Now,

$$N^3(\boldsymbol{x}_1, \boldsymbol{x}_2, \boldsymbol{x}_3) \equiv N^1(\boldsymbol{x}_1)N^1(\boldsymbol{x}_2)N^1(\boldsymbol{x}_3)g^3(\boldsymbol{x}_1, \boldsymbol{x}_2, \boldsymbol{x}_3) \tag{4.14}$$

where g^3 is the *triplet correlation* function. If the distance among particle triplets in amorphous materials is large, g^3 tends towards the unity. In a homogeneous system,

$$N^3 = (N_0)^3 g^3 \tag{4.15}$$

If we use the previous definitions and results we immediately find that

$$\int N^3 \mathrm{d}\boldsymbol{x}_3 = (N-2)N^2(\boldsymbol{x}_1, \boldsymbol{x}_2) \tag{4.16}$$

$$\int\int N^3 \mathrm{d}\boldsymbol{x}_3 \mathrm{d}\boldsymbol{x}_2 = (N-1)(N-2)N^1(\boldsymbol{x}_1) \tag{4.17}$$

$$\int\int\int N^3 \mathrm{d}\boldsymbol{x}_3 \mathrm{d}\boldsymbol{x}_2 \mathrm{d}\boldsymbol{x}_1 = N(N-1)(N-2). \tag{4.18}$$

When we repeat the procedure we find that the result of integrating $N^{\mathrm{N}}(\boldsymbol{x}_1 ... \boldsymbol{x}_n)$ is $N!$, namely the number of ways N distinguishable particles can be located in the same number of distinct positions.

Each *higher* order correlation function supplies us with better structural information about the system under examination as compared to the corresponding lower order function. In particular, if we take (4.16) as an example, we will notice that an *infinite* number of *triplet* particle arrangements is compatible with the *same pair* distribution function. This result immediately leads us to understand how badly detailed the information contained in $g(x)$

is as regards a disordered structure. Unfortunately, $g(x)$ often represents the maximum amount of information that can be obtained from the experiment.

The very problem of obtaining from the experiment, particularly from diffraction, a univocal picture of the arrangements of atoms, has led to the development of *structural models.*

The average number of particles found in a homogeneous isotropic spherical shell with radius $\boldsymbol{x}$ and thickness $d\boldsymbol{x}$, centred on a particle chosen as the origin 0, may be expressed as

$$\int_0 \int_{shell} N^1(\boldsymbol{x}_1) N^1(\boldsymbol{x}_2) g^2(\boldsymbol{x}_1, \boldsymbol{x}_2) \mathrm{d}\boldsymbol{x}_1 \mathrm{d}\boldsymbol{x}_2$$

$$= \int_{shell} N^1(\boldsymbol{x}_2) \int_0 \delta(\boldsymbol{x}_1 - 0) g^2(\boldsymbol{x}_1, \boldsymbol{x}_2) \mathrm{d}\boldsymbol{x}_1 \mathrm{d}\boldsymbol{x}_2 \qquad (4.19)$$

since $N^1(\boldsymbol{x}_1) = \delta(\boldsymbol{x}_1 - 0)$, provided, of course, that origin 0 has been determined.

Now, considering the integration over the shell, where the integrand is

$$N^1(\boldsymbol{x}_2) g^2(\boldsymbol{x}_2, 0) \mathrm{d}\boldsymbol{x}_2$$

and if we take $\boldsymbol{x}_2 = \boldsymbol{x}$ as the only significant variable, we obtain

$$\int_{shell} N^1(\boldsymbol{x}) g^2(\boldsymbol{x}) \mathrm{d}\boldsymbol{x}.$$

Furthermore, since the system is homogeneous,

$$N^1(\boldsymbol{x}) = N_0 = \text{constant}$$

and for isotropy we consider $|\boldsymbol{x}|$. Given this hypothesis, we obtain

$$\int_{shell} N^1(x) g^2(x) \mathrm{d}x = \int_{R_\mathrm{m}}^{R_\mathrm{M}} 4\pi x^2 N_0 g(x) \mathrm{d}x \qquad (4.20)$$

where R_m and R_M are respectively the minimum and the maximum radii of the shell. If a shell has infinitesimal thickness, the pair distribution density (4.20) reduces to

$$4\pi x^2 N_0 g(x) \mathrm{d}x.$$

If we now substitute the density function $\varrho(x)$ for $g(x)$, to indicate the atomic pair correlation function, then

$$4\pi x^2 N_0 g(x) = 4\pi x^2 \varrho(x) \equiv J(x) \qquad (4.21)$$

Literature usually refers to r instead of x; $J(x)$ is called the *Radial Distribution Function* (RDF).

Figure 4.2 schematically shows the trend in density $\varrho(x)$ in a disordered system of identical particles, starting from an arbitrarily chosen atom. It

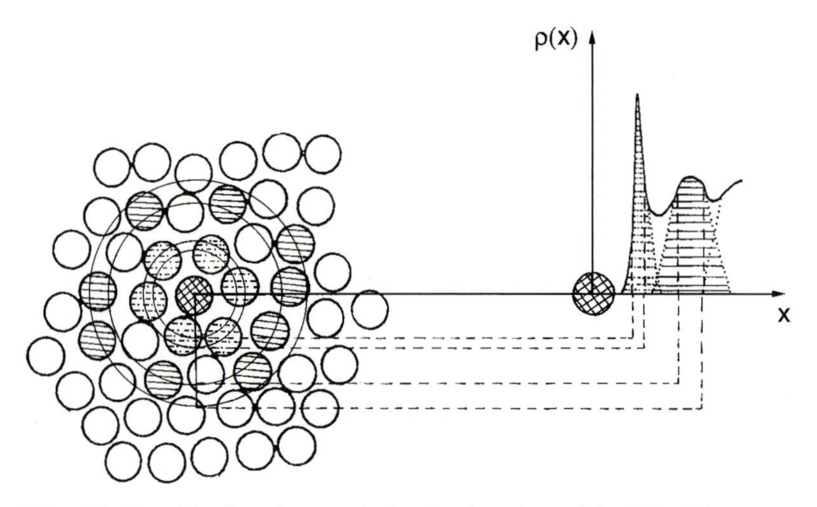

Fig. 4.2. Trend in the microscopic density function, $\varrho(x)$, defined for a structurally disordered system of identical atoms, starting from an arbitrary origin atom. For distances x less than the average interatomic distance between first neighbours, x_1, $\varrho(x)$ is null

will be noticed that where, by definition, no interatomic correlation occurs, namely for values of x below the average interatomic distance between first neighbours, x_1, $\varrho(x)$ is zero. At the other end of the spectrum, for large values of x, where the material may be considered homogeneous, the absolute value of $\varrho(x)$ coincides with the average value for the density of the material, ϱ_0. In the region between these two extremes, in a glassy system, the density function, when represented as a function of x, has an *oscillating* trend. The x values that correspond to the peaks in function ϱ are interpreted as the average interatomic distances. The radial distribution function, in turn, has a null value below the first peak, centred on x_1, like the analogous peak of $\varrho(x)$. Furthermore, as can be seen in Fig. 4.1, $J(x)$ oscillates around the parabolic curve for the trend in the average density $4\pi x^2 \varrho_0$.

Attribution of any precise physical meaning to $J(x)$ has limited meaning, as does any procedure to deduce structural parameters of a non-crystalline system, starting from the bare analysis of the radial distribution function. First of all, $J(x)$ is a one-dimensional representation of a three-dimensional structure; as such, the information contained in it is a spatial average, and is only valid the more isotropic the system is. Moreover, from the RDF definition (4.20), the integral of a given peak gives us the effective coordination number, Z, for that atomic shell. The first peak is the best defined one. In non-crystalline materials, the width χ_a^2 is due to static disorder (topological and/or chemical), and corresponds to a distribution of the interatomic bond lengths, χ_d^2 and to thermal disorder χ_t^2

$$\chi_\mathrm{a}^2 = \chi_\mathrm{d}^2 + \chi_\mathrm{t}^2. \tag{4.22}$$

The position of the second peak in the $J(x)$ function gives the average interatomic distance between second neighbours x_2. According to Fig. 4.1, once both parameters are known, we can calculate the bond angle γ

$$\frac{\gamma}{2} = \arcsin\left(\frac{x_2}{2x_1}\right) \tag{4.23}$$

The severe limitations to the radial distribution function are already highlighted at the level of the second peak. In an amorphous system, the second peak is consistently broader than the first peak. This can be attributed, at least in covalently bonded solids, to a change in the bond angles of up to 1/10 of the equilibrium value. This contribution has to be added to the contribution from thermal disorder. More realistically, we have to consider that, for an amorphous solid, the area right below the second peak depends also on correlations of order higher than the second. The role of these contributions is highlighted by the fact that the second $J(x)$ peak does not fall to zero on the side of increasing X values.

Taking the $J(x)$ trend as a whole as x increases, we notice that as the order of the considered peak of the radial distribution function increases, the relative weight of the higher order pair correlations contributing to that peak also increases, thus making any simple geometric interpretation of the characteristic properties of $J(x)$ impossible right from the third peak. All the problems discussed so far make it necessary to introduce structure modelling.

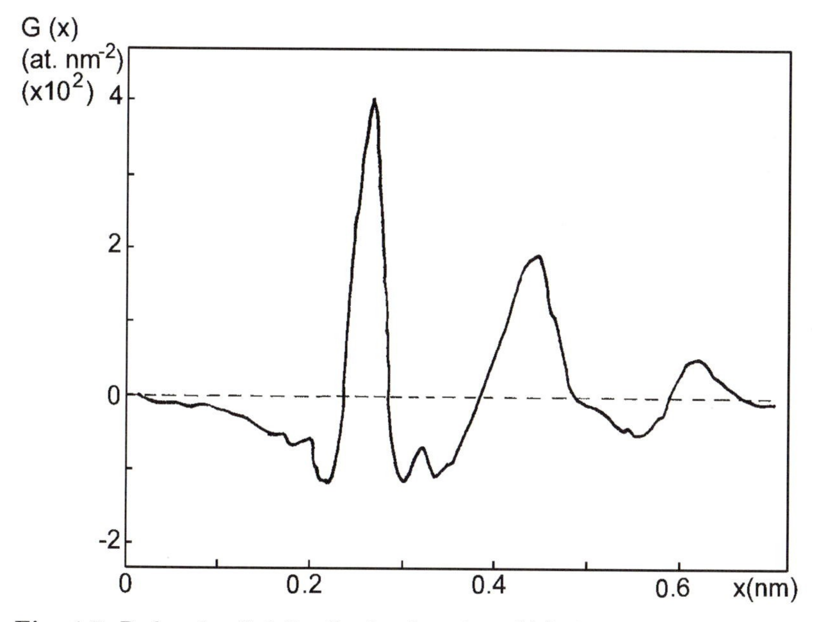

Fig. 4.3. Reduced radial distribution function, $G(x)$, for a structurally disordered condensed system

The *reduced* radial distribution function (RRDF) is often used without changing the supplied information:

$$G(x) = 4\pi x[\varrho(x) - \varrho_0] = \frac{J(x)}{x} - 4\pi x\varrho_0. \tag{4.24}$$

The function $G(x)$ (Fig. 4.3) is said to be "reduced" because it oscillates around zero instead of being an increasing, or decreasing, function of x, like $J(x)$ is. In the region of radial values below x_1, where $J(x)$ is zero, $G(x)$ decreases linearly like $-4\pi x\varrho_0$.

$G(x)$ for large values of x is asymptotic to zero, thus indicating a lack in correlation; the information on the average density of the material is deduced from the slope of $G(x)$ in the origin.

4.3 Experimental Techniques: Diffraction

Diffraction effects are associated with any kind of wave propagation when the waves meet an obstacle whose size can be compared with the incident radiation wavelength. When the scattering centres are atoms organised in condensed systems, *X-rays*, *electrons* and *neutrons*, are probes whose wavelength may be in the order of atomic sizes.

X-ray diffraction is commonly used in structural analysis owing to the ease by which a collimated beam of adequate intensity is obtained. In the case of scattering from a single crystal, X-rays with wavelength λ, elastically scattered by families of atoms lying on regularly spaced planes, give rise to constructive interference whenever the Bragg equation holds

$$2d\sin\theta = n\lambda \tag{4.25}$$

where d is the interplanar spacing and n an integer. We only observe beams of diffracted X-rays for values of λ and of the scattering angle 2θ such that

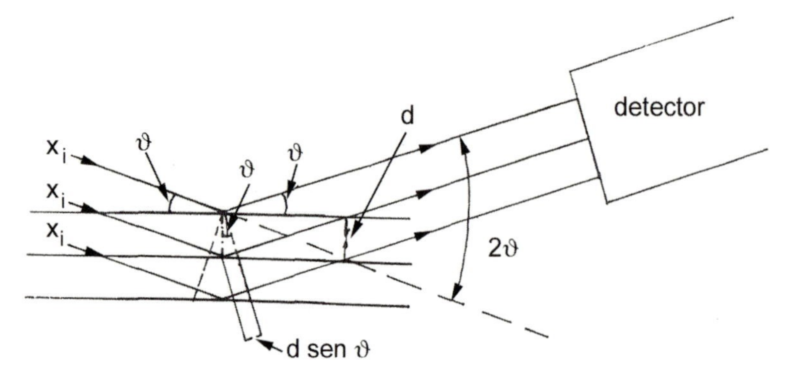

Fig. 4.4. Typical geometry in a diffraction experiment; three atomic layers are represented

(4.25) is fulfilled, as shown in Fig. 4.4. Using a sensitive screen placed on a plane normal to the direction of the monochromatic incident beam, we observe a series of sharp spots distributed in a complicated array, though highly regular in shape. From the study on the position of the spots and their variation in intensity we can determine the structure of the crystal.

Collections of crystals oriented in a random manner, just like in the polycrystalline systems, and subject to X-ray diffraction, produce typical concentric circles, with sharp edges, of varying intensities from each other. Lastly, in the case of an *amorphous* material, we observe *diffuse halos*.

When there is no structural periodicity the Bragg law cannot be used to interpret the experimental data and the diffraction conditions from a disordered structure have to be obtained from the very beginning.

We shall specifically examine the case of X-ray diffraction from an amorphous sample. The radiation is scattered by the electrons in the atoms of the material. The coherent scattering due to a single electron constitutes the fundamental process. We suppose that the X-ray wavelength is far different from the values that correspond to the X-ray absorption edges for the examined material. In turn, the electron bonding energy is very low compared to the X-ray photon energy; thus, a fairly simple classical representation of the interaction gives us results that are equivalent to the results obtained from a complete quantum analysis.

We shall start by taking a single electron and consider it as a *point* charge. The Thomson relation describes its behaviour as a scattering source. We shall assume hereafter that the scattered waves may be considered plane waves, described by the pertinent wavevectors $\boldsymbol{k}$.

As far as coherent scattering is concerned, where the radiation wavelength is conserved, and using an unpolarised beam, the differential scattering cross-section is

$$\mathrm{d}\sigma/\mathrm{d}\Omega = F^2(1+\cos^2\theta)/2. \tag{4.26}$$

The angular factor $(1+\cos^2\theta)/2$, indicated as Θ, is called the polarisation factor. The corresponding polar diagram is represented in Fig. 4.5. The explicit form for the term F in (4.26) is

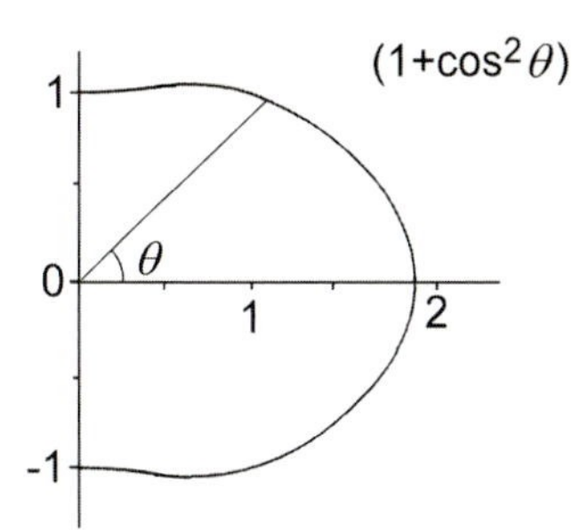

Fig. 4.5. Polar diagram of the angular factor $(1+\cos^2\theta)$

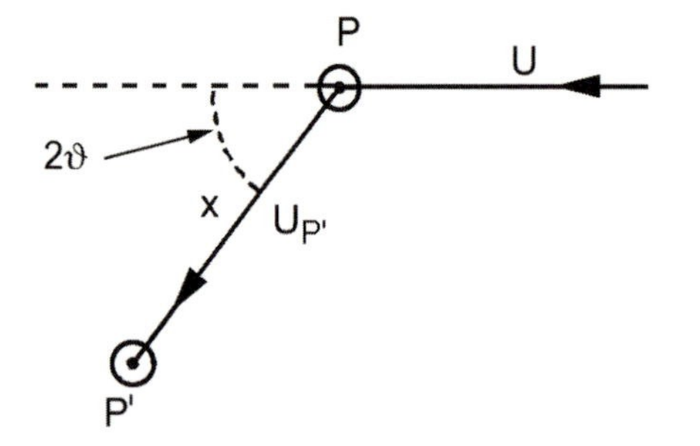

Fig. 4.6. Scattering geometry for radiation of intensity U scattered from a point-like electron located in P

$$F = (e^2/4\pi\varepsilon_0 m_e c^2)$$

where e and m_e are the electron charge and mass, ε_0 is the dielectric constant of the vacuum and c the speed of light in the vacuum.

Given this, we suppose that an electron located at point P will cause radiation of intensity U to be scattered through an angle 2θ (Fig. 4.6). The intensity of the scattered radiation observed at P', which is distant X from P, is

$$U_{P'} = (U/X^2)F^2\Theta. \tag{4.27}$$

If, more realistically, we now want the electron charge to be distributed uniformly within a volume $\mathrm{d}V$, as $\varrho\mathrm{d}V$, and not concentrated at one point, then the classical scattering induced by the "extended electron" will be $\varrho\mathrm{d}V$ times the scattering amplitude from the point-like electron. In order to consider the difference in optical path between scattered rays, originating from different points of the charge distribution at distance $\boldsymbol{x}$ from an arbitrary origin (e.g. the centre of symmetry for the system, which, in the case of an atom, coincides with the centre of the atom), the scattering intensity becomes

$$(U/X^2)F^2\Theta = S_e S_e^* = |S_e|^2 \tag{4.28}$$

where the scattering factor S_e is given by

$$S_e = \int \varrho(\boldsymbol{x}) \exp\left[(2\pi \mathrm{i}/\lambda)(\widehat{\boldsymbol{\xi}} \cdot \boldsymbol{x})\right] \mathrm{d}V. \tag{4.29}$$

In (4.29), the vector difference between the two vectors $\boldsymbol{k}_2$ and $\boldsymbol{k}_1$ for, respectively, the scattered radiation and the incident radiation, is $\boldsymbol{\xi} = (\boldsymbol{k}_2 - \boldsymbol{k}_1)$, and $\widehat{\boldsymbol{\xi}}$ is the versor, $\boldsymbol{\xi}/|\boldsymbol{\xi}|$. Since the wavelength is conserved in coherent scattering, then $\boldsymbol{k}_2$ and $\boldsymbol{k}_1$ have the same magnitude, $(2\pi/\lambda)$. Referring to Fig. 4.7, and using trigonometric relations, we obtain the wavevector magnitude $|\boldsymbol{\xi}|$, where

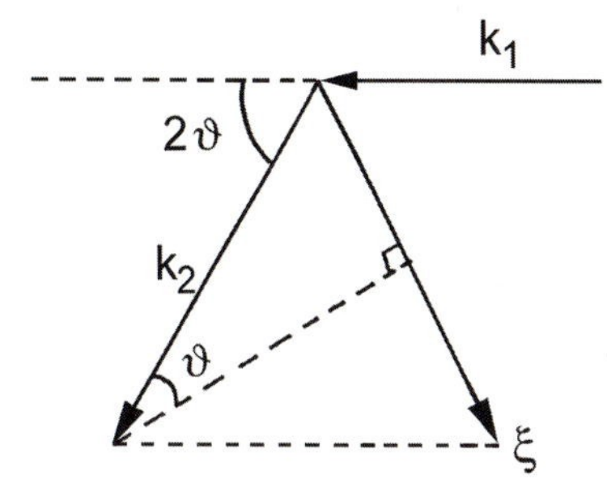

Fig. 4.7. Geometrical illustration to define the scattering wavevector $\boldsymbol{\xi}$

$$|\boldsymbol{\xi}| = 2(2\pi/\lambda)\sin\theta.$$

If we give this scattering wavevector the usual symbol $\boldsymbol{k}$, then we obtain in vectorial form

$$\boldsymbol{k} = \frac{4\pi}{\lambda}\sin\theta\,\widehat{\boldsymbol{\xi}}. \tag{4.30}$$

Neutron scattering literature refers to $\boldsymbol{Q}$ as the scattering vector. For simplicity, we shall use the symbol $\boldsymbol{k}$ even when we discuss the results of neutron scattering experiments.

We adopt the simplified, though widely applied, hypotheses that the symmetry is spherical, thus $\varrho(\boldsymbol{x}) = \varrho(x)$ and the origin coincides with the origin of $\boldsymbol{x}$. If we choose a set of three spherical coordinates x, $\eta(0 \le \eta \le 2\pi)$, $\varphi(0 \le \varphi \le \pi)$, and we integrate with respect to η, then the scattering factor for a single electron, S_e, becomes

$$S_\mathrm{e} = \int\limits_{x=0}^{\infty}\int\limits_{\varphi=0}^{\pi} \varrho(x)\exp\left[\mathrm{i}kx\cos\varphi\right] 2\pi x^2 \sin\varphi\,\mathrm{d}\varphi\,\mathrm{d}x. \tag{4.31}$$

If we then integrate this over φ we obtain

$$S_\mathrm{e} = \int\limits_{0}^{\infty} 4\pi x^2 \varrho(x)\frac{\sin kx}{kx}\mathrm{d}x. \tag{4.32}$$

For the general case where the atom contains i electrons we simply assume that the electron distribution has spherical symmetry. This is true for heavy atoms where the fraction of valence electrons, involved in non-spherically symmetric chemical bonding, is small. As such, the atomic scattering factor, or *form factor*, is simply the summation of the individual amplitudes

$$S = \sum_i S_{\mathrm{e},i} = \sum_i \int_0^\infty 4\pi x^2 \varrho_i(x)\frac{\sin kx}{kx}\mathrm{d}x \tag{4.33}$$

and the total scattered intensity U_t is

$$U_{\mathrm{t}} = (U/X^2)F^2\Theta \times SS^* = |S|^2 \times (U/X^2)F^2\Theta \,. \tag{4.34}$$

When k is small, the value of $[(\sin kx)/(kx)]$ is roughly around unity and the form factor (4.33) gives us the number of electrons in the atom. S tends to the atomic number Z when k, and thus $(\sin\theta)/\lambda$ tend to zero. The form factor dependence on k is strong, and it is a characteristic feature of each element. This dependence is tabulated in the literature for all atomic species. As an example, the typical trends for oxygen and hydrogen is given in Fig. 4.8.

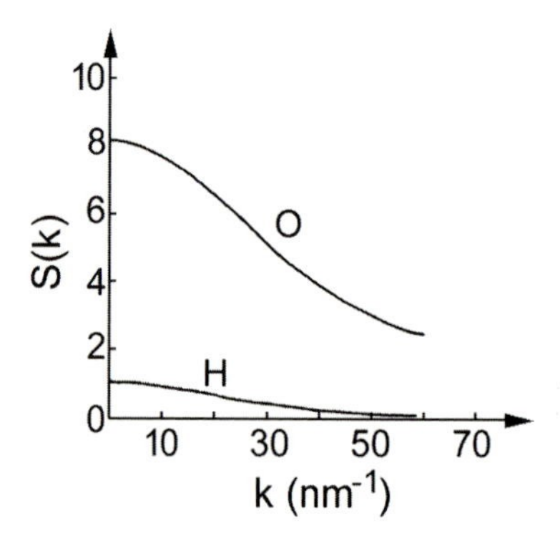

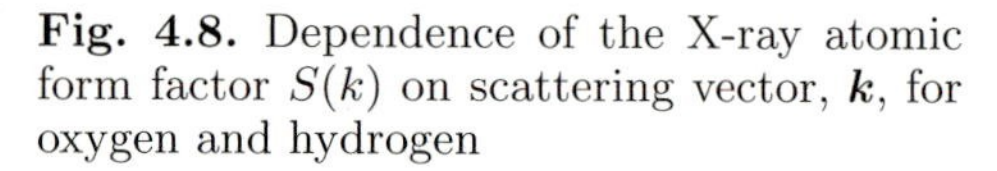
Fig. 4.8. Dependence of the X-ray atomic form factor $S(k)$ on scattering vector, $\boldsymbol{k}$, for oxygen and hydrogen

For a *distribution* of n atoms, the scattering intensity is given by the summation of the scattering amplitudes from each of the n atoms in the $\boldsymbol{x}_n$ positions, multiplied by their complex conjugates, thus

$$(U/X^2)F^2\Theta = \sum_n S_n \exp\left[(2\pi\mathrm{i}/\lambda)(\widehat{\boldsymbol{\xi}} \cdot \boldsymbol{x}_n)\right] \times \sum_m S_m \exp\left[(2\pi\mathrm{i}/\lambda)(\widehat{\boldsymbol{\xi}} \cdot \boldsymbol{x}_m)\right] . \tag{4.35}$$

In (4.35), which is valid for both crystalline and amorphous materials, we assume that the form factors have real values. In crystals, the translational periodicity implies that for any two atomic position vectors, $\boldsymbol{x}_n$ and $\boldsymbol{x}_m$, there is a reciprocal lattice vector $\boldsymbol{k}$ where $\boldsymbol{k} = \boldsymbol{x}_n - \boldsymbol{x}_m$. In this case, the sums in (4.35) constitute a geometrical progression that can be evaluated. In an amorphous system, with no simple relation between atomic position vectors, the sums are left as they are and we can only re-write equation (4.35) in a more compact manner, with the notation $(\boldsymbol{x}_n - \boldsymbol{x}_m) = \boldsymbol{x}_{nm}$

$$(U/X^2)F^2\Theta = \sum_n \sum_m S_n S_m \exp\left[(2\pi\mathrm{i}/\lambda)(\widehat{\boldsymbol{\xi}} \cdot \boldsymbol{x}_{nm})\right] . \tag{4.36}$$

Figure 4.9 schematically shows the geometrical relation between $\boldsymbol{x}_{nm}$ and $\widehat{\boldsymbol{\xi}}$. If we can assume that the amorphous solid is *isotropic*, the $\boldsymbol{x}_{nm}$ vector spans all the orientations with the very same probability, thus it is possible

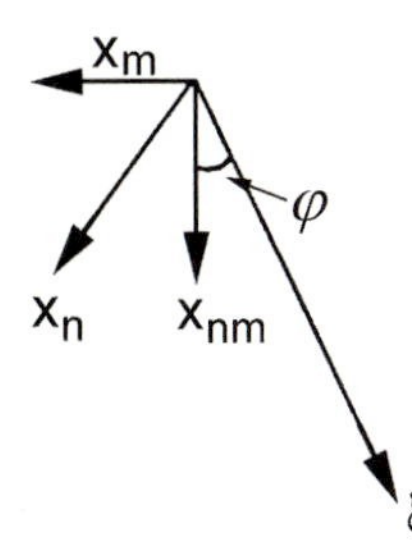

Fig. 4.9. Geometrical relation between ξ and $\boldsymbol{x}_{nm}$, defined as the difference between the two position vectors $\boldsymbol{x}_n$ and $\boldsymbol{x}_m$

to calculate the average of the exponential term in (4.36), for the intensity of the scattered radiation

$$\left\langle \exp\left[(2\pi\mathrm{i}/\lambda)(\widehat{\boldsymbol{\xi}} \cdot \boldsymbol{x}_{nm})\right]\right\rangle = \frac{\sin kx_{nm}}{kx_{nm}}. \tag{4.37}$$

When we substitute the result in (4.37) for (4.36), we obtain, for a random distribution of atoms that scatter the radiation, the following relation

$$(U/X^2)F^2\Theta = \sum_n \sum_m S_n S_m \frac{\sin kx_{nm}}{kx_{nm}}. \tag{4.38}$$

Equation (4.38) is often called the Debye equation and is the simplest form, under very general conditions, to express the angle and the wavelength dependence (see (4.30)) of the average scattered intensity from a completely random spatial distribution of identical atoms.

It is now necessary to specify what *kind* of atom we are considering in the analysis. In the simple case of an elemental system, the Debye equation can be re-written as

$$(U/X^2)F^2\Theta = \sum_n S^2 + \sum_n \sum_{m\neq n} S^2 \frac{\sin kx_{nm}}{kx_{nm}}. \tag{4.39}$$

When we introduce the density function $\varrho_n(x_{nm})$ for the origin atom n, still on the assumption of spherical symmetry, and by integrating over the sample volume, we obtain

$$(U/X^2)F^2\Theta = \sum_n S^2 + \sum_n S^2 \int \varrho_n(x_{nm}) \frac{\sin kx_{nm}}{kx_{nm}} \mathrm{d}V_m. \tag{4.40}$$

The microscopic *density* function, averaged out on the n atoms of the sample is

$$\langle \varrho_n(x_{nm})\rangle = \varrho(x) \tag{4.41}$$

If the macroscopic average density is ϱ_0, when we add and then we subtract a term in ϱ_0, we obtain

$$
\begin{aligned}
(U/X^2)F^2\Theta = \\
&\sum_n S^2 + \sum_n S^2 \int 4\pi x^2 \left[\varrho(x) - \varrho_0\right] \frac{\sin kx}{kx} \mathrm{d}x \\
&+ \sum_n S^2 \int 4\pi x^2 \varrho_0 \frac{\sin kx}{kx} \mathrm{d}x .
\end{aligned} \tag{4.42}
$$

The sum over n gives us N, the total number of atoms in the sample. The quantity $[\varrho(x) - \varrho_0]$ tends to zero for distances greater than the interatomic distances relative to the atoms within the first three-four coordination shells. This confirms that the lack in long range order, and thus the lack in strong correlations that extend over great distances, make the density function $\varrho(x)$ converge to the average macroscopic density of material ϱ_0, for sufficiently large values of x.

In integrating the second term of (4.42), the contribution of the scattering centres which are neighbours with each other and between them and the origin, is dominant, whereas the third term represents the interaction between atoms that are a long distance apart, and thus consists in small angle scattering. If L indicates the sample size, then the integral is usually limited to values $|\boldsymbol{k}| < (2\pi/L)$ of the scattering vector. Given L values are in a few millimetre range and the X-ray wavelength in a few tenths of a nanometre range, the third term of (4.42) includes X-ray scattering at very small angles, so that the primary (transmitted) beam obscures the scattered beams.

Equation (4.42) thus reduces to the *total scattering intensity*, $I(k)$, so

$$
\begin{aligned}
I(k) &= (U/X^2)F^2\Theta \\
&= NS^2 + NS^2 \int_0^\infty 4\pi x^2 \left[\varrho(x) - \varrho_0\right] \frac{\sin kx}{kx} \mathrm{d}x .
\end{aligned} \tag{4.43}
$$

The upper integration limit indicates that the typical size of the sample is much greater than the atomic size. In Fig. 4.10 $I(k)$ is the curve we obtain directly from the experiment after we have subtracted the incoherent scattering contributions, such as those due to Compton scattering.

Currently, energy dispersive detectors are available that can measure the coherent contribution to scattered intensity, $I'(k)$ only. In the case of X-ray diffraction $I(k)$ has an oscillating trend around the curve of the squared

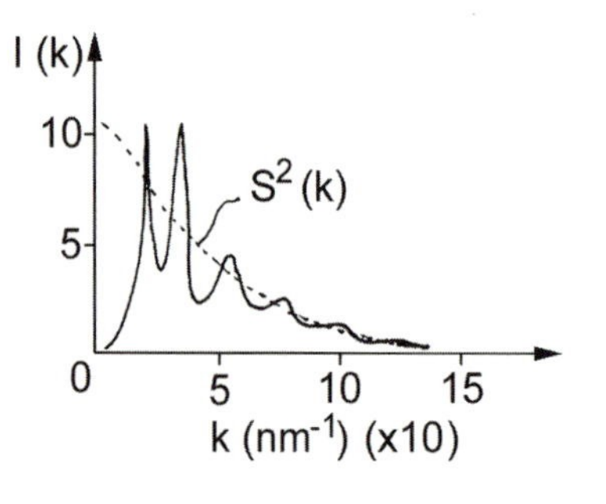

Fig. 4.10. Qualitative trend in the total scattering intensity, $I(k)$, as a function of the scattering vector k, for an amorphous film (adapted from [4.1])

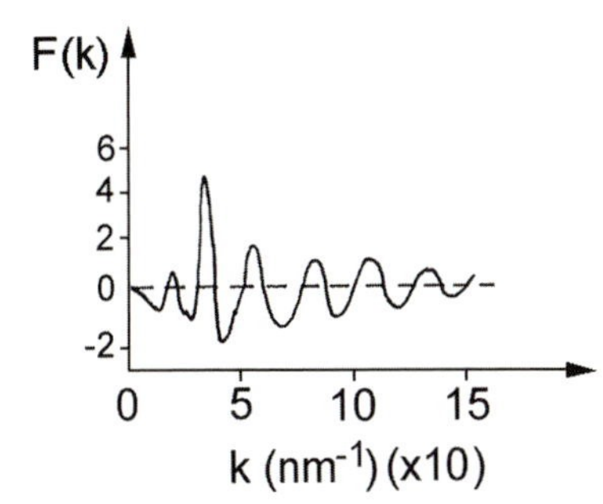

Fig. 4.11. Qualitative trend in the reduced scattering intensity, $F(k)$, as a function of the scattering vector k, for an amorphous film (adapted from [4.1])

atomic scattering intensity, S^2, which strongly depends on the value of the scattering vector, k. This property is characteristic of X-rays, whereas, in the case of neutron diffraction, the scattering amplitude is independent from k unless there is a magnetic scattering contribution from unpaired electrons.

A further simplification of (4.43) is obtained by introducing the *reduced scattering intensity* $F(k)$

$$F(k) = k \left\{ \left[\frac{(U/X^2)\, F^2 \Theta}{N} - S^2 \right] \Big/ S^2 \right\}. \tag{4.44}$$

$F(k)$, whose schematic trend is given in Fig. 4.11, oscillates around zero instead of being an increasing, or decreasing, function of k. If we express the first part of (4.43) as a function of $F(k)$, then using (4.44) we obtain

$$(U/X^2) F^2 \Theta = \left(\frac{S^2 F(k)}{k} + S^2 \right) N. \tag{4.45}$$

When we substitute (4.45) in (4.43) and we use the definition for $G(x)$ provided in (4.24), then

$$F(k) = \int_0^\infty G(x) \sin(kx)\; \mathrm{d}x. \tag{4.46}$$

This equation gives us the link between a quantity that is *directly* obtainable experimentally, such as $F(k)$, and a function that describes the structure of the disordered system in real space, namely $G(x)$.

Given its structure, (4.46) can be Fourier transformed, giving, upon inversion

$$G(k) = (2\pi)^{-1/2} \int_0^\infty F(k) \sin(kx)\; \mathrm{d}k. \tag{4.47}$$

Equation (4.47), from which we obtain the correlation function in real space, in principle requires experimental data for infinite values of the scattering vector k. Since, in practice, it is impossible to fulfil the condition, this is a major difficulty when analysing experimental data. However, the analysis of the reduced scattering intensity, $F(k)$, often supplies us with more information than we can obtain from the radial distribution function when we compare the structure of amorphous systems that are similar to each other. This typically occurs with samples of the same material but with slightly different stoichiometries. The reason for this is clearer when we introduce the *structure factor* $\mathfrak{S}(k)$ which is given as

$$\mathfrak{S}(k) = [(U/X^2)F^2\Theta]/NS^2 \tag{4.48}$$

The dividend of $\mathfrak{S}(k)$ is defined in terms of the scalar product $(\boldsymbol{k} \cdot \boldsymbol{x})$ and includes the k dependence. The structure factor is *independent* of the kind of radiation used, due to the normalisation by the S^2 factor, and oscillates about an average unity value. If we now express $F(k)$ in terms of the structure factor we obtain

$$F(k) = k[\mathfrak{S}(k) - 1]. \tag{4.49}$$

The relevance of high k values in $F(k)$ is emphasised in (4.49). When analysing the data for $\mathfrak{S}(k)$, errors could be made that affect both the function ordinate and the abscissa. The first does not alter the structure or the characteristic properties of $G(x)$ (see (4.24)), but it affects the estimate of the coordination number Z. In amorphous metallic systems the latter is defined to an accuracy of a few percent. The resolution for $G(x)$, $\Delta x = (2\pi/\left|\boldsymbol{k}_{\max}\right|)$, which is determined from the position of the abscissa starting from the experimental value of $k_{\max}$, is usually between 0.08 nm, when we use CuK_α radiation, and 0.01 nm for high values of $k_{\max}$, typically achieved with X-rays from a synchrotron, with $\lambda \simeq 10^{-2}$ nm and $k_{\max} \simeq 600\ \mathrm{nm}^{-1}$.

The procedure used to convert data in the reciprocal space into correlation functions in real space becomes more and more delicate as the interval of accessible experimental values for k becomes narrower. In practice, let us consider (4.30), by which the diffusion vector is defined as $|\boldsymbol{k}| = (4\pi \sin\theta)/\lambda$. The highest value for k is obtained when $\theta = 90°$, and this depends critically on the wavelength λ of the X-rays used. For Mo K_α radiation, $\lambda = 0.071$ nm, which corresponds to $|\boldsymbol{k}_{\max}| = 177\ \mathrm{nm}^{-1}$ (for Cu K_α we obtain $|\boldsymbol{k}_{\max}| = 81.5\ \mathrm{nm}^{-1}$).

The transformation from reciprocal space to real space would only be perfect if (see (4.46)) we knew $F(k)$ for an infinite set of k values. In practice, the data is interrupted at a finite $k_{\max}$ value; as such we have to multiply $F(k)$ by a truncation function $T(k)$, which is defined in such a way as to obtain a unit value for $k < k_{\max}$ and null for $k > k_{\max}$; the result is

$$\begin{aligned} G'(x) &= (2\pi)^{-1/2} \int_0^\infty F(k) \sin(kx)\ \mathrm{d}k \\ &\simeq (2\pi)^{-1/2} \int_0^\infty F(k)T(k) \sin(kx)\ \mathrm{d}k. \end{aligned} \tag{4.50}$$

Obviously, when using this procedure we introduce truncation errors in the Fourier transform, namely in $G(x)$. The weight of these errors is less the higher the value of $k_{\max}$, i.e. the less the wavelength of the X-rays used.

When examining the limitations to the procedure used to define $G(x)$, we note that the curves for $F(k)$ are obtained directly from the measured data, and as such are not affected by any spurious effects of the Fourier transformation procedure. The drawback to observing the correlation curves in reciprocal space is that we cannot immediately derive from these curves any structural correlation in real space. We understand this problem when we study the trend in the contributions to $F(k)$ from correlations of different order, and compare them with the contributions from the very same correlations to $G(x)$.

With reference to an ideal elemental disordered system, at low temperature, and if the interatomic correlations are confined to the first correlation shell (see Fig. 4.2), with correlation length x_1, then $G(x)$ gives us a *unique* peak, centred on x_1; for simplicity we shall presume this is a lorentzian peak, so that

$$G(x) = \frac{1}{4\pi}\left[\frac{1}{\chi_1^2 + (-x - x_1)^2}\right]^{1/2}$$

where the peak width, χ_1^2, is due to static disorder alone. Thus, after Fourier transforming, $F(k)$, (or $\mathfrak{S}(k)$), is a dampened sine curve,

$$F(k) = \exp\left[-k\chi_1\right] \sin kx_1$$

where χ_1 is the damping degree. Using the relation $k_1 = (2\pi/x_1)$ we obtain, from the period k_1 of the sine curve, the distance x_1 between first neighbours.

Using the same assumption made for the first coordination shell, we could deal with the effect of another shell of higher order, with bond length x_2. The contribution from this structural correlation to $F(k)$ would be a sine curve with k_2 and χ_2 parameters. When the values for k are small, $F(k)$ would have a complicated structure, as a result of the mixing of the two contributions. On the other hand, the width of the coordination shells increases with their order, namely $\chi_1^2 > \chi_2^2$ for $i > j$ (see Fig. 4.2), and the contribution to the reduced scattering intensity of increasingly ill defined structural coordinations would be progressively lowered. Based on this hypothesis, for large values of k, the structure of $F(k)$ should depend only on the correlation between first neighbours and would thus be a simple sine curve of period $(2\pi/x_1)$, from which we could deduce the bond length x_1.

One notable example of structural correlation which can be interpreted with the above hypothesis is given by the transition metal-metalloid amorphous alloys where the structure factor $\mathfrak{S}(k)$ has a first, sharp peak centred around 20–30 nm^{-1}, as shown in Fig. 4.12. On the other hand, for "large" distances, i.e. greater than approximately double the correlation distance between first neighbours, the radial distribution function shows oscillations that

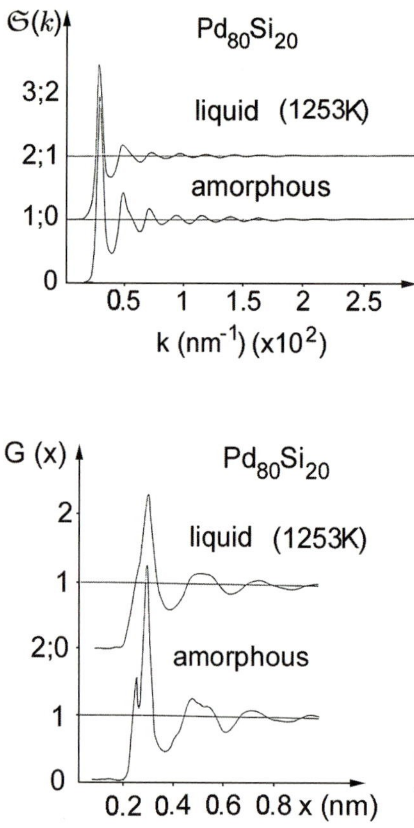

Fig. 4.12. Trend in the structure factor $\mathfrak{S}(k)$ for amorphous and liquid $Pd_{80}Si_{20}$ (adapted from [4.2])

Fig. 4.13. Trend in the reduced radial distribution function, $G(x)$, for amorphous and liquid $Pd_{80}Si_{20}$ (adapted from [4.2])

are almost periodic. These correspond to the characteristic repetition distance of the diameter of hard spheres, typical of quasi-random packing of spheres. As such, we obtain an initial approximation of the structure (Fig. 4.13). Generally speaking, on the other hand, the first sharp peak in $F(k)$ or in $\mathfrak{S}(k)$ *does not* correspond to any simple structural correlation in real space; indeed, we cannot associate its position solely to the distance between first neighbours.

The above scattering analysis particularly refers to X-rays; if the structure of non-crystalline materials is probed by electron scattering, thin samples are irradiated in situ in an electron microscope with an electron beam of energy between 50 and 100 keV, which corresponds to wavelengths around 5×10^{-3} nm.

The experimental geometry requires us to keep the electron wavelength constant. The scattering angle is varied by deflecting the electron beam with a scanning coil.

The *form factor* for electron scattering is

$$S_{e^-}(k) = (2m_e e^2/\hbar^2)\left\{[Z - S_x(k)]/k^2\right\} \tag{4.51}$$

where m_e and e are respectively the electron rest mass and charge, and $S_x(k)$ the corresponding form factor for X-ray scattering.

Coulomb interaction makes the scattering much stronger than when using X-rays. This implies that multiple scattering contributes significantly to the signal if the thickness of the sample exceeds a few tens of nanometres. Moreover, a strong background signal is inevitable due to electron energy-loss anelastic processes.

The need to use very thin samples, usually obtained by mechanical and chemical procedures, is a considerable practical problem; we could cause modifications in the structure of the thin film compared to the thicker samples we start with. Particularly, in the case of experiments on vitrified materials, the relaxation processes that occur in a structure which is in any case metastable, are highly dependent on the internal stresses that are inevitably induced in the thin sample, and can greatly affect the structure observed under the microscope.

The validity of the structural analysis of disordered systems so far examined is limited to *monatomic* systems. However, this is of limited practical use since most of the amorphous systems are polyatomic metallic alloys, or compounds, which need a more complicated treatment.

We shall now examine the simplest case of a binary compound XY. Obviously a single correlation function in real space is insufficient to adequately describe the structure. We require three *partial* functions to respectively treat the $X-X$, $Y-Y$ and $X-Y$ pair correlations. Thus it is necessary to carry out three different diffraction experiments to achieve, in a non-ambiguous way, three functions for the considered atomic pairs. Depending on the nature of the compound components, X-ray scattering, neutron scattering and electron scattering can be jointly performed on the same sample. If we use neutron diffraction we can examine a number of isotopically substituted samples, all with the same stoichiometry, though with different neutron scattering lengths, b. In any case we have to perform three scattering experiments, each with different structure factors, or scattering lengths, to obtain the three partial correlation functions.

We are thus talking about a strategy which is realistic in the case of binary compounds, but cannot be used for more complicated systems because a compound with n components requires $n(n+1)/2$ partial functions. The number of possible distinct pairs in a system with n atomic species is $\binom{n}{2}$ to which we must add the number of homologous pairs $X-X$, $Y-Y$...., which are n. As such the number of radial distribution functions is

$$\binom{n}{2} + n = \frac{n!}{(n-2)!2} + n = \frac{n\,(n-1)}{2} + n = \frac{n\,(n+1)}{2}$$

So, for materials with more than two components the only way we can carry out a possible structural analysis is to adopt a technique that specifically examines the chemical *short range* structure. The best way of doing this is usually to use EXAFS spectroscopy (see Sect. 4.4).

In scattering experiments on multi-component systems the drawbacks to the information obtained from experiments on elemental systems persist; indeed, these drawbacks are even worse for samples with a complicated composition. The information, both regarding the spatial atomic arrangement and the chemical correlation, on the system is obtained indirectly. We can obtain a good definition of the bond length between first neighbours but often doubts persist even about the coordination number for second neighbours, which can be determined by Fourier transforming the intensity data (see (4.47)).

For a polyatomic system with n components (4.43), which is for monatomic systems, is formally generalised as

$$\begin{aligned}
\left(U/NX^2\right) F^2\Theta = & \sum_{i=1}^{n} c_i S_i^2 \\
& + \sum_{i=1}^{n}\sum_{j=1}^{n} c_i c_j S_i S_j \int_0^\infty 4\pi x^2 \left[\frac{\varrho_{ij}(x)}{c_j}\right]\left(\frac{\sin kx}{kx}\right) \mathrm{d}x \\
& - \left[\sum_{i=1}^{n} c_i S_i\right]^2 \int_0^\infty 4\pi x^2 \varrho_0 \left(\frac{\sin kx}{kx}\right) \mathrm{d}x
\end{aligned} \tag{4.52}$$

where c_i and S_i are the atomic fraction and the scattering factor for element i, and $\varrho_{ij}(x)$ is the number density of j atoms that, in the unit volume, are found at x distance from any i species atom.

The reduced scattering intensity is thus rewritten as

$$\begin{aligned}
F(k) &= k\frac{\frac{(U/X^2)F^2\Theta}{N} - \langle S^2\rangle}{\langle S\rangle^2} \\
&= \int_0^\infty 4\pi x^2 \left[\varrho(x) - \varrho_0\right]\left(\frac{\sin kx}{kx}\right) \mathrm{d}x.
\end{aligned} \tag{4.53}$$

In this equation we use the functions

$$\langle S\rangle = \sum_{i=1}^{n} c_i S_i; \quad \langle S^2\rangle = \sum_{i=1}^{n} c_i S_i^2; \quad \varrho(x) = \sum_{ij} p_{ij}\varrho_{ij}(x)/c_j. \tag{4.54}$$

The weighting factors p_{ij} are given as $p_{ij} = c_i c_j S_i S_j / \langle S\rangle^2$.

The similarity between (4.53) and (4.44) for elemental systems is formal; the scattering factors in (4.53) are *weighted* and $\varrho(x)$ depends on the very scattering factors through the weight factors p_{ij}.

It is useful to introduce the *partial interference functions*, $I_{ij}(k)$, which are widely used in the literature, as

$$I_{jk}(k) - 1 = \int_0^\infty 4\pi x^2 \left[\left(\frac{\varrho_{ij}(x)}{c_j}\right) - \varrho_0\right]\left(\frac{\sin kx}{kx}\right) \mathrm{d}x \tag{4.55}$$

where the symmetry relation $I_{ij} = I_{ji}$ is fulfilled since $\varrho_{ij}/c_j = \varrho_{ji}/c_i$.

Thus, in the case of practical interest of a binary compound whose components are given as 1 and 2, we can write the equation for the reduced scattering intensity as

$$\begin{aligned}\left(U/NX^2\right) F^2\Theta - \left\langle S^2\right\rangle = c_1^2 S_1^2 \left[I_{11}(k) - 1\right] \\ +c_2^2 S_2^2 \left[I_{22}(k) - 1\right] \\ +2c_1 c_2 S_1 S_2 \left[I_{12}(k) - 1\right] .\end{aligned} \tag{4.56}$$

The equations thus far given constitute the Faber–Ziman formalism.

Still referring to the partial interference functions (4.55), Bhatia and Thornton introduced three new correlation functions for binary compounds; the number-number, $S_{NN}(k)$, the concentration-concentration, $S_{CC}(k)$, and the number-concentration, $S_{NC}(k)$. These three functions are interesting because of their physical meaning.

The number-number correlation function is

$$S_{NN}(k) = c_1^2 I_{11}(k) + c_2^2 I_{22}(k) + 2c_1 c_2 I_{12}(k) \tag{4.57}$$

It oscillates about the unity where k tends toward infinity, and represents *topological SRO*. When we transform $S_{NN}(k)$ using the Fourier transform we obtain the number-number distribution function in real space

$$(2\pi)^{-1/2} \int_0^\infty k \left[S_{NN}(k) - 1\right] \sin kx \, \mathrm{d}k = 4\pi x \left[\varrho_{NN}(x) - \varrho_0\right] \tag{4.58}$$

There is a formally simple relation between the number-number distribution function and the corresponding partial quantity in the Faber–Ziman formalism

$$\varrho_{NN}(x) = c_1 \varrho_1(x) + c_2 \varrho_2(x) \tag{4.59}$$

where $\varrho_j(x) = \sum_{i=1}^{2} p_{ij}(x)$ respectively, for $j = 1$ and $j = 2$.

The concentration-concentration correlation function $S_{CC}(k)$ is

$$S_{CC}(k) = c_1 c_2 \left\{1 + c_1 c_2 \left[I_{11}(k) + I_{22}(k) - 2I_{12}(k)\right]\right\} .$$

$S_{CC}(k)$ describes the *chemical SRO* and, for a random distribution of atom species 1 and 2, $S_{CC}(k) = c_1 c_2$. If not, $S_{CC}(k)$ has an oscillating trend about this value and tends to it when k tends to infinity. Considering the quantity $S_{CC}(0) - c_1 c_2$, if it has a positive value then the atoms of each of the two species tend to coordinate with first neighbours of the same species; thus *clustering* results. If, though, $S_{CC}(0) - c_1 c_2$ has a negative value then it is likely that the atoms of a species prefer to coordinate with atoms of the other species; thus we have *chemical SRO*.

Again, using the Fourier transform we obtain, for $\varrho_{CC}(x)$,

$$(2\pi)^{-1/2} \int_0^\infty k \left\{ \left[S_{CC}(k) - c_1 c_2 \right] / c_1 c_2 \right\} \sin kx \, \mathrm{d}k = 4\pi x \varrho_{CC}(x)$$

whose relation with the Faber–Ziman correlation functions is given by

$$\varrho_{CC}(x) = c_2 \varrho_1(x) + c_1 \varrho_2(x) - \varrho_{12}(x)/c_2 \tag{4.60}$$

where the definitions already given for $S_{NN}(k)$ are used for ϱ_1 and ϱ_2.

We rewrite (4.60) as

$$\varrho_{CC}(x) = \left[c_2 \varrho_1(x) + c_1 \varrho_2(x) \right] \alpha(x)$$

where we have introduced the Warren–Cowley generalised chemical short range order parameter

$$\alpha(x) = 1 - \frac{\varrho_{12}(x)}{c_2 \left[c_2 \varrho_1(x) + c_1 \varrho_2(x) \right]} .$$

The radial correlation function for concentration $4\pi x^2 \varrho_{CC}(x)$ is modulated about zero. Negative peaks, or minima, correspond to distances where unlike atom pairs prevail ($CSRO$), whereas positive peaks, or maxima, correspond to distances where like atom pairs prevail.

Since for both $\varrho_{NN}(x)$ and $4\pi x^2 \varrho_{NN}(x)$ the first peak only is usually well defined we can rewrite the chemical short range order parameter with explicit reference to the number of atoms in the first coordination shell,

$$\alpha_1 = 1 - \frac{N_{12}}{c_2(c_2 N_1 + c_1 N_2)} . \tag{4.61}$$

Here N_{12} is the number of atoms of species 2 first neighbours of a species 1 atom, so

$$N_i = \sum_{j=1}^{n} N_{ij}; \qquad N_{ij} = \int_{x_\mathrm{b}}^{x_\mathrm{t}} 4\pi x^2 \varrho_{ij}(x) \mathrm{d}x$$

where x_b and x_t are respectively the values for the bottom and top radii of the first coordination shell. Only when $\varrho_{NC}(x) = 0$, can (4.61) be reduced to the Warren–Cowley $CSRO$ parameter.

The term $\varrho_{NC}(x)$ gives us the correlation between the density fluctuations and the concentration fluctuations; $\varrho_{NC}(x)$ has a null value for $\varrho_1(x) = \varrho_2(x)$, a condition that occurs in a substitutional binary alloy with atoms of identical sizes. In this case, then, $S_{NC}(k)$ is obviously also null,

$$S_{NC}(k) = c_1 c_2 \left\{ \left[c_1 I_{11}(k) + c_2 I_{12}(k) \right] - \left[c_1 I_{21}(k) + c_2 I_{22}(k) \right] \right\} . \tag{4.62}$$

The Fourier transform of $S_{NC}(k)$ is $\varrho_{NC}(x)$, thus

$$(2\pi)^{-1/2} \int_0^\infty k S_{NC}(k) \sin kx \, dk = 4\pi x \varrho_{NC}(x). \tag{4.63}$$

Generally speaking, $\varrho_{NC}(x)$ has an oscillating trend about zero and is null as x tends towards infinity. The link with the Faber–Ziman correlation function is

$$\varrho_{NC}(x) = c_1 c_2 [\varrho_1(x) - \varrho_2(x)]. \tag{4.64}$$

When we invert this equation we obtain the reduced scattering intensity (see (4.62)). The procedure is formally simple; only if the factors p_{ij}, and thus S_i and S_j, all have the *same* dependence on, or are independent from k, no approximations are introduced in the information on the real space correlations. The second condition ($S_i = b_i$), holds for neutrons whereas none of the conditions are generally satisfied for X-rays since S_i are functions of k. Using the Bhatia–Thornton partial functions, again for a binary compound, the *scattering intensity* is given as

$$\frac{(U/X^2)\, F^2 \Theta}{N} = \langle S \rangle^2 S_{NN}(k) + 2\left[S_1(k) - S_2(k)\right] \langle S \rangle S_{NC}(k) + + \left[\left(\langle S^2 \rangle - \langle S \rangle^2\right)/c_1 c_2\right] S_{CC}(k). \tag{4.65}$$

As an example, we shall now examine the results from the structural analysis of a disordered binary system. We shall study the amorphous alloy $Ni_{50}Nb_{50}$ prepared in the form of ribbons using ultra-fast quenching from the melt.

Both neutron scattering experiments, using the isotopic substitution method, and X-ray scattering were performed. The analysis shows up both the Faber–Ziman (Fig. 4.14, part (a)) and the Bhatia–Thornton (Fig. 4.14, part (b)) partial structure factors. On examining part (a) of Fig. 4.14, it is clear that the $S_{\text{Ni-Nb}}$ and $S_{\text{Ni-Ni}}$ contributions are very similar to each other, apart from the pre-peak of around 19 nm^{-1}, in $S_{\text{Ni-Ni}}$ which is present in NiNb alloys, even with differing compositions. The same peak is found also in $S_{\text{Ni-Nb}}$ where, though, it only occurs in the spectra for alloys that are sub-stoichiometric in niobium.

The pre-peak is an index of chemical ordering in the amorphous alloys; the preference for hetero-coordination between first neighbours implies there is greater probability for homo-coordination in the second coordination shell. This gives rise to a larger period in the structure in the physical space, which corresponds to the pre-peak in the reciprocal space. The asymmetry of this ordering feature, which is more pronounced for nickel atoms than for niobium atoms, can be explained by considering the atom sizes. The nickel atoms, which are smaller, are less likely to be first neighbours with each other than are the niobium atoms.

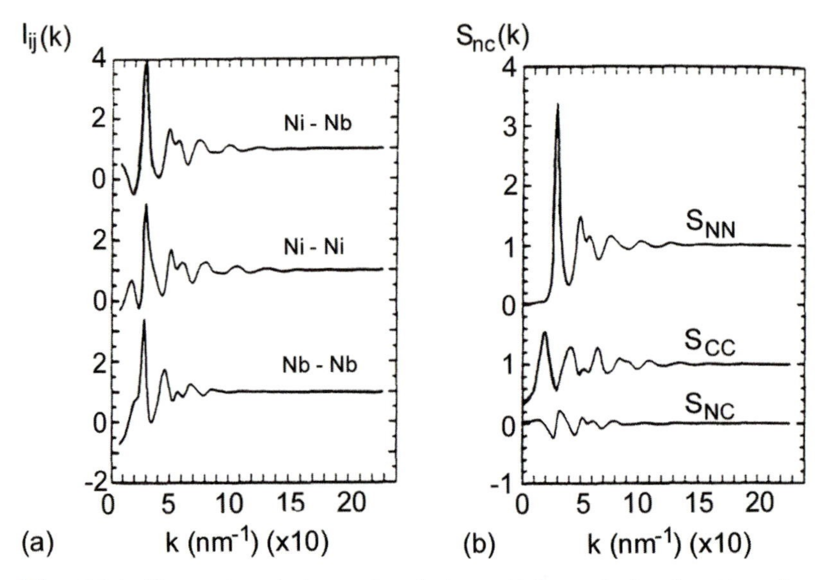

Fig. 4.14. Comparison between trends in partial correlation functions for amorphous NiNb: (**a**) according to Faber–Ziman, $I_{ij}(k)$; (**b**) according to Bhatia–Thornton, $S_{nc}(k)$ (adapted from [4.3])

Part (b) of Fig. 4.14 clearly shows the sharp peak for S_{NN}, which is due to the topological arrangement of the atoms. Even S_{CC}, which is due to chemical ordering, has rather strong oscillations. This confirms the results obtained from the Faber–Ziman partial functions, whereas the oscillations in S_{NC} are due to the size difference between Ni and Nb and are, in fact, independent of the specific chemical short range order.

From the reduced Faber–Ziman partial radial distribution functions, as shown in part (a) of Fig. 4.15, we extract the average distances between pairs of Ni atoms (0.25 nm), Nb atoms (0.302 nm) and Ni–Nb atoms (0.264 nm), as well as the relative coordination numbers (respectively $5.0, 7.5, 7.4$). We confirm the observation that the Ni atoms, smaller in size, can be first neighbours with each other, though with less efficiency than the Nb atoms, which are larger. We are here dealing with a phenomenon that is common to many amorphous metallic systems, whether they are made of transition-transition metals or of transition metals with metalloids.

Lastly, part (b) of Fig. 4.15 shows the reduced Bhatia–Thorntorn partial pair distribution functions. The presence of consistent topological order is revealed by the trend in G_{NN} (where $\varrho N(x) = [G_{NN}(x)/4\pi x] + \varrho_0$), whereas the maxima in $G_{CC}(x)$ give us distances where correlations between pairs of like atoms (respectively Ni–Ni or Nb–Nb) predominate, and the minima correspond to distances where the correlations between pairs of Ni and Nb atoms predominate.

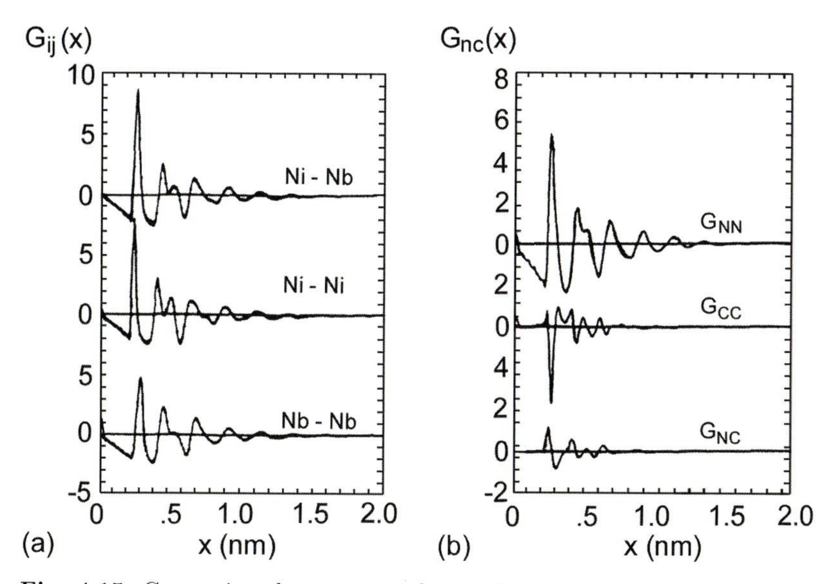

Fig. 4.15. Comparison between trends in reduced partial pair distribution functions for amorphous NiNb: (**a**) according to Faber–Ziman $G_{ij}(x)$; (**b**) according to Bhatia–Thornton, $G_{nc}(x)$ (adapted from [4.3])

4.4 Experimental Techniques: EXAFS

When X-rays pass through a material they can either be scattered or absorbed. The probabilities for each process depend on both the energy of the radiation and the nature of the target material.

As the X-ray energy increases, in any material we will observe sudden jumps in the trend of the absorption coefficient. This is due to the radiation-matter interaction; when the radiation is energetic enough, it can extract electrons from specific electron shells. For example, the emission of a 1*s* electron corresponds to the K absorption edge. In the case of an isolated atom, the change in the absorption coefficient is a monotonic function where the photon energy decreases smoothly, except for the mentioned sudden jumps. In condensed matter we observe a characteristic oscillating behaviour, from the high energy side of an absorption edge, as shown in Fig. 4.16. This structure is defined as the Extended X-ray Absorption Fine Structure, or EXAFS. The analysis of this pattern gives us detailed information on the *local* structure about a given atom in a crystalline or disordered environment.

Experimentally, we collect an *absorption* spectrum around an X-ray absorption edge for a particular atom. The experiments are usually carried out in transmission; depending on the energy of the photons, the incident intensity I_i of the X-ray beam is measured together with the intensity I_t transmitted through a sheet with a thickness of s

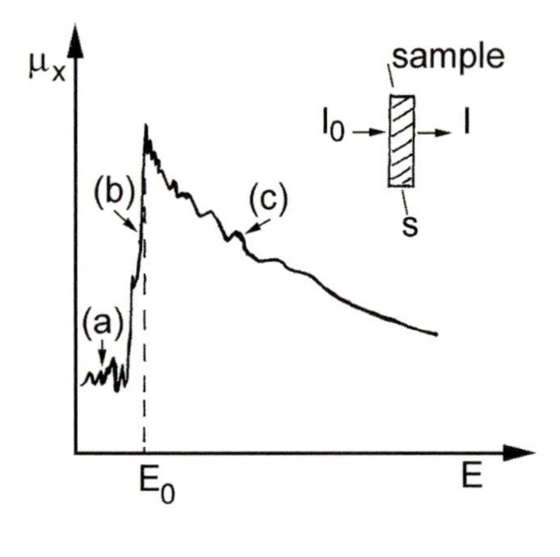

Fig. 4.16. Typical trend in X-ray absorption coefficient, μ_x, as a function of photon energy E. The highlighted regions are (a) the pre-edge region, (b) the absorption edge, (c) the region beyond the edge, with XANES and EXAFS oscillations. The experiment geometry is also schematised

$$I_{\mathrm{t}} = I_{\mathrm{i}} \exp\left[-\mu_x \cdot s\right]. \tag{4.66}$$

If we subtract the background signal μ_0 from the measured absorption coefficient, μ_x, we obtain the EXAFS signal amplitude

$$\chi(E) = \left(\mu_x - \mu_0\right)/\mu_0. \tag{4.67}$$

E represents the photoelectron energy: this value is not known exactly because the reference level E_0, which is the difference between the binding energy for the internal shell excited electron and the vacuum cannot be accurately determined. The background absorption, μ_0, is difficult to measure and is taken as an approximation, using a polynomial expression.

When we consider the scattering factor $S(k)$, we note that it is a function of both the change in the scattering wavevector and of the radiation frequency ω,

$$S(k,\omega) = S_0(k) + S_1(k,\omega) + \mathrm{i}S_2(k,\omega) \tag{4.68}$$

The first term in the sum is the form factor (see (4.33)), namely the Fourier transform of the electron density, and it does not depend on the frequency, whereas the other terms represent the so-called dispersion corrections; these corrections are important only if the frequency is very close to an absorption edge for the atom species under examination.

An X-ray absorption spectrum may be divided into three regions: in the pre-edge region we generally observe a monotonic reduction in the X-ray absorption coefficient, μ_x, as a function of the photon energy, E.

The absorption edge region follows, where μ_x increases suddenly at the energy E_0.

Lastly, in the post-edge region, atoms organised in a condensed structure give rise to the Extended X-ray Absorption Fine Structure, or EXAFS; this is a set of oscillations of μ_x, with energy approximately between 70 eV and 800 eV beyond the edge. On the other hand, if the μ_x oscillation energy is closer than about 70 eV to the edge value, then we call the spectrum X-ray Absorption Near Edge Structure, or XANES. Besides the acronym XANES, this kind of spectroscopy is also called NEXAFS, i. e. Near Edge X-ray Absorption Fine Structure.

The *pre*-edge is caused by photo-excitation of electrons of the internal shells to low energy bonded states of the ionised atom, or to resonant continuum states. We are here talking about an absorption spectroscopy, which can in principle highlight the local coordination of transition metal atoms, or of rare earths in an amorphous matrix. In fact, depending both on the local coordination of the metallic atoms, which act as impurities in this process, and on the kinds of atoms these metallic atoms are coordinated with, the relative occupation for levels d or f changes; they also shift in energy. In practice, however, it is only possible to distinguish between tetrahedral and octahedral coordinations, but no other structural parameters can be identified.

The fine structure beyond the edge, whether EXAFS or XANES, is not caused by direct absorption processes but by a sort of internal diffraction.

During the absorption process the photon is treated as a classical electromagnetic field, and the electron as a quantum particle. If the photon wavelength is large as compared to the spatial extension of the excited internal shell, then the absorption coefficient, as a function of the energy, $\mu_x(E)$, can be calculated using the Fermi golden rule, with a time dependent perturbative approach

$$\mu_x(E) = (4\pi^2 N e^2 \omega / c) \left| \langle e \left| E \right| c \rangle \right|^2 D_{E_\mathrm{F}} \tag{4.69}$$

where N is the number of atoms for a given species in the sample, ω is the X-ray frequency, $|c\rangle$ and $|e\rangle$ are the wave functions for the starting inner shell atomic level and for the final state of the photoelectron, E is the electric dipole transition and D_{E_F} is the final density of states at the Fermi level.

Since the photon field is spatially uniform in this theory, we make use of a scalar potential that is proportional to distance x, with the X-ray beam polarised in the x direction.

For a free atom, the final state is a free electron beyond the region affected by the atomic potential

$$e\rangle = |0\rangle \simeq \exp\left[\mathrm{i}\delta_L\right]\ h_L(\boldsymbol{k}x) + \text{conjugate complex}$$

where $\exp\left[\mathrm{i}\delta_L\right]$ is the shift caused by the atomic potential and h_L is an output spherical wave with angular momentum L.

The oscillating trend of μ_x, observed for an atom in a bonded structure, may be attributed to the matrix element, or to the density of the states in (4.69). D_{E_F} is the free electron density, if the photon energy is above the threshold; this energy changes monotonically, so the oscillating trend can only be caused by the matrix element. This is a reasonable hypothesis since both the output wave and the backscattered wave contribute to the wavefunction of the final state $|e\rangle$; the interference between these two waves modulates the

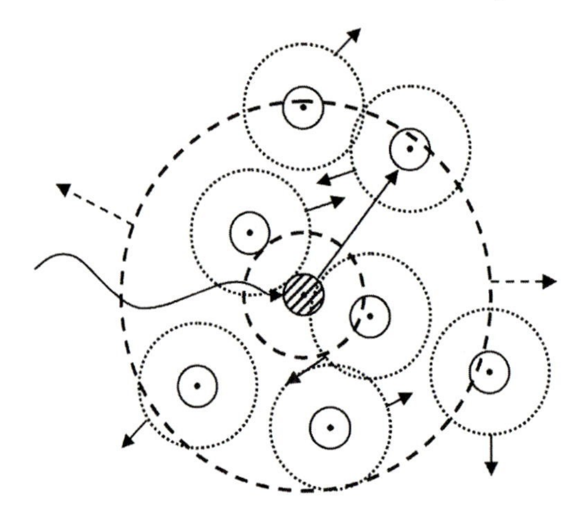

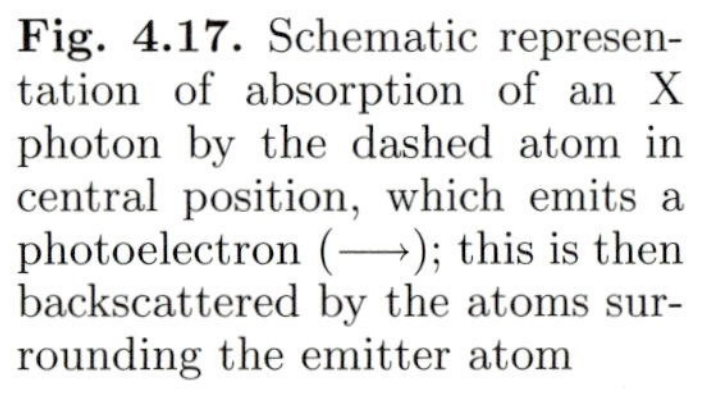

Fig. 4.17. Schematic representation of absorption of an X photon by the dashed atom in central position, which emits a photoelectron (⟶); this is then backscattered by the atoms surrounding the emitter atom

matrix element. In fact, owing to the other atoms around the absorbing atom, the output photoelectron is *backscattered* and its associated wave interferes with the output waves. This process modifies the matrix elements that trace the absorption, as schematised in Fig. 4.17.

The more the X-ray photon energy exceeds the absorption threshold the more the energy of the produced photoelectrons increases, and the lower the corresponding wavelength. When the local interatomic distance equals an integer, or a semi-integer multiple of the photoelectron wavelength, constructive interference occurs. The periodic oscillations in X-ray absorption,

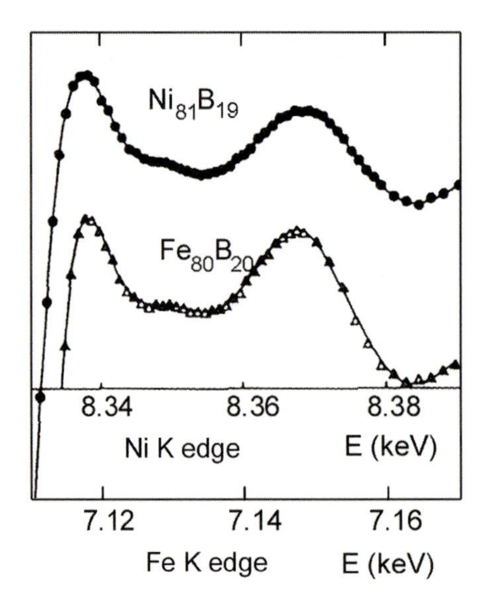

Fig. 4.18. Experimental XANES spectra around the K edges of nickel and iron, in $Ni_{81}B_{19}$ and in $Fe_{80}B_{20}$ (adapted from [4.4])

observed above the absorption edge, are the result of alternating constructive and destructive interference conditions, and make up the fine structure.

The EXAFS part of the absorption spectrum is caused by the scattering of *two* atoms, the absorbing one and the backscattered one. If multiple scattering dominates, namely *more* atoms are involved in the backscattering of an electron, then before such electron returns to the absorption site XANES oscillations will occur. Hence, both EXAFS and conventional diffraction give us information on the atomic pair correlations, whereas XANES gives us information on triplet (or higher order) correlations.

In the structurally disordered systems partial decoupling between correlations of different order occurs, and any knowledge of higher order correlations adds significant structural information (see Sect. 4.2). However, pair correlations and higher order correlations are not completely independent of each other since the need to densely fill the space with atoms restricts the choice of atomic arrangements. Furthermore, given the nature of the disordered structures, higher order correlations cannot be associated with the presence of a specific structural element; we have to consider the statistic weight of the various structural elements.

XANES spectrum analysis is based on comparing the specific features of the experimental spectrum and those of simulated spectra; these are obtained by ad hoc algorithms that calculate multiple scattering from various trial geometric structures whose parameters are changed.

For example, the XANES spectra measured for the amorphous systems Fe_{80} B_{20} and Ni_{81} B_{19} are characterised by being very similar to each other and by their having a small, though significant, "hump" (see Fig. 4.18). The simulated spectra, starting from different structural models in which Fe and Ni are taken as being equivalent, were compared with experimental spectra. Only by introducing triangular structures Ni–B–Ni into a specific model, with a Ni–B distance of 0.2 nm and a Ni–Ni distance of 0.25 nm, is it possible to reconstruct the characteristic "hump" (see Fig. 4.19).

To calculate the EXAFS amplitude in an analytical form is easier. We assume that not only are the electron wavefronts planar, but so are the wavefronts generated by backscattering electron shells, as schematised in Fig. 4.17. This is reasonable for energies in the typical range of the EXAFS spectra (at least 70 eV above the absorption edge value). We can thus study the experimental results with an analytical equation, avoiding any comparison of the data with simulated spectra.

We choose as our reference atom and coordinate origin an atom of a known species (X), whose absorption edge value is known; in Fig. 4.17 this atom is shown in a central position. The number of unlike atoms (Y) in the shell between x and $x + dx$ from the origin is

$$n_{\mathrm{Y}}(x) = \varrho_{\mathrm{Y}} g(x) 4\pi x^2 dx$$

where ϱ_{Y} is the number density of the Y atoms and $g(x)$ the pair correlation function relative to the Y atoms, with reference to origin (X). The EXAFS

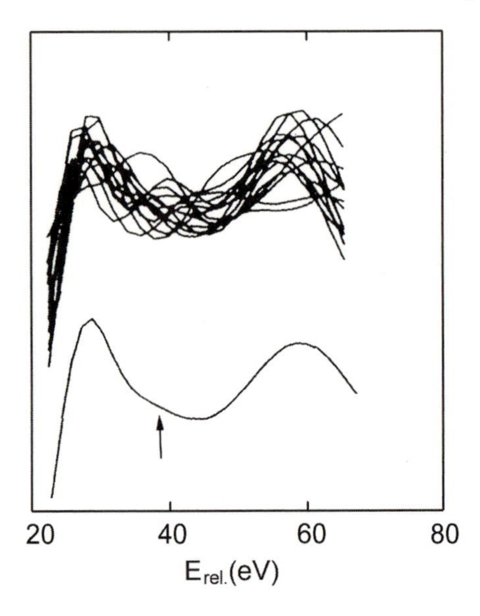

Fig. 4.19. Simulation of the essential features in the experimental XANES spectra in Fig. 4.18, as obtained using a model structure which includes metal-metalloid three-body correlations (adapted from [4.4])

function is

$$\chi_X(k) = \sum_Y \int_0^\infty \frac{|S_Y(\pi)|}{kx^2} \{\sin(2kx + 2\delta_L + \psi_Y) \times \exp\left[-2x/\lambda_Y(k)\right] \varrho_Y g(x) 4\pi x^2\} \, dx \tag{4.70}$$

where $|S_Y(\pi)|$ is the backscattered amplitude from the Y atoms in the considered shell and $\lambda_Y(k)$ the mean free path of the *elastically* scattered electrons, the only ones that contribute to the signal. This is the result of the total interference process about each absorbing atom. The phase shifts δ_L, and ψ_Y take into account the scattering from atomic potentials.

If there are N scatterer atoms at a fixed distance x_1 from the origin, then $g(x)$ is given by

$$g(x) = \frac{N}{4\pi x_1^2 \varrho} \delta(x - x_1)$$

and the EXAFS function becomes

$$\chi(k) = -\frac{N\,|S(\pi)|}{kx_1^2} \exp\left[-2x_1/\lambda(k)\right] \sin(2kx_1 + 2\delta_L + \psi). \tag{4.71}$$

In this case we obtain x_1 from the period of the oscillations and N from their amplitude, whereas the $S(\pi)$ dependence on energy, i.e. on wavevector, allows us to identify the species of scattering atom. Now, if the distances of the N scattering atoms show Gaussian distribution, as a result of e.g. thermal vibrations, then

$$g(x) = \frac{N}{4\pi x_1^2 \varrho} \frac{1}{\sqrt{2\pi}\chi_r} \exp\left[-(x - x_1)^2/2\chi_r^2\right]$$

and so

$$\chi(k) = -\frac{N\,|S(\pi)|}{kx_1^2} \exp\left[-2x_1/\lambda(k)\right] \exp\left[-2k^2\chi_r^2\right] \times \sin(2kx_1 + 2\delta_L + \psi). \quad (4.72)$$

With respect to (4.71), (4.72) is reduced by the Debye–Waller factor in the second exponential term. The standard deviation of the Gaussian distribution, χ_r, is determined by interpolation starting from the experimental data. χ_r^2 is not the usual mean squared amplitude of an atom, but the mean squared fluctuation of the *relative* positions of the central atom and the backscattering atoms. These fluctuations may have a dynamic, namely thermal, origin, or a static one when they are caused by structural disorder.

The Fourier transform of $\chi(k)$ shows peaks at distances that correspond to the coordination shells for the first neighbours of the reference atom and this gives us an approximate picture of the structure. The peaks are displaced with respect to the exact values of x_1 due to the dependence of $(2\delta_L + \psi)$ on the wavevector.

EXAFS is a complicated oscillating function; it is made up of the sum of as many sine curves as there are atomic shells. All such curves have the same structure as (4.72) and each bears a $2kx_j$ period different from the others. The amplitude of each sine curve gives us the number of neighbours in the considered shell, modified by an envelope due to the scattering amplitude, to the Debye–Waller damping and to the damping due to the electron mean free path. However, since $\lambda(k)$ is limited, and χ_r^2 increases with the shell order, only a few shells with low order contribute significantly to the amplitude.

In practice, we can deduce the coordination number N, the interatomic spacing x and the mean square deviation χ_r^2, for the first two coordination shells surrounding the absorbing atom.

One example of how the EXAFS technique is used is the investigation on the local coordination of the compounds Dy Fe_2 and Tb Fe_2, both crystalline and amorphous. The typical structure around each constituent atom has been identified separately by the respective EXAFS spectra since the K absorption edge of iron, at 7.1 keV, is well separated from the $L_{2,3}$ edges of dysprosium and terbium, which are at more than 8 keV. The surroundings of the iron atoms are not modified by the crystal-amorphous transition whereas the surroundings of the rare earth atoms do change drastically.

In the crystalline state, X-ray diffraction indicates that Dy–Fe and Tb–Fe distances are 0.3036 nm and 0.3046 nm respectively; these values decrease for materials in the glassy state, down to 0.264 nm and 0.270 nm, as obtained from the EXAFS measurements. Such distances are very close to the sum of the covalent radii of the constituent metals. Furthermore, the number of rare earth atoms coordinated among themselves changes from 12, for both

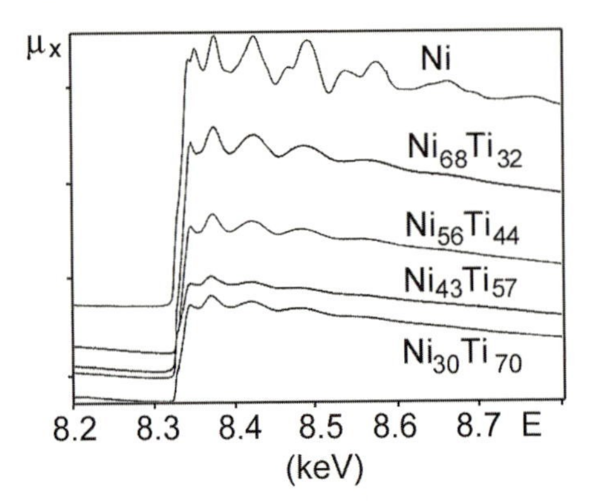

Fig. 4.20. EXAFS absorption spectra around the K edge of nickel. The compositions of amorphous alloys are reported (courtesy of M. Jaouen)

crystalline alloys, to 7.1 ± 1 for Dy and 8.4 ± 1.8 for Tb. Lastly, in the metallic glasses we observe a shift of about -1.5 eV in the position of the L_3 edge, as compared to the same edge in the crystal. This means that the initial state of the rare earth in the amorphous material is less ionised than in the crystal. When we consider all these observations together, we understand that the glass transition for this family of alloys implies an increase in their degree of covalence.

Another example of the EXAFS analysis is the study of the local order in amorphous $Ni_xTi_{(1-x)}$, using various alloys with stoichiometries $x = 68; 56; 43; 30$ at.% respectively. The experiment was carried out at the absorption edges K of Ti and Ni respectively, to examine how nickel influences the K edge titanium and vice versa. Figure 4.20 shows the trend of the coefficient μ_x around the K edge of Ni in alloys progressively enriched in Ti, whereas Fig. 4.21 shows a comparison between X-ray absorption spectra in crystalline and amorphous $Ni_{30}Ti_{70}$. The EXAFS signal at the K edge of Ti (Fig. 4.22) is more influenced by the changes in sample stoichiometry than does the signal

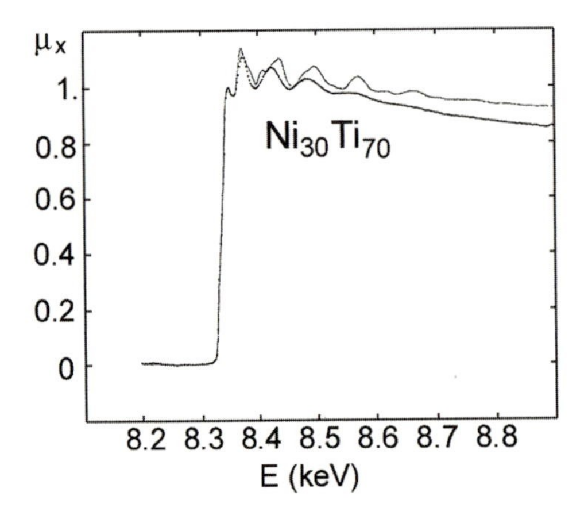

Fig. 4.21. Comparison between absorption in crystalline (—) and amorphous (....) $Ni_{30}Ti_{70}$ (courtesy of M. Jaouen)

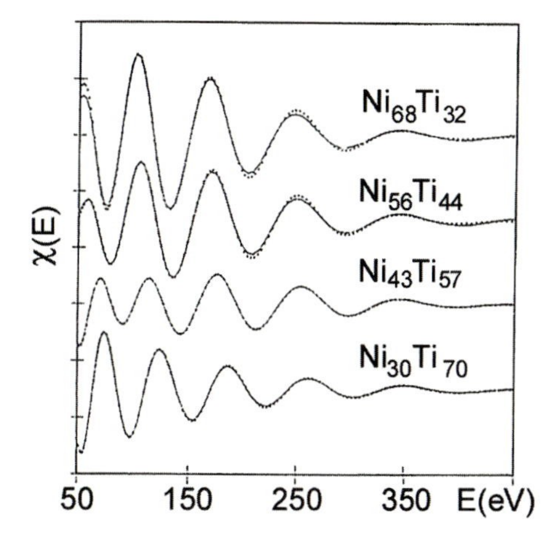

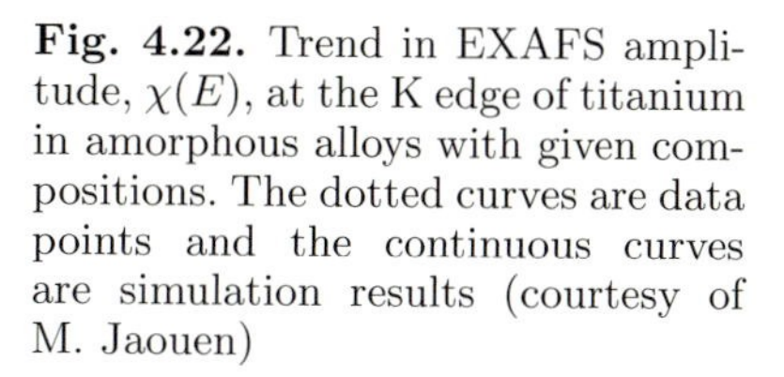

Fig. 4.22. Trend in EXAFS amplitude, $\chi(E)$, at the K edge of titanium in amorphous alloys with given compositions. The dotted curves are data points and the continuous curves are simulation results (courtesy of M. Jaouen)

at the K edge of Ni (Fig. 4.23). In both cases, the strong signal damping as the energy increases indicates that there is considerable topological disorder; the weight of this disorder varies in the neighbourhood of each atomic site, even when we consider the same component of the alloy.

Referring to X-ray diffraction data, it was possible to optimise the *average* interatomic distances and the average coordination numbers were obtained; these are almost the same $(11 - 11.5)$ in all the alloys. However, there is a strong tendency towards chemical short range order since the coordination number for the nickel atoms moves from 13 to 10.5 as the nickel content grows in the alloys, whereas the corresponding values for titanium increase from 10.5 to 12.5. This means that unlike atoms will more likely pair up around a given species of atom. Furthermore, if we consider that the value of static topological disorder is the same for all the atomic shells (thermal disorder is limited since the experiments were performed at $T = 78$ K), then a local structure is formed with a higher degree of chemical organisation in

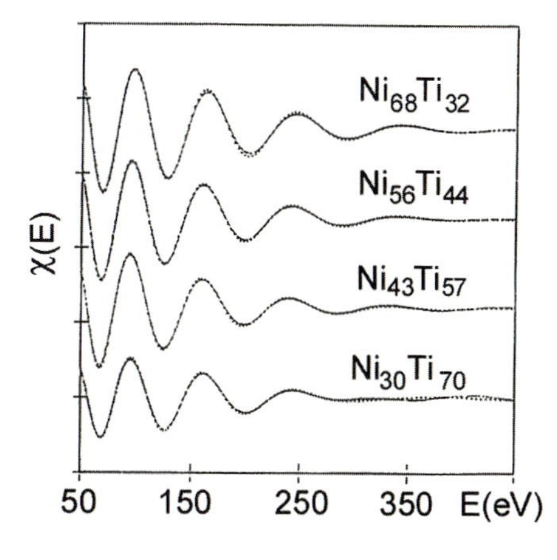

Fig. 4.23. Trend in EXAFS amplitude, $\chi(E)$, at the K edge of nickel in amorphous alloys with given compositions. The dotted curves are data points and the continuous curves are simulation results (courtesy of M. Jaouen)

the surroundings of the minority atomic species. The average interatomic distances are in excellent agreement with the interatomic distances deduced from X-ray diffraction. These distances increase with the titanium content in the alloy, but remain constantly less than the distances calculated for a disordered solid solution. This is further proof that chemical short range order is realised where the shorter Ti–Ni bonds are favoured over the longer Ti–Ti bonds. We can easily correlate this result with the values of the atomic radii, the nickel atoms being smaller than the titanium atoms.

Lastly, the increase in the distance between pairs of titanium atoms as the concentrations of these atoms decreases confirms the clear tendency to chemical ordering. In alloys with very little titanium, on the one hand, Ti–Ni bonds are formed; on the other the Ti–Ti bonds are stretched, compared to the bond length in NiTi alloys with high titanium content.

4.5 Experimental Techniques: Mössbauer Spectroscopy

Mössbauer spectroscopy is a nuclear spectroscopy based on the emission of γ-rays associated with any transition between nuclear energy levels. The sensitivity of the method is based on the observation that the γ-ray emission, due to radioactive decay of atoms of a suitable source embedded in a solid may be *recoil-free*, namely without energy loss.

In general, the energy partitioning between an emitted γ photon and the recoiling source is

$$E_\gamma = h\nu = E_0 - E_{\mathrm{r}}. \tag{4.73}$$

The energy E_γ of the γ photon is less than the difference, E_0, between the excited energy level and the ground state energy since a fraction, E_{r} of the energy is absorbed by the nucleus as the recoil energy. In a non-relativistic scheme,

$$E_{\mathrm{r}} = p_{\mathrm{n}}^2/2m_{\mathrm{n}} = (h\nu)^2/(2m_{\mathrm{n}}c^2)$$

where p_{n} and m_{n} are the momentum and the rest mass of the nucleus. Likewise, the absorption energy E_γ must be grater than E_0 by the amount E_{r}. The resonant excitation of a nucleus of the same species, via absorption of a γ photon, requires that the photon energy is

$$E_\gamma = E_0 + 2E_{\mathrm{r}} \tag{4.74}$$

thus being greater than E_0. With a γ photon of energy $E = h\nu \simeq 10$ keV and with mass number A = 50, we obtain $2E_r \simeq 2 \times 10^{-3}$ eV, which is much greater than the line width ΔE_0 (typically $(\Delta E_0/E_0) \simeq 10^{-10} - 10^{-15}$). It is thus impossible to observe resonant nuclear fluorescence, namely a *free* nucleus cannot absorb γ rays emitted from another free nucleus of the same species. In 1958 R. Mössbauer discovered that radioactive nuclei, bonded into suitable crystalline lattices, showed resonant nuclear florescence since both the emission and the absorption of γ photons occur without any recoil energy losses because the solid, in a classical picture, recoils as a whole of infinite mass. The nucleus mass, in turn, tends towards infinity, the recoil energy towards zero and the γ photons energy to E_0.

From a quantum point of view, the probability that a γ photon will be emitted, or absorbed, without a phonon being emitted at the same time, namely without energy transfer to the crystal, is finite.

In normal conditions, the atomic nucleus is in the ground state. The exact level of this state can be slightly perturbed due to the molecular environment of the nucleus itself. The environment is determined by both the kind of nearest neighbour atoms and by the spatial distribution of these atoms, namely the structure of the condensed system. From a quantum point of view, there is a finite probability that the wave function of the electrons surrounding the nucleus may extend *inside* the nucleus volume. The interaction between these electrons and the nuclear matter, particularly the protons, modifies the nuclear energy states. The degree of modification varies slightly when the nucleus is embedded in atomic structures that are different from each other, which determine different specific features of the electron environment.

The efficiency of the transmission of γ rays through matter is high; however, if the energy of γ radiation from the source is equal to the excitation energy of the lowest excited level of the absorbing nuclei, then the radiation is resonantly absorbed. After an interval of about 10^{-9}s, the γ rays are isotropically re-emitted. The absorber behaves just like a uniform scattering source and a significant fraction of the radiation incident on this absorber is not transmitted and thus is not detected.

Since the chemical perturbation of nuclear energy levels is weak, we can modify the frequency of the incident γ rays by Doppler effect. The relative motion of the source with respect to the absorber at very low Doppler velocities, v_D, is used. Compared to the relative stationary conditions, if the source is moved towards the absorber there is an increase in the γ ray frequency, whereas if the source is moved away from the absorber there is a lowering in the frequency.

Typical changes in energy values between levels, caused by modifications in the surrounding chemical environment are, for ^{57}Fe and for ^{119}Sn (two commonly used nuclei, together with ^{121}Sb, ^{129}I, ^{169}Tm, just to cite some of the about forty possible Mössbauer sources) around 5×10^{-8}eV, which correspond to Doppler velocities in the range of one millimetre per second.

A Mössbauer spectrum is characterised by position, intensity and width of the absorption lines. The shorter the mean life of the excited nuclear state, the wider the observed line is.

If the atoms of the source and the absorber were embedded in the very same chemical and structural environment, then the resonance line would be observed at zero Doppler velocity.

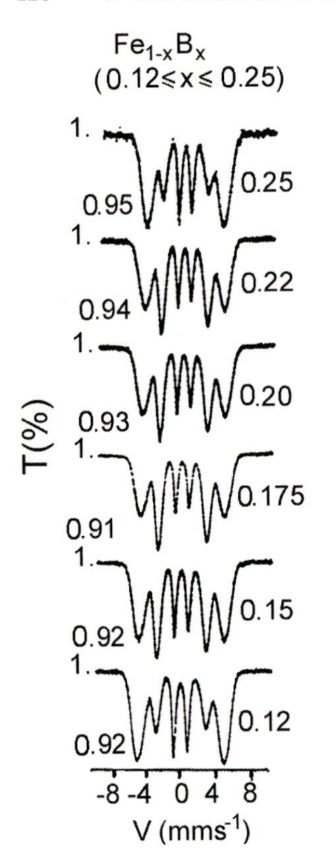

Fig. 4.24. Hyperfine Mössbauer spectra for six amorphous FeB alloys with given compositions (adapted from [4.5])

A number of interactions may perturb the resonance line. Among them, the *isomeric*, or chemical, shift, δ, is due to the density of the s electrons at the nucleus being not null. This density shifts the relative energy of the fundamental and excited nuclear levels, and thus the energy of the resonance line, without lifting the degeneracy.

So,

$$\begin{aligned} \delta &= E_{0,\mathrm{a}} - E_{0,\mathrm{e}} \\ &\simeq (2/3)\pi Z e^2 \left[\langle R_{\mathrm{e}}^2\rangle - \langle R_{\mathrm{g}}^2\rangle\right] \left\{|\psi_{\mathrm{a}}(0)|^2 - |\psi_{\mathrm{e}}(0)|^2\right\} \end{aligned} \tag{4.75}$$

where indices a and e refer to the target (absorber) and the source (emitter), R_{e} and R_{g} are the nuclear radii of the excited and ground states, and the electronic wave functions $\psi(0)$ are calculated at the centre of the nucleus.

The isomeric shift can supply us with structural information since modifications in the atomic arrangement in the neighbourhood of the resonant nucleus cause modifications in the electron charge density.

Furthermore, an electric field gradient can exist at the position of the nucleus, being caused by the change in density of the bonded electrons, which is not isotropic. Such a gradient may cause *quadrupole* interaction with the

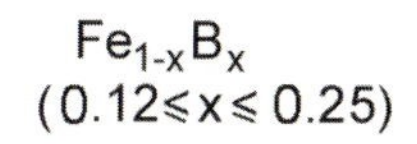

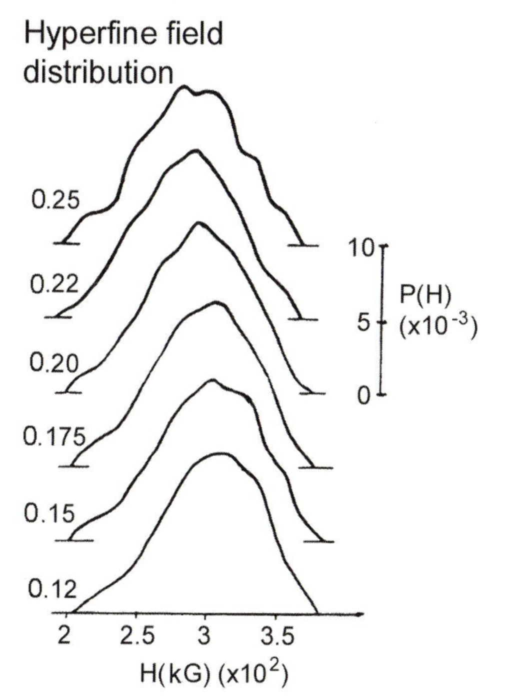

Fig. 4.25. Calculated distributions of hyperfine fields, starting from five discrete values for the magnetic fields (adapted from [4.5])

nuclear energy levels, which give rise to a nuclear quadrupole momentum. Geometrically speaking, this momentum is generated when we can schematically describe the electric charge distribution by two opposite facing dipoles, set a distance d apart.

For example, in the case of ^{57}Fe, the nuclear ground state ($I = 1/2$) does not change but the degenerate excited level ($I = 3/2$), is split into two, and corresponds to $I = \pm 1/2$ and $I = \pm 3/2$; the resonance line is split into a "quadrupole doublet".

Finally it is possible for a magnetic field, that may be internal, to completely lift the degeneracy both of the ground state and of the excited level of the nucleus, thus creating a characteristic structure called the "magnetic hyperfine spectrum". In the case of ^{57}Fe this magnetic hyperfine spectrum consists of six lines, which correspond to the possible transitions allowed because of the selection rule $\Delta m = 0, \pm 1$. Only if the internal fields are distributed isotropically will the intensity ratio between the hyperfine lines be $3:2:1:1:2:3$.

Not only is the Mössbauer spectroscopy useful in structural analysis, it is also helpful in studying the magnetic properties of metallic glasses.

An illustrative case concerns the glassy magnetic alloy $Fe_{(1-x)}B_x$ with $12 \leq x \leq 25$ at. %; Fig. 4.24 reports the Mössbauer spectra and Fig. 4.25 the calculated distributions of hyperfine fields. Starting from the profile of the six hyperfine lines in Fig. 4.24 we observe that a single hyperfine spectrum is not sufficient to reproduce the characteristic properties of the broad experimental lines. There is, however, a distribution of internal magnetic hyperfine fields $p(H_i)$. In principle, this distribution may be due to various structural causes, among which topological, or compositional, fluctuations. On the assumption that $p(H_i)$ may be correctly represented with five discrete values for the magnetic fields, then the histogram distribution has been modified to obtain the best fit to the experimental spectrum. The resulting distribution $p(H_i)$ in Fig. 4.25 is very similar to the $p(n)$ distribution for the coordination numbers n for first neighbours, typical of the model for the metallic glass structure which consists in a dense random packing of hard spheres. Despite the success this model has had, it has been observed that the hyperfine field at the iron nucleus is specifically sensitive to the number of *metalloid* atoms being first neighbours of the iron atom.

In principle the analysis of the degree of quadrupole splitting might give us some structural information about non-magnetic glassy alloys. Unfortunately this analysis cannot be performed because it would require a general theory of the tensor of the electric field gradient which is presently lacking.

4.6 Experimental Techniques: Vibrational Spectroscopies

Vibrational spectroscopies have long been reliable and widely used analytical methods supplying us with a great deal of useful information to specify the molecular structure of organic and inorganic compounds, revealing both microscopic and macroscopic details of materials. These techniques are based on the interaction between electromagnetic radiation and matter, and allow us to determine the *vibrational* frequencies of the molecules we are interested in.

For simplicity, we shall first examine the motion of an isolated molecule, as decomposed into translations, rotations and vibrations. The position of a point in space is given by three coordinates; three degrees of freedom are associated with this point in space. A molecule having M atoms has $3M$ coordinates; three of these coordinates give us the translations where the molecule motion as a whole is described by the change in the coordinates of its centre of gravity. A further three degrees of freedom (two for linear molecules) give us the rotations about the molecule's centre of gravity. Lastly, the remaining $(3M-6)$ or $(3M-5)$ pertain to the vibrations, in which bond distances and/or angles change in a periodic manner.

The number of molecular degrees of freedom coincides with that of normal, or fundamental, modes of vibration. Normally, each of these modes is not localised at a specific molecular bond; it involves several atoms. For this reason the vibrational problem is not studied using Cartesian coordinates, but internal coordinates Q that describe the atom motions in terms of changes in bond angles and bond lengths, as well as torsion (or dihedral) angles (see Sect. 4.8). In a normal mode of vibration all the atoms vibrate in phase; furthermore, it is possible to decompose any vibration into the superposition of a number of normal modes. The set of normal modes of vibration thus is a basis for an irreducible representation of the vibrational motion of the system.

In general, each normal mode of vibration corresponds to a vibrational frequency ν. In some cases two distinct vibrational modes have the same frequency, and are said to be degenerate.

The normal modes of vibration are classified based on their type (Fig. 4.26) and on their symmetry. These motions are either symmetric or antisymmetric, depending on whether the molecule symmetry is preserved, or not, during the vibration. For example, the normal modes of vibration of the CS_2 molecule are schematised in Fig. 4.27, classified by type and symmetry. Modes ν_s and ν_{as} are not degenerate, whereas the δ_s mode is twice degenerate, since the linear molecule can oscillate both in the plane of the sheet and in the plane normal to the plane of the sheet.

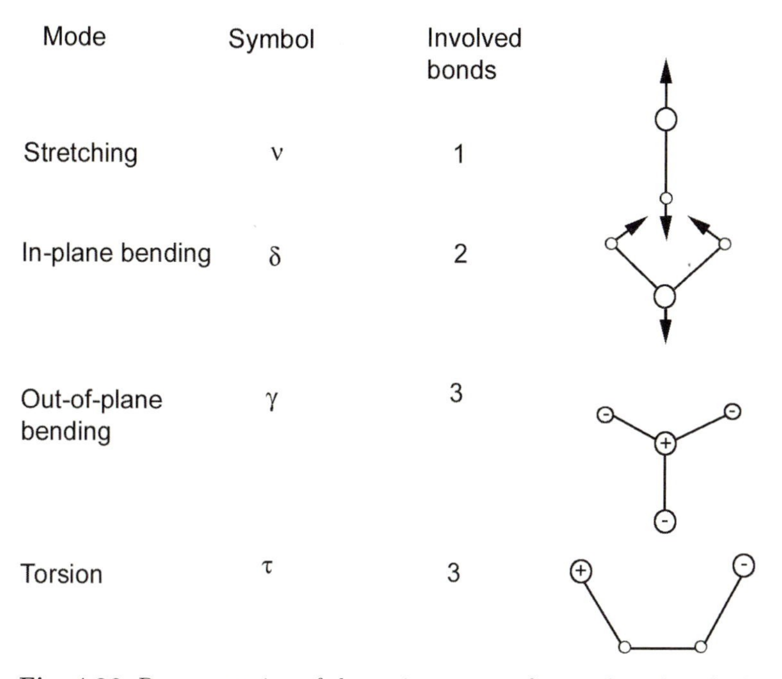

Fig. 4.26. Representation of the various types of normal modes of vibration

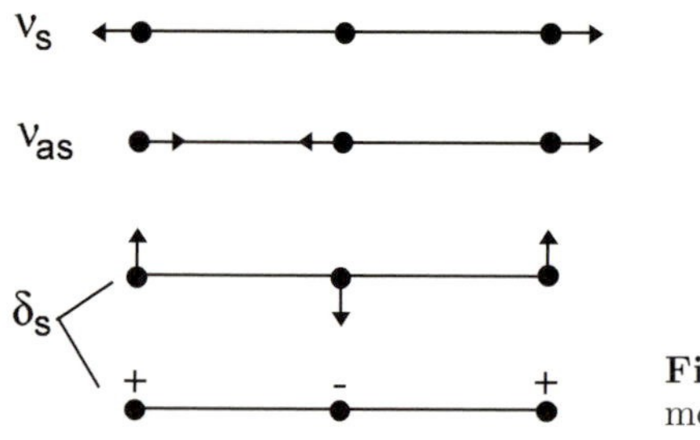

Fig. 4.27. Normal vibrations of the molecule CS_2

The symmetry of each molecule tells us which of the $(3M-6)$, or $(3M-5)$ modes of vibration are infrared (IR) or Raman active. An external electromagnetic field will distort the electron charge distribution around the molecule nuclei. This induces an electric dipole moment $\boldsymbol{\mu}$, which is proportional to the electric field $\boldsymbol{E}$

$$\boldsymbol{\mu} = \alpha \boldsymbol{E},$$

where α is the molecule polarisability and, in a semi-classic representation of radiation-matter interaction, tells us how much the electron charge distribution can be deformed. The molecule has dipole moment $\boldsymbol{\mu}$ only if the centre of the electron charge distribution does not spatially coincide with the centre of positive charge. For two elemental charges $+e$ and $-e$, separated by a distance $\boldsymbol{x}, \boldsymbol{\mu} = e\boldsymbol{x}$. A vibration is IR active only if a non-zero variation in the electric dipole moment $\boldsymbol{\mu}, (\partial\boldsymbol{\mu}/\partial Q) \neq 0$, is associated with it. The molecule does not have to have a permanent dipole moment to be IR active, but the considered fundamental vibration must induce a change in dipole moment.

Similarly, a vibration is Raman active only if it alters the polarisability $\boldsymbol{\alpha}$ of the molecule, namely $(\partial\boldsymbol{\alpha}/\partial Q) \neq 0$.

The structure of the molecule greatly affects the characteristic properties of the vibrational spectra associated with that molecule. The geometry of the molecule determines the symmetry of the fundamental vibrations, which in turn cause the changes in $\boldsymbol{\mu}$, and in $\boldsymbol{\alpha}$.

In general, the IR and Raman spectra of a system are complementary to each other, in that some fundamental vibrations that do not cause a change in the dipole moment may be associated with a change in polarisability, and vice versa. It may well occur that a mode may provoke the contemporary change in both $\boldsymbol{\mu}$ and in $\boldsymbol{\alpha}$, or, conversely, that that mode is neither IR nor Raman active. For this reason often only the use of both kinds of spectroscopy can give us a clear picture of the vibrational behaviour of a system. Referring back to Fig. 4.27, the normal mode of vibration ν_s induces modifications in the polarisability, and is Raman active. While the permanent dipole moment of CS_2 is zero, modes ν_{as} and δ_s are IR active since $(\partial\boldsymbol{\mu}/\partial Q)$ is not null for both.

When we represent the vibrational spectra it is common practice to make use of the wave number ν', which is defined as the reciprocal of wavelength

λ and is given in cm^{-1}; the relation between ν', ν and λ is

$$\nu'[\text{cm}^{-1}] = \frac{\nu[\text{s}^{-1}]}{c[\text{cm}\,\text{s}^{-1}]} = \frac{10^{-4}}{\lambda[\mu\text{m}]}$$

IR Spectroscopy. The IR spectrum of a sample is measured by exposing the target to polychromatic radiation. The spectra are collected in transmission, and represent the transmittance as a function of the radiation wavelength; in the case of supported films the signal is corrected for the substrate absorption.

The absorption coefficient, A, is obtained from the equation

$$T = \frac{(1-R)\exp[-At]}{1-R^2\exp[-2At]} \tag{4.76}$$

where T is the percent transmittance, R the reflectivity and t the thickness of the film. Equation (4.76) strictly applies to samples that are not supported by a substrate; however, it can often be used for samples with a suitable film-substrate geometrical configuration.

The vibrational frequencies are obtained as the absolute absorbed radiation frequencies over the wavelength interval under examination; the intensity of the sample beam is compared with a suitable reference beam.

The IR bands are originated when (I) a particular chemical bond is stretched (ν), or (II) the in-plane bond angle is either opened or closed (*planar bending* band δ) while bond lengths are unchanged, or (III) an atom vibrates through a plane defined by three surrounding atoms (*out-of-plane bending* band γ) or, lastly, (IV), a molecular torsion occurred (band τ), so that the dihedral angle, between two planes which share a bond, varies (Fig. 4.26). In general we observe that the frequencies associated with the four kinds of vibrational band described are progressively lower, based on the sequence: stretching, in-plane bending, out-of-plane bending and torsion.

When electromagnetic radiation of energy E_p impinges onto the molecule, then a fundamental mode of vibration is excited, provided E_p is equal, for example, to the energy difference between the first excited vibrational level, E_1, of the molecule and the fundamental vibrational level, E_0. A resonant transition of the system from the vibrational ground state E_0, to the excited level, E_1, occurs. Thus,

$$E_\text{p} = h\nu_\text{p} = h\nu_\text{v} = E_1 - E_0$$

where E_p and ν_p are the energy and the frequency of the photon and ν_v is the vibrational eigenfrequency of the molecule.

When we schematise the molecule as a harmonic quantum oscillator, the sequence of its energy eigenvalues is given by $E_n = h\nu_\text{v}(n+1/2)$ where n is an integer, zero included. The fundamental transitions are characterised by the selection rule $\Delta n = +1$. We also observe contributions, called overtones, that correspond to $\Delta n = +2; +3...$; the transition probability associated with these overtones decreases progressively as the energy difference between the initial state and the final state increases; the intensity of the respective absorption bands decreases accordingly.

The above analysis of IR spectroscopy refers to an isolated molecule, whereas in a condensed system, such as a film, even if it is structurally ordered, the molecules interact and this leads to band shift and/or broadening as well as to the possible appearance of new bands.

Raman Spectroscopy. If we irradiate a sample with an intense beam of monochromatic radiation (laser), with a frequency ν_{L}, that cannot be absorbed by the molecules under examination, and whose energy lies between the typical energy for IR spectroscopy and the energy of the electronic transitions, then a fraction of the incident beam will be scattered isotropically. In the scattered light spectrum, called the Raman spectrum, frequencies ν_{jn} are observed that are in relation to the vibrational frequencies of the molecule.

We consider that the molecule occupies the electronic ground state and that the photon energy $E_{\mathrm{L}} = h\nu_{\mathrm{L}}$ is insufficient to cause transition to the first excited electronic state. Any transition will involve the fundamental and the excited vibrational states of the electronic ground state. The scattering process is represented starting from the photon-molecule collisions, as schematically shown in Fig. 4.28. Both the photon and molecule energies are unchanged if the collision is elastic, so that the frequency of the scattered photon, ν_{s}, is equal to ν_{L}. This type of diffusion, called Rayleigh scattering, gives rise to an intense line in the spectrum when the molecule, which has been excited by the photon, undergoes transition from the fundamental vibrational state to a *virtual* state v, with higher energy, and then immediately decays to the fundamental state.

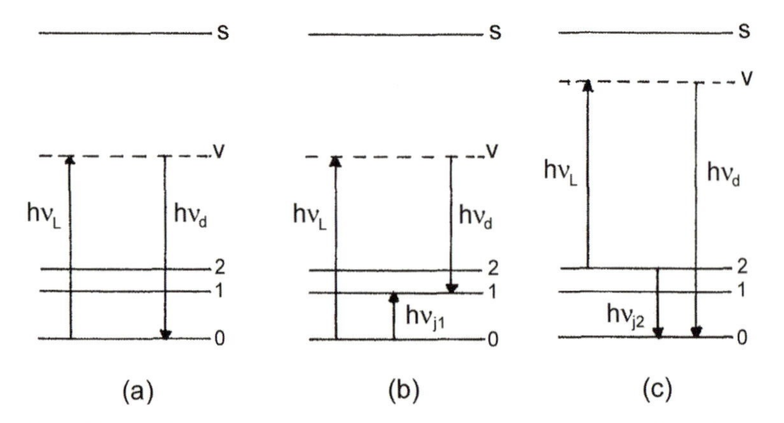

Fig. 4.28. Schematic representation of the transitions associated to (**a**) Rayleigh scattering; (**b**) Stokes scattering; (**c**) anti-Stokes scattering. s: stationary level; v: virtual level; 0, 1, 2: fundamental and excited vibrational levels

We must not confuse the concept of virtual level with stationary energy level s, which was introduced when we examined absorption.

If the molecule decays, for example, to the first excited vibrational state, which is associated with frequency ν_{j1}, it will absorb a fraction of the photon energy, and so the photon frequency, after the collision, reduces to $\nu_{\mathrm{S}} = \nu_{\mathrm{L}} - \nu_{j,1}$, and we will observe a line in the spectrum called the Stokes line. The absorption of the photon $h\nu_{\mathrm{L}}$ and the emission of the photon $h\nu_{\mathrm{s}}$ occur *simultaneously* and cannot be separated from each other in time.

Similarly, the molecule may initially be in an excited vibrational level of frequency $\nu_{j,2}$; the molecule will lose energy and return to the fundamental vibrational level, whereas the scattered photon will have a frequency $\nu_{\mathrm{S}} = \nu_{\mathrm{L}} + \nu_{j,2}$. The corresponding Raman anti-Stokes line is normally much less intense than the corresponding Stokes line since, at room temperature, most of the molecules are observed in the fundamental vibrational state.

This model allows us to explain why the intensity of the Stokes line I_{S}, is higher than the anti-Stokes lines, I_{aS}. The population of the fundamental vibrational level is by far greater, for optical phonons at moderate temperatures, than that of the excited vibrational levels, and the likelihood of the system being in an excited vibrational state is much less than it being in the fundamental state. When we consider that the ratio between the populations depends on the Bose–Einstein statistics,

$$\frac{I_{\mathrm{S}}}{I_{\mathrm{aS}}} = \frac{\left(\nu_L - \nu_j(\boldsymbol{k}_j)\right)^4}{\left(\nu_L + \nu_j(\boldsymbol{k}_j)\right)^4} \exp\left[h\nu_j(\boldsymbol{k}_j)/k_B T\right] \tag{4.77}$$

then the $I_{\mathrm{S}}/I_{\mathrm{aS}}$ ratio in equation (4.77) is certainly greater than one, as has been experimentally observed.

The experimental geometry essentially includes the laser source, a scattering region where the sample is illuminated by a parallel beam of light, and a detector that measures the intensity scattered at a finite angle; in many cases the detector is positioned normal to the direction of the incident beam.

In the Raman spectrum the scattered light intensity is usually displayed as a function of the Stokes, or anti-Stokes frequency (shift). The former should be negative, however it is conventionally considered positive.

Raman scattering in molecules is completely analogous to what occurs in a crystal, though we must remember that molecular normal modes of vibration correspond to collective vibrational motions (normal modes) for the ions that occupy the lattice sites in the crystal.

Understanding of Raman scattering is easy in terms of a qualitative, quantum representation, when we consider an ideal crystal, which constitutes the immediate reference for us to study structurally disordered systems. Photon-crystal elastic collision gives rise to Rayleigh scattering, whereas Raman scattering corresponds to an anelastic interaction where the photon destroys, anti-Stokes, or creates, Stokes, one or more lattice phonons. In any case, a photon in the incident beam, with a frequency ν_{L} and wavevector $\boldsymbol{k}_{\mathrm{L}}$, is destroyed and both a photon of the scattered beam with $(\nu_\mathrm{s}, \boldsymbol{k}_\mathrm{s})$ and a phonon, with a frequency ν_j and wavevector $\boldsymbol{k}_j$, are created. The scattering process is thus second order since it requires two interactions of the radiation field with the scattering system (destruction of the incident photon and creation of the scattered photon).

The energy and the momentum are conserved between the initial and final states. For Rayleigh scattering,

$$\nu_\mathrm{L} = \nu_\mathrm{s} \tag{4.78}$$

$$\boldsymbol{k}_\mathrm{L} = \boldsymbol{k}_\mathrm{s} \tag{4.79}$$

and for Raman scattering

$$\nu_\mathrm{L} = \nu_\mathrm{s} \pm \nu_j(\boldsymbol{k}_j) \tag{4.80}$$

$$\boldsymbol{k}_\mathrm{L} = \boldsymbol{k}_\mathrm{s} \pm \boldsymbol{k}_j . \tag{4.81}$$

For Raman scattering, normally, ν_L is much higher than $\nu_j(\boldsymbol{k}_j)$, so ν_L is nearly equal to ν_s (typical values are ν_L about $2 \times 10^4 \mathrm{cm}^{-1}$, ν_j about $10^3 \mathrm{cm}^{-1}$; these values for ν_j correspond to phonon frequencies in the optical branch). Furthermore, we choose frequencies where, in practice, there is no dispersion in the refraction index n, given the small difference between ν_L and ν_s, so $n(\nu_\mathrm{L}) \simeq n(\nu_\mathrm{s}) = n$. Since $\boldsymbol{k}_\mathrm{L}$ and $\boldsymbol{k}_\mathrm{s}$ are defined *within* the crystal,$\boldsymbol{k}_\mathrm{s}$ $= \nu_\mathrm{L} n(\nu_\mathrm{L})/c \simeq \boldsymbol{k}_\mathrm{s} = \nu_\mathrm{s} n(\nu_\mathrm{s})/c$. Furthermore, $\boldsymbol{k}_\mathrm{L}$ and $\boldsymbol{k}_\mathrm{s}$ are much smaller than the wavevector at the Brillouin Zone boundary, so (see (4.79)) $\boldsymbol{k}_j$, in turn, is very small. This means that in first order Raman scattering only the optical phonons at the centre point of the Brillouin zone can be excited. This condition coincides with the selection rule

$$\boldsymbol{k}_j \simeq 0 \tag{4.82}$$

where, in practice, the parity sign is assumed.

When a photon, with $h\nu_\mathrm{L}$ energy, in the visible or the ultraviolet regions of the spectrum interacts with a crystal, it perturbs the electron wave functions since only the electrons are light enough to follow the rapidly changing electric field caused by the photon. The system wave functions become linear combinations of all the possible wave functions of the perturbed crystal with time dependent coefficients. Formally, we assume that the crystal reaches a non-stationary energy level with higher energy. This is the already mentioned virtual level which, in the classical description, corresponds to a forced oscillation of the electrons at frequency ν_L of the incident radiation. In a quantum approach, the virtual level is essential in modelling the perturbation process (see Fig. 4.28).

Spectroscopy of Amorphous Systems. If a solid is structurally disordered, then the absence of long range order coincides with the lack of large enough crystallites to produce well defined diffraction peaks, or spots. There are two general criteria we can use to analyse the vibrational spectrum of an amorphous solid: the first is that the general characteristic features of the spectrum are similar to the corresponding crystalline system, if the short range order (see Sect. 4.7) does not change, as is observed in most materials; the second is that the local spatial arrangement of bonds in amorphous systems can activate vibrational modes, which are forbidden by the extended symmetry in the crystal; this gives rise to the appearance of particular spectral features. For example, non-polar crystals, with covalent bonds, do not have static dipole moment, and so each induced dipole moment can only depend on a dynamic effect which induces atomic displacements that cause the bonds to either stretch or to contract. If we express the bond compression C_{ij} as

$$C_{ij} = (\boldsymbol{r}_i - \boldsymbol{r}_j) \cdot \boldsymbol{x}_{ij} \tag{4.83}$$

where $\boldsymbol{r}_i$ and $\boldsymbol{r}_j$ are the displacement vectors and $\boldsymbol{x}_{ij}$ is a unit vector that connects the i and the j sites, then the dipole moment $\boldsymbol{\mu}$ is

$$\boldsymbol{\mu} = 2\sum_i \left(\sum_j C_{ij}\right)\left(\sum_l \boldsymbol{x}_{il}\right) . \tag{4.84}$$

For example, the induced moment $\boldsymbol{\mu}$ in crystals with tetrahedral symmetry is zero, since $\sum_l \boldsymbol{x}_{il} = 0$; for this reason crystalline silicon and germanium are not IR active, nor are any of the non-polar cubic crystals with two atoms per primitive cell. Though this selection rule is strictly valid for crystals, it is relaxed by lattice disorder.

As an alternative to representing the amorphous system as a crystal that has lost its extended symmetry, we can represent the system as the extended network of chemical bonds of the continuous random network (CRN) model (see Sect. 4.7), which is widely adopted for covalent solids.

The IR spectra are interpreted by omitting the contribution from all the bonds and by considering the solid as being made up of *molecular units*; these units are then dealt with as if they were all separate from each other. The internal motions in each unit are similar to the fundamental vibrations of an isolated molecule. The "molecular vibrations" of the various units are influenced by the potential generated by the surrounding molecular units. This is similar to what we observe in the lattice modes of crystalline solids. These vibrations are caused by the relative motions of the various molecular units. Here we are talking about vibrations with wave numbers typically below 800 cm^{-1}, thus frequencies lower than the frequencies of the internal modes. The difference between a crystal and an amorphous system is that the bands in the amorphous system are broad, whereas in the crystal they are sharp and well resolved. The shift and broadening of the IR bands reflect the influence of the various local chemical surroundings of each molecular unit. The effect of structural disorder is that, in principle, a different chemical environment surrounds each molecular unit, and induces a specific shift in each fundamental vibrational frequency. The large number of such contributions, which differ from each other, arising from the various structural units, gives the broad absorption bands observed experimentally.

If the amorphous system contains particular molecular units whose vibrations are excited independently from the surrounding matrix, then the IR spectroscopy is an excellent method for investigating short range structural order. In the case of terminator atoms whose bond coordination is different from that of most atoms of the solid, we observe very strong effects. If the *terminator* atoms are lighter than the other atomic species they give rise to localised "impurity" high frequency modes; these are a characteristic feature of the behaviour of hydrogen atoms in an amorphous matrix.

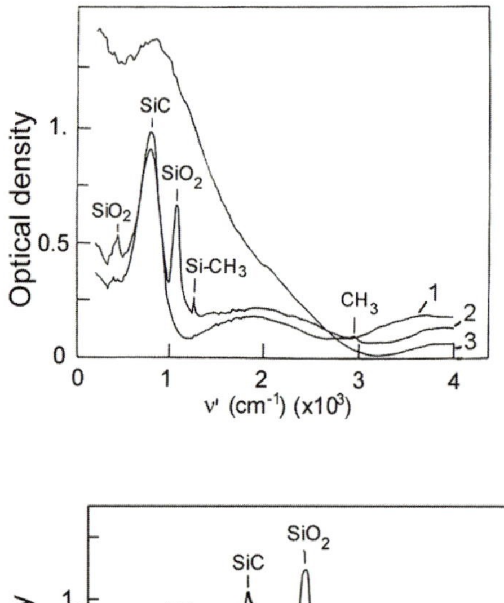

Fig. 4.29. Optical density IR spectra of amorphous SiC films treated differently; (1) as deposited film; (2) film heated in air at 1273 K for 2.5 hours; (3) film heated in vacuum up to 1673 K (adapted from [4.6])

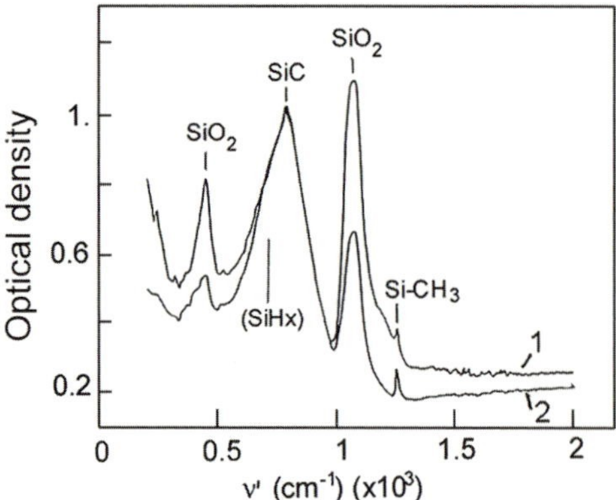

Fig. 4.30. Magnification of the spectral region up to 2×10^3 cm^{-1} for (1) film heated in air at 1273 K for 2.5 hours; (2) film heated in air at 1273 K for 15 hours (adapted from [4.6])

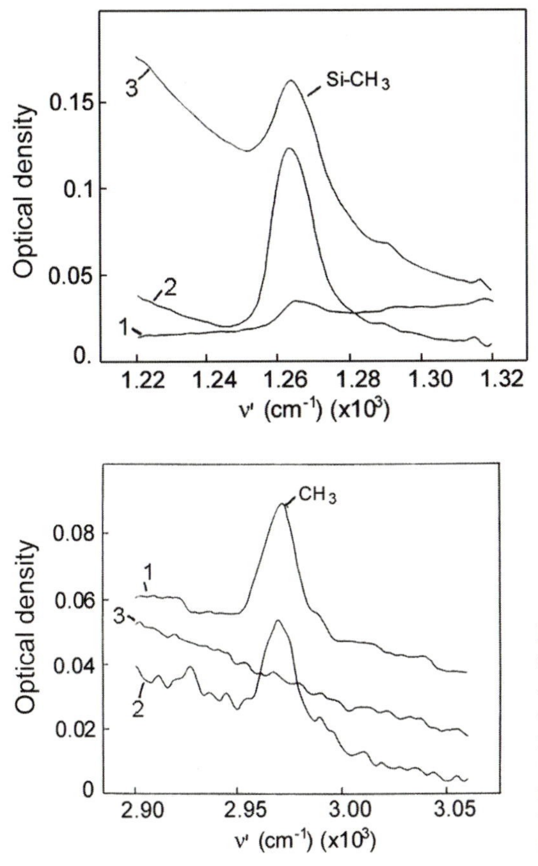

Fig. 4.31. Magnification of the spectral region including the umbrella-shaped deformation mode of group CH_3, coordinated like Si–CH_3, (1) as deposited film; (2) film heated in air at 1273 K for 2.5 hours; (3) film heated in air at 1273 K for 15 hours (adapted from [4.6])

Fig. 4.32. Magnification of the spectral region including the asymmetrical stretching mode of group CH_3, (1) film treated in air at 1273 K for 2.5 hours; (2) film heated in air at 1273 K for 15 hours; (3) film heated in vacuum up to 1673 K (adapted from [4.6])

One exemplary study on the structure of amorphous materials demonstrating the potential of IR spectroscopy concerns the vibrational behaviour of thin amorphous films of silicon carbide, $Si_{45}C_{55}$, annealed in different ways. The samples were prepared by radio frequency sputtering, starting from sintered cathodes, with SiC composition, and accidentally incorporating about 5 at. % hydrogen. Neither the hydrogen bond state nor its evolution as a function of the temperature and the modifications in the chemical environment are known. From a comparison of the characteristic features of the IR absorption spectra of a deposited film with the features of films that have been subjected to annealing in various environments (air, or vacuum) at different temperatures and for different times, as shown in Figs. 4.29 to 4.32, we observe that:

a) in freshly deposited films there is only a band centred at 795 cm^{-1}, attributed to SiC. This band is broad, indicating structural disorder in the material. The presence of H in the films was not revealed by IR spectroscopy (see Fig. 4.29);

b) heat treatment in air at 1273 K for various times induced SiO_2 formation (Fig. 4.30) and the coordination of hydrogen, that was incorporated during film formation, into structural CH_3 groups, both free and bonded like Si–CH_3. We clearly observe a band at 2970 cm^{-1}, caused by the asymmetrical stretching of CH_3, and a band centred at about 1267 cm^{-1} caused by the umbrella-like deformation of the Si–CH_3 group shown in Fig. 4.31. The intensity of the latter increases as the heat treatment continues, whereas the intensity of the band at 2970 cm^{-1} decreases with the heat treatment (see Fig. 4.32). Hydrogen is not released from the films, not even at 1273 K; this is presumably because the surface layer of SiO_2 is an efficient barrier to any hydrogen desorption;

c) by heating the films under vacuum to 1673 K, where the re-crystallisation process for the SiC matrix starts, the hydrogen contained in the samples under various coordination is released; this is evident from Fig. 4.29, curve 3 and Fig. 4.31, curve 3.

The characteristic features of the IR spectrum at this high temperature are very different from the features of the spectra recorded at lower temperatures; this is an indication of dramatic structural changes, due to hydrogen release, which induces strong deterioration in the surface structure of the samples. Indeed, hydrogen evaporation from the films seems to be a necessary condition to trigger the amorphous solid-crystal transition, both as regards the SiC matrix and the excess carbon that is present right from the beginning in the samples. This transition seems to be associated to the formation of a highly defective and/or highly non-uniform structure.

Raman Spectroscopy in Amorphous Systems. The effect of structural disorder on the Raman spectrum of a solid is examined by considering that the unit cell in an amorphous solid has infinite size, and thus the Brillouin zone reduces down to a single point, with $\boldsymbol{k} = 0$. This means that $\boldsymbol{k}$ becomes a meaningless index of the vibrational modes. The only quantity that is useful to describe the phonon modes in amorphous solids, which is defined also for crystals, is the vibrational density of states (DOS), namely the number of states per unit frequency interval.

This is because when the crystal-amorphous solid transition occurs the progressive breakdown in the Raman selection rules makes all the phonons into the Brillouin zone able to contribute to the scattering process. We can see this effect by introducing a correlation length ξ, which is characteristic of the spatial extension of a normal mode. Given a perfect crystal, $\xi = \infty$, which means that the mode under examination is a true phonon, with a definite wavevector $\boldsymbol{k}$. It is reasonable to consider that the space-time correlation function for the mode with index j is proportional to

$$\mathcal{F} = \exp\left[\mathrm{i}\boldsymbol{k}_j \cdot \boldsymbol{r}\right] \exp\left[-r/\xi\right] \tag{4.85}$$

where the exponential damping term mixes the $\boldsymbol{k}_j$ states, which are all distinct in the crystal. Since the Fourier transform of $\mathcal{F}$ is proportional to the

Raman scattered intensity for mode j, then if we sum over all the modes we obtain, for the scattered intensity (Stokes component)

$$I(\omega) = \sum_{l} a_l \omega^{-1} \left[1 + n(\omega, T)\right] g_l(\omega) \tag{4.86}$$

where a_l is the coupling constant for band l of the vibrational states, $g_l(\omega)$ is the phonon density of states for the same band and $n(\omega, T)$ is the phonon occupation number

$$n(\omega, T) = \frac{1}{\exp\left[\hbar\omega/k_B T\right] - 1} \tag{4.87}$$

From (4.86) we see that the Raman spectrum of an amorphous material gives us a representation of the vibrational density of states (unless there are particular visibility effects due to the Raman coupling coefficient) since the intensity $I(\nu)$, when we also consider factor ν^4 in (4.77), is proportional to the material DOS. In many materials, including amorphous silicon, germanium and carbon, the DOS is similar to the DOS of the corresponding crystal, convoluted with a Gaussian which takes into account the phonon mean life in the amorphous material.

One example of structural analysis carried out using IR and Raman spectroscopy together regards the study of the structure of thin films of boron nitride, BN. In a number of ways the system is similar to carbon and it is expected to have many applications. Apart from being chemically inert in aggressive environments, and stable up to very high temperatures, BN crystallises in phases that have several very different properties from each other: at room temperature the material has a hexagonal structure (h-BN), it is sp^2 hybridised and similar to graphite. At very high pressures and temperatures we can synthesise the cubic phase (c-BN), sp^3 hybridised, similar to diamond both as regards the electronic and the mechanical properties. In particular, c-BN (45 GPa) is the second hardest material after diamond (around 96 GPa). It is very difficult and expensive to synthesise the massive material in the cubic phase, similarly to the synthesis of artificial diamond. For this reason, attention turned to the low pressure and temperature deposition by way of a number of techniques of c-BN films on suitable substrates. The greatest difficulties encountered are due to two factors: if the stoichiometric ratio between boron and nitrogen, 1 : 1, is not maintained, h-BN is nearly always deposited. Furthermore, the nucleation of c-BN occurs only if we can locally achieve very high pressure, for example by bombarding the surface of the film with ions with well defined energy, so that locally the favoured coordination is sp^3 instead of sp^2. This implies that in the film and at the interface with the substrate, high (up to 10 GPa) internal stresses of a compressive nature are present; such stresses can induce poor film adhesion to the substrate. Deposition onto substrates that are kept at high temperature, around 600 – 800 K, favours relaxation of most internal stresses; however, these annealings make it very unlikely BN can be used to coat many technologically important materials, which undergo heavy deterioration at the considered temperatures. In particular, the use of c-BN for electronic applications (c-BN can easily be doped, both p with beryllium and n with silicon) is precluded.

We can more easily obtain at room temperature films with a mixed coordination $sp^2 - sp^3$, with a variable percentage of tetrahedral bonds, up to $70 - 80\%$, and intermediate hardness (about 30 GPa). These films are almost always structurally disordered; since the fraction of material with "valuable" coordination sp^3 depends heavily on the choice of process parameters, irrespective of the specific preparation method adopted, then a structural analysis is necessary that can recognise the presence, and the possible abundance, of the two coordinations.

The vibrational spectroscopies, used together, allow us to identify the sp^2 and sp^3 coordinations, as well as the hexagonal and cubic crystalline phases.

The IR absorption spectrum of h-BN is characterised by two peaks; the weaker at 783 cm^{-1} is caused by the out-of-plane bending of the B-N-B group, whereas the most intense, at 1370 cm^{-1}, is associated with the in-plane stretching of the B-N group. The cubic phase has a single peak at 1065 cm^{-1}, and is interpreted as the transverse optical phonon (TO).

Several films were deposited at low temperature (around 350 K) by magnetron sputtering of a sintered BN target in an atmosphere made up of a mixture of Ar 97 at.% - N_2 3 at.%, keeping the substrate (Si(100)) polarised in radio-frequency. The IR spectra highlight a strong dependence of the BN

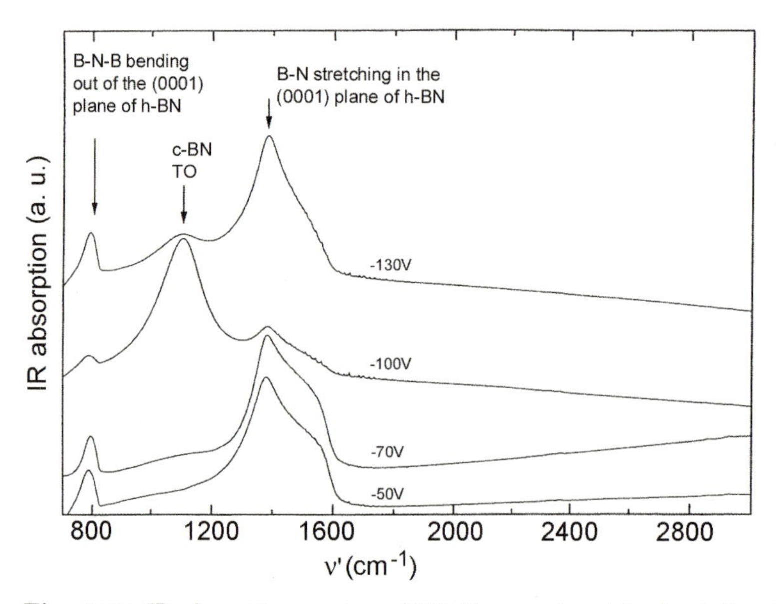

Fig. 4.33. IR absorption spectra of BN films produced in Ar + N_2 atmosphere with the substrate polarised at various voltages (adapted from [4.7])

coordination on the power applied to the target (between 100 and 200 W) and on the bias voltage at the substrate (between −50 and −130 V). Figure 4.33 shows how the fraction of sp^3 coordinated material in four films deposited at 150 W varies from 0% (−50 V) to 11% (−70 V), 90% (−100 V) and 30% (−130 V). Glancing angle X-ray diffraction patterns from purely sp^2 coordinated films show the (1000) peak of h-BN; from peak width at half maximum correlation lengths of 3.2 nm, in the hexagonal planes, and 1.6 nm in the direction normal to the hexagonal planes, are deduced. In the pattern of the film with the highest fraction of sp^3 coordination the (111) peak of c-BN is evident with a correlation length of 1.5 nm. Raman spectroscopy has been used to confirm the structural information on films obtained using IR spectroscopy and X-ray diffraction.

The first order Raman spectrum of crystalline BN, whether hexagonal or cubic, exhibits characteristic peaks at $\boldsymbol{k} = 0$, which are associated with optical phonons. h-BN has two active modes, one at 51.8 cm^{-1} and the other at 1366 cm^{-1}, whereas c-BN has two modes, the transverse optical (TO) at 1056 cm^{-1} and the longitudinal optical (LO) at 1306 cm^{-1}.

The Raman cross-section for BN is small, particularly for c-BN; this makes these spectroscopy measurements critical even for single crystals. Given the amplitude of the optical gap (5.8 eV in h-BN, more than 6.4 eV in c-BN), the material is transparent in the visible region; since the Ar^+ laser works at a single frequency at 514.5 nm wavelength, it is necessary to polarise the scattered beam so as to depress the intensity of the silicon peak at about 970 cm^{-1}. Figure 4.34 and Fig. 4.35 show the Raman spectra for h-BN and c-BN; Gaussian interpolations are also reported and the arrows indicate the

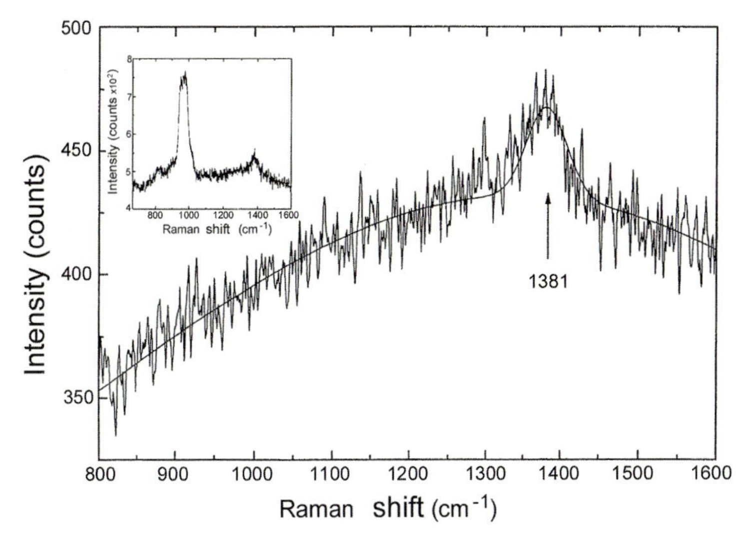

Fig. 4.34. Raman spectrum for an h-BN film; the complete spectrum is shown in the inset (adapted from [4.7])

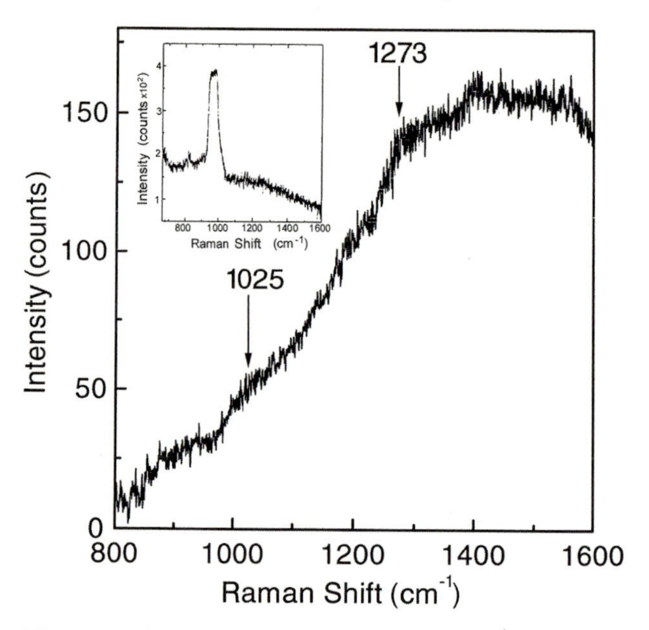

Fig. 4.35. Raman spectrum for a c-BN film; the complete spectrum is shown in the inset (adapted from [4.8])

maxima. In both figures the insets represent the complete spectra and highlight the silicon peak. Just as in other experiments on polycrystalline BN films with nanometre sizes, we notice that the peaks degenerate in bands, which result particularly broad and unstructured for c-BN. Taking into account phonon confinement due to the small crystal size, as deduced from X-ray diffraction, we can attribute the peak at 1381 cm^{-1} (Fig. 4.34) to h-BN. This is the 1366 cm^{-1} peak, blue shifted due to phonon confinement. The two broad maxima at 1025 cm^{-1} and 1273 cm^{-1} (Fig. 4.35) correspond respectively to the TO c-BN peak, shifted with respect to its position at 1056 cm^{-1} in single crystals, and to a peak that is found only in c-BN films due to the nanometre size of the crystallites.

In conclusion, from the global examination of the experimental methods used in the structural analysis, particularly as regards short range order, of glassy materials, it is clear that determining the structure of amorphous solids is a somewhat complex matter and that none of the methods are in reality sufficient if used alone. In the more favourable case of *elemental* systems, the conventional diffractometric methods may well be sufficient enough to determine short range order and, together with the development of structural models, medium range order (see Sects. 4.7 and 4.8).

On the other hand, the study of *multi-component* systems is much harder to undertake. The diffraction techniques, and techniques directly related to them such as anomalous scattering, isotopic substitution and neutron magnetic scattering, can be used in practice only for binary compounds, which need three pair correlation functions.

For those systems with more than two components, EXAFS is the only reliable tool for direct structural investigation; the advantage of being sensitive to the average local order around the atomic species that absorb the X-ray photons is, unfortunately, offset by the absence of sensitivity to atomic correlations beyond the second coordination shell, and by the difficulty in quantifying the coordination numbers. The limitations to XANES spectroscopy, which in principle shows up three body coordinations, are connected to the difficulties in analysing spherical waves and the need to compare results with simulated spectra obtained from trial structures.

Mössbauer and vibrational spectroscopies supply us with relatively indirect structural information which usually requires backing up with results from other structural probes.

4.7 Short Range Order

Our degree of knowledge about the nature of the amorphous state is often influenced by an unconscious bias, namely that the amorphous state must correspond to a set of properties that is *radically* different to those properties that refer to the well established models for the two extreme cases: total order, typical of the crystalline state, and complete disorder, associated with a dilute gas.

The dividing line between crystalline and glassy materials is given by the lack, in seconds, of translational symmetry that can be detected using interferometric methods. The notion of crystal is connected to the use of X-ray diffraction to detect structural order, which extends over not less than 1.5–2 nm. On the other hand, many physical properties that are determined by *short* distance atomic organisation, from the structure of hyperfine fields, as manifested by Mössbauer spectroscopy, to the magnetic properties that are determined, in alloys, by the nature of the first neighbour atoms of the magnetic atoms, to the EXAFS features often exhibit surprising similarities both in crystals and in amorphous solids with the same composition.

It is not possible, thus, to give a unique definition of short range order (SRO) in non-crystalline materials; indeed, SRO totally depends on the specific features of the kind of chemical bonds in the system, in particular on the bond *directionality*; precisely this directionality sets constraints to the symmetry of the atomic arrangements.

It is clear how important local atomic coordination is, and we are inclined to define, for a macroscopic set of identical atoms that are densely packed in a disordered way, the correct size and symmetry of those "structural building blocks" that are mainly responsible for the properties and stability of the glassy state. These structural building blocks play the same role as the unit cell in a crystal; from a chemist's point of view, the factors that determine stability in an amorphous system and in a crystal are the same: concentration of valence electrons, atomic size, chemical affinity.

In general, we speak about Topological Short Range Order (TSRO) when we consider only the sizes of the atoms, which are packed together as dense as possible, whereas Chemical Short Range Order (CSRO) is determined by the chemical nature of the various atoms. CSRO can easily be correlated with the (meta)stability of amorphous solids up to a specific temperature, thus with the ease by which a material is vitrified (see Sect. 3.3) and thus, from the operational point of view, with the cooling rate required for the liquid to turn amorphous.

The trend in the crystallisation process, starting from the liquid state, is given by interatomic attractive forces. The resulting atom arrangement has two properties, first valence electron energy is minimum, and second, the steric constraints that correspond to the point symmetry for one of the 14 Bravais lattices are fully met.

The stability of a given atomic configuration depends on the characteristic properties of the initial atomic orbitals which, after hybridisation, give rise to new atomic orbitals with the required point symmetry. The van der Waals forces are radial and can thus stabilise any densely packed atomic configuration. The ionic forces are radial, too; however, ions with the same charge sign must organise themselves in space as far apart as possible. The metallic bond typical for the simple metals involves almost free conduction electrons and does not imply any restraints on the resulting atomic arrangement. On the other hand, among the strongly bound d electrons, those with t_{2g} symmetry preferably stabilise the tetrahedral configurations in the fcc and hcp structures, whereas those with e_g symmetry stabilise the less dense cubic configurations in the bcc structure. Lastly, the sp^3 configuration corresponds to bonds with the tetrahedral symmetry of the diamond lattice.

Quite arbitrarily, we can examine the non-crystalline systems with covalent, strongly directional bonds, where the chemical factors are dominant in realising short range order, separately from the metallic and ionic systems where the non-directional chemical bonds imply that geometric considerations prevail in the study of topological order. Such a study is often associated with the problem to densely pack spherical atoms, whether hard or not, and, at the same time, inhibit any crystal structures forming.

We shall start with the most complicated description of short range order in those metallic systems, both elemental, where topological factors largely predominate, and alloys, either with other metals or with metalloids. In these systems, in addition to topological factors, chemical factors are required to achieve local order.

It is reasonable to think that a crystal forms when structurally identical small elements with a simple structure, that include the significant chemical bonds in the solid, join together by way of simple crystallographic operations such as translations, rotations, reflections, and interpenetrating growth.

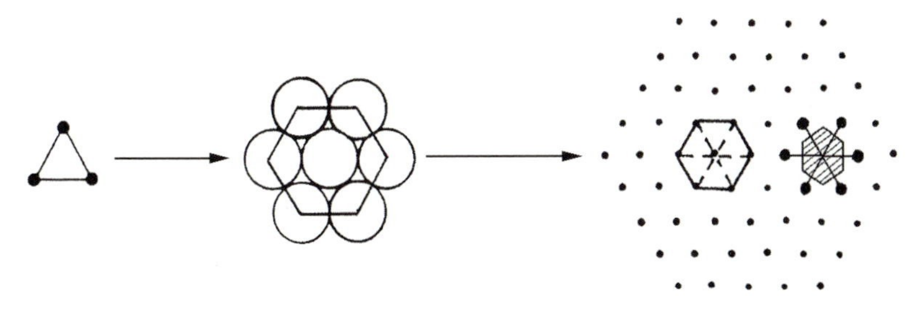

Fig. 4.36. Construction of a Voronoi polygon for ordered and regular tiling of the plane with triangles

The vast family of the Frank–Kasper inter-metallic phases is a typical example of this approach. The structural building blocks do not coincide, but they are part of the crystal unit cell, which is very large and meets the conditions of one of the 230 space groups so that the crystal with infinite extension can form by way of pure translations. The structural building block is, on the other hand, a *tetrahedron* and allows high density atom packings.

Before we proceed to discuss the meaning and implications of this structural unit, we have to better examine the simpler problem of how the most dense packing is achieved in a plane. The greatest packing efficiency for identical disc-shaped atoms corresponds to an equilateral triangle as our structural building block (see Chap. 1). We can describe the densest possible tiling, in statistical geometric terms, by using the *Voronoi polygons* (Voronoi polyhedra, in three dimensions). Just like the Wigner–Seitz and Brillouin cells, by definition each of these cells has a single atom located at its centre and is made up of the intersections between the segments that perpendicularly bisect the "bonds" that join the central atom to all its first neighbours.

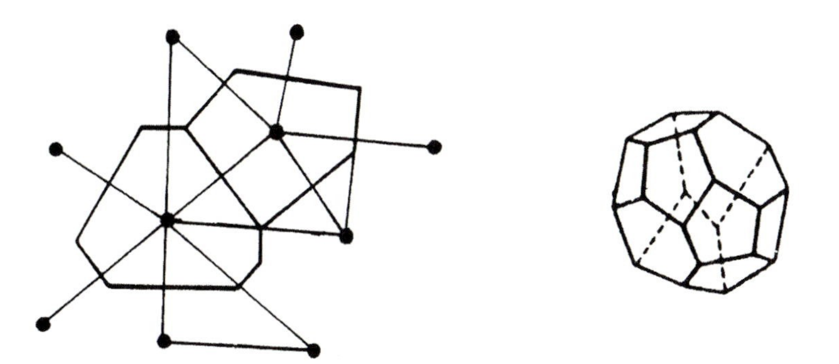

Fig. 4.37. Voronoi polygons for a disordered tiling and a Voronoi polyhedron for a disordered solid: notice the high number of pentagonal faces; the number of sides to a face corresponds to the number of tetrahedra packed around the "bond" bisected by that face.

As regards tiling a plane with equilateral triangles (Fig. 4.36), the Voronoi polygons are all regular hexagons. In general, in the case of disordered atomic distributions, we obtain polygons with a variable number of sides.

Analogous to the above example, the Voronoi polyhedra are the smallest convex polyhedra that contain a single central atom. They are formed by the intersection among the planes that normally bisect the "bonds" which connect the atom chosen as the central atom to its first neighbours. The faces of the Voronoi polyhedra are irregular polygons, very often *pentagons*, as exemplified in Fig. 4.37. The number of sides for a given face is equal to the number of packed tetrahedra about the "bond" that is bisected by that face. There being pentagonal faces implies there is a consistent degree of icosahedral order.

From a three-dimension point of view, the densest packing of identical atoms is given by the tetrahedral arrangement of four hard spheres, each of which is in contact with the other three. The packing efficiency, namely the fraction of volume filled by matter, is around 78%, whereas, for an octahedral structural unit, this fraction is around 72%.

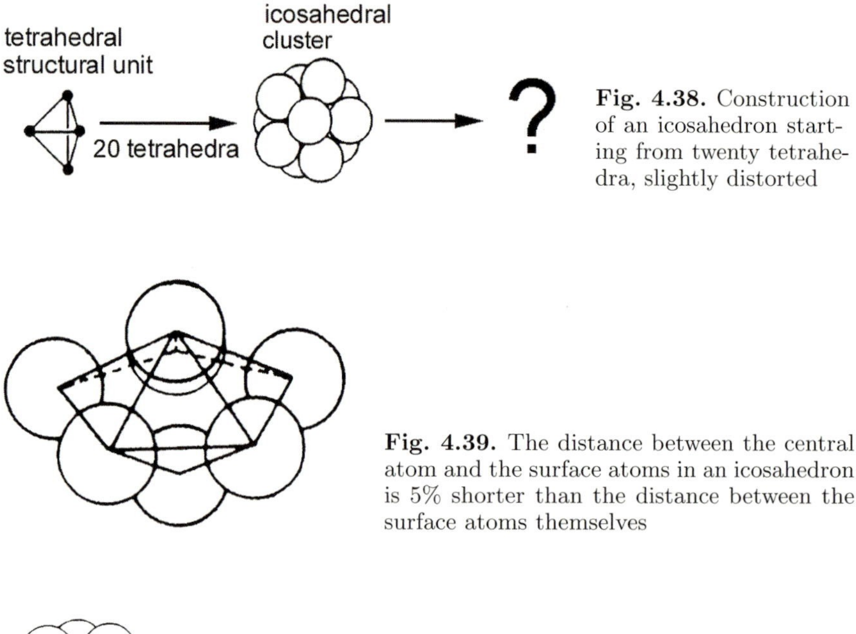

Fig. 4.38. Construction of an icosahedron starting from twenty tetrahedra, slightly distorted

Fig. 4.39. The distance between the central atom and the surface atoms in an icosahedron is 5% shorter than the distance between the surface atoms themselves

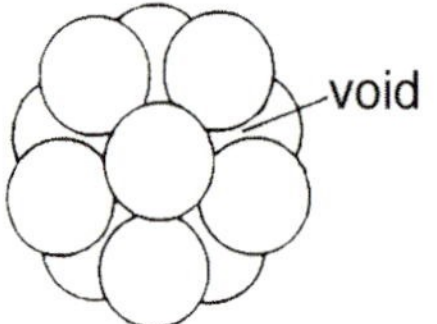

Fig. 4.40. Surface atoms in an icosahedron do not touch each other

If we start with 20 tetrahedral building blocks, as in Fig. 4.38, we obtain, with small distortions, an icosahedron. Similarly to the hexagon, the surface of the icosahedron is divided into regular triangles: six in the hexagon and twenty in the icosahedron (see Chap. 1). However, unlike the hexagon in the plane, the icosahedra, with their six five-fold rotation axes, cannot be arranged in a simple and periodic manner to fill the space. The fact that it is impossible for perfect tetrahedra to fill the space already emerges from the icosahedron geometry where the distance between the surface atoms is 5% greater than the distance between surface atoms and the central atom (Fig. 4.39). If we try to bring the surface atoms closer together, without affecting the distance between them and the central atom, then the small interatomic surface voids in the initial atom arrangement (Fig. 4.40), coalesce to form a single fracture that opens up in some points of the surface, as shown in Fig. 4.41. This fracture constitutes an excess "free volume", even though it is insufficient to host a thirteenth surface atom. Similarly, if we try to construct a structural unit that is simpler than the icosahedron, still using the tetrahedra, we immediately come across insurmountable problems in trying to occupy the available space. In the simplest of cases (see Fig. 4.42), if we densely assemble the largest number possible of tetrahedra (five) about

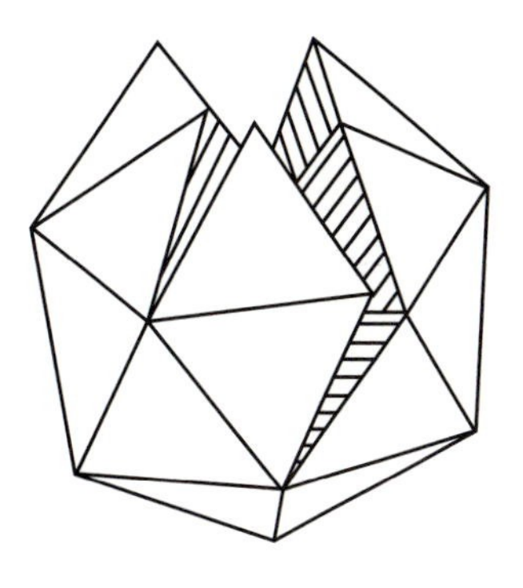

Fig. 4.41. In an icosahedron, if we force contact between surface atoms a fracture is caused at a point on the surface

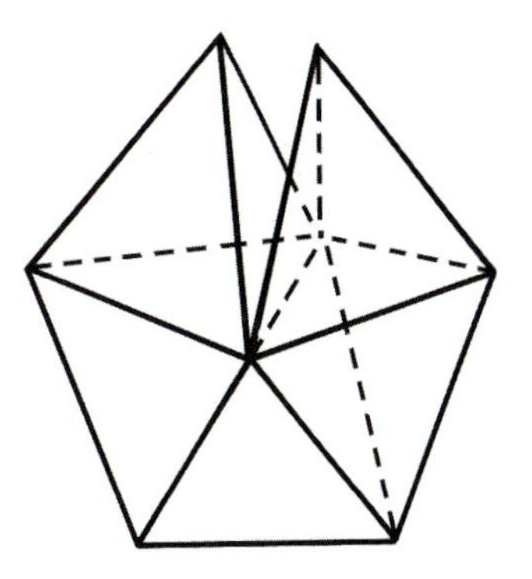

Fig. 4.42. When five tetrahedra are joined around a common edge a fracture is caused, subtended by an angle of $7°35'$

a common edge, we observe an unoccupied "segment" of space subtended by an angle of $7°35'$. The only possible way of filling the space with such units is to introduce line defects, namely to distort the structure, at the expense of a considerable increase in the free energy of the system.

The definitive incompatibility between tetrahedral local order and a simple scheme of extended tetrahedra packing is called *frustration*. Surface atoms are said to be frustrated because the positions of the atoms surrounding the empty "segment" are not, at the same time, points of minimum energy of the pair interactions with all the first neighbours. Yet, the tetrahedron seems to be the most natural structural building block in studying short range order and transition to long range order in condensed phases made up of hard-sphere atoms. As a matter of fact, nature adopts various strategies to overcome the unavoidable frustration.

The cube is the only regular polyhedron that allows us to fill the space fully when it is used alone. In general, in order to accomplish this we need at least *two* kinds of polyhedra, packed together using well defined proportions and sequences, so that we can obtain the maximum global density. This information is contained in the space group of a crystal; however, it is lost in the liquid.

Among the crystals, those with the simple fcc or hcp structures, with highest packing densities (Fig. 4.43) can be constructed by alternating an octahedron with every two tetrahedra. As such we obtain a 74.05% packing efficiency. In the fcc structure, the tetrahedra are packed about the octahedra so that they share edges and vertices, but do not have common faces. In the hcp structure, the tetrahedra have common vertices and faces, but never with more than *one* tetrahedron. The bcc structure is made up of considerably *distorted* tetrahedra.

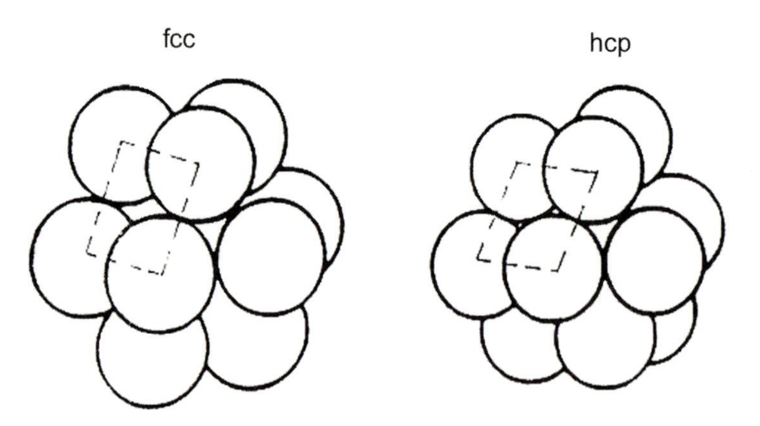

Fig. 4.43. Elemental clusters of spherical atoms, packed according to the fcc and hcp schemes (see also Fig. 1.20)

To increase the overall packing density, a number of tetrahedra must share their faces, and this can be achieved by piling them along a *spiral* or by coordinating 12 atoms about a central atom in order to form a icosahedron, which is the first coordination shell of a regular tetrahedral arrangement.

Each atom is in contact with the 12 first neighbours also in the fcc and hcp crystals; the difference between this local elementary structure and the icosahedral structure is that in the fcc (hcp) cluster the 12 first neighbours are not distributed uniformly on the surface of the central atom they are in contact with, whereas when the central atom is icosahedrally coordinated, the 12 first neighbours are distributed symmetrically on it, but *are not* in contact with each other, as shown in Fig. 4.40. The icosahedron surface is divided into twenty regular triangles. If we take the central atom as our reference atom, then each of the 12 first neighbours is coordinated on the surface of the central atom with five atoms whose centres make up a regular pentagon.

In 1952, Frank compared the energies of a 13-atom icosahedral cluster with a similar fcc cluster; both clusters were relaxed in Lennard–Jones $\mathcal{U}(x)$ potential.

This is a particular case of the potential

$$U(x) = \frac{n\varepsilon}{n-6}\left(\frac{n}{6}\right)^{6/(n-6)}\left\{\left(\frac{\sigma}{x}\right)^{n} - \left(\frac{\sigma}{x}\right)^{6}\right\}$$

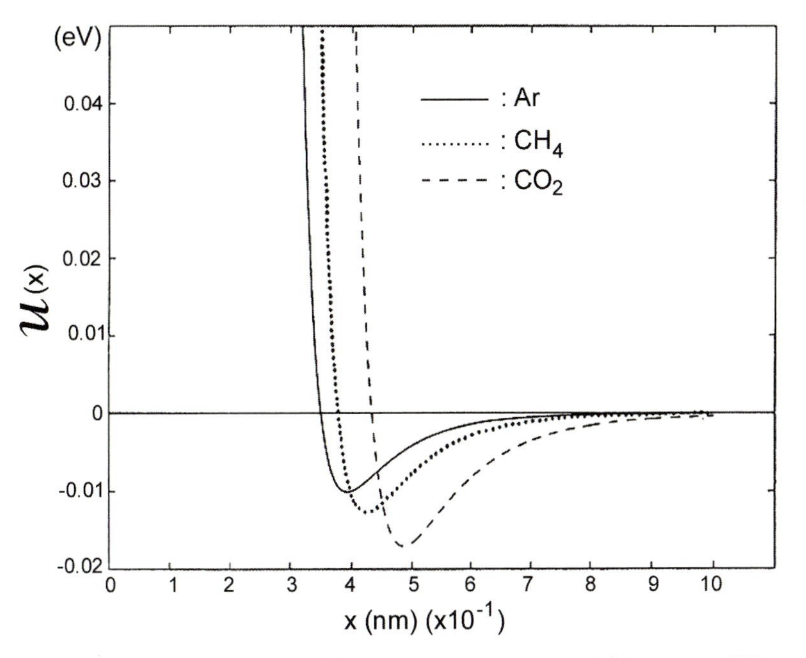

Fig. 4.44. Trend in Lennard–Jones interatomic potential in argon, CH_4 and CO_2

where σ is the distance at which $U(x) = 0$ and ε the depth of the potential well. When exponential n, which usually varies between 9 and 15, equals 12, $U(x)$ is called the Lennard–Jones potential.

$$\mathcal{U}(x) = 4\varepsilon \left\{ \left(\frac{\sigma}{x}\right)^{12} - \left(\frac{\sigma}{x}\right)^{6} \right\}$$

The trend of $\mathcal{U}(x)$ is given in Fig. 4.44. The surprising discovery from the analysis of clusters was that the energy of the icosahedral structure is 8.4% less than that of the cubic structure, although the latter is made up of tetrahedra and octahedra. The considerable increase in local stability of the icosahedral cluster is not sufficient, though, to remove the frustration in the *extended* structure.

The crystalline state is made compatible with an extended tetrahedral order by way of slightly distorting the tetrahedra. The result is that the crystal has a regular arrangement of $-72°$ disclination lines (see Chap. 5) which corresponds to removing a tiny portion of the material in a medium showing extended icosahedral order.

The frustration associated with obtaining perfect extended tetrahedral structures is partially relaxed by mixing together atoms with two different sizes. The typical example is given by the Frank–Kasper phases where the triangles on the surface of each icosahedron form ideal tetrahedra with the central atom, provided the diameter of the central atom is less than about 10% of the external atoms.

Frank and Kasper suggested the use of four different polyhedra with normal coordination, that respectively coordinate $12, 14, 15$ and 16 atoms (Fig. 4.45) with different sizes. These polyhedra are made up of tetrahedra only, and are characterised by having very high packing efficiency. Even though the local density of the tetrahedra is greater than the average crystal density it cannot be kept for steric reasons. In three-dimensional crystalline lattices, the polyhedra with normal coordination are located in skeletons and generally give rise to layered structures. The Frank–Kasper phases, of which the Laves phases (see Chap. 1) are an example, are described as being sequences of high density atomic layers tiled with alternating hexagons, pentagons and triangles and stacked in such a way as to form, between the layers, tetrahedral interstices only.

In the C15 structure with XY_2 composition, whose packing efficiency we have already discussed (see Chap. 1), the interatomic distances Y–Y are often less than the distances observed in the Y elemental lattice, whereas the opposite is true for the X–X distances. As such, we can then suppose that the small Y_4 tetrahedra determine the structural stability of the crystal. If we substitute Cu with Zn in $MgCu_2$, we notice that the form of the skeleton, in which the Y (Cu and/or Zn) atoms are packed, depends on the number of valence electrons per atom, e/a. In particular, where $e/a = 1.8$, we observe a structural transition from the C15 phase to the C36 hexagonal phase (type $MgNi_2$) and where $e/a = 2$ a transition occurs to the still hexagonal C14 phase, (type $MgZn_2$), as shown in Fig. 4.46.

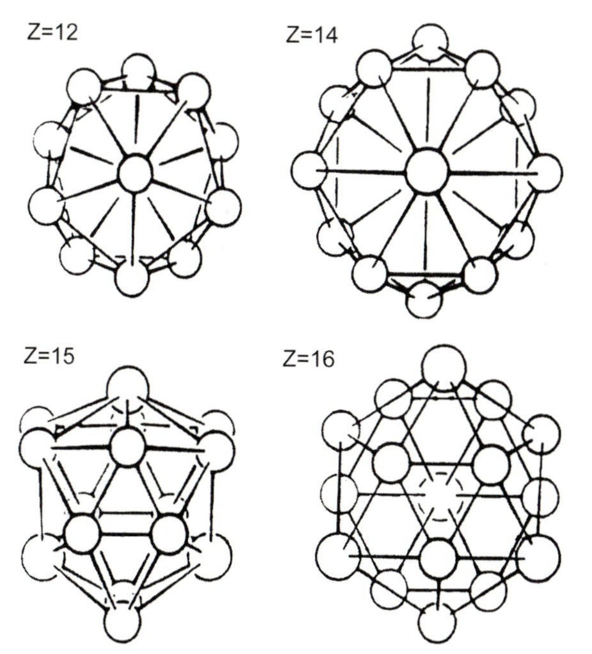

Fig. 4.45. The four normal coordination polyhedra according to Frank and Kasper consist only of tetrahedra and respectively coordinate 12, 14, 15 and 16 atoms

This sequence throws light on the role played by the conduction electrons, in the band states that form when X–X overlapping occurs, in stabilising long range order. The short range order is determined by the Y_4 tetrahedra, which are formed because of the directional atomic orbitals involving d or p electrons.

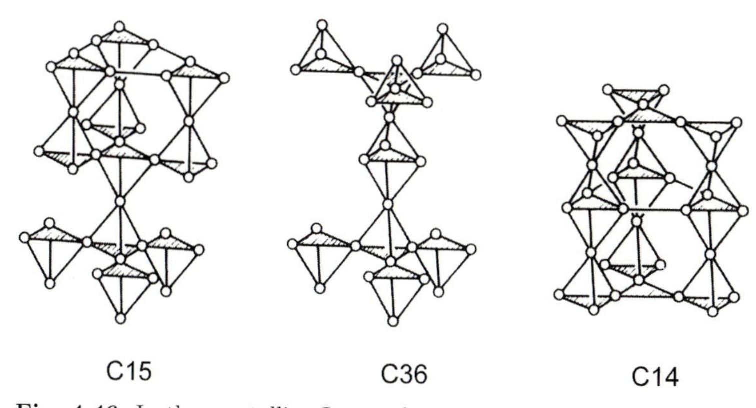

Fig. 4.46. In the crystalline Laves phases with composition XY_2, the relative arrangement of Y_4 tetrahedra changes in structures C15, C36 and C14 as the average number of electrons per atom, e/a, increases

The shape of the elementary structural unit depends critically on the concentration, size and electron configuration of the constituent atoms; it seems reasonable to assume that the structural building blocks for a given alloy are essentially the same in the crystalline, glassy and liquid states. They realise various *relative arrangements* of the above structural units depending on the degree of the atomic mobility typical for the phase under examination.

The same material in the liquid state is not subject to the steric constraints the crystal is; tetrahedra can form in the liquid in unlimited number and arrange themselves into icosahedral coordination. Local polytetrahedral order in the icosahedron differs from the order of tetrahedra and octahedra distribution in the fcc crystal due to the *orientational* part of the coordination. This is not identified in diffraction experiments, due to the orientational average in the liquid, yet, it is the specific characteristic of the polytetrahedral (icosahedral) order, according to the models and simulations for this structure.

As the atomic mobility typical of a liquid at high temperature is reduced towards the crystallisation temperature, new structural elements, e.g. octahedra, form and are packed in the correct proportions. The experimental proof that the short range order in a liquid is *not* the same as that in the corresponding crystal is given by the supercooling of small droplets of mercury to temperatures around $(2/3)T_{\mathrm{m}}$. This proves that atoms with the same coordination number and coordination distance, thus with similar atomic volume (as is observed in a material both in the crystalline and in the liquid phases), may have different short range order.

According to Frank, the reason why large supercooling takes place in liquid metals, where the central interatomic potential dominates, is precisely the presence of icosahedral (polytetrahedral) order.

The liquid metals are supercooled because in order to substitute a local stable configuration (icosahedral) with one of the extended configurations typical of a crystal, with higher energy, we have to overcome a high energy barrier.

The fact that the solidification is blocked in the glassy, disordered state implies that there are *constraints* which impede the structural building blocks arranging themselves according to appropriate crystalline order. It is just because both the elementary structural building blocks and the bond energies coincide in the amorphous solid and in the crystal that these constraints are only steric in origin, and must thus be explained in terms of the specific nature of the chemical short range order.

The free energy in a liquid alloy is dominated by the mixing entropy, in particular at high temperature. Whether the structural building blocks survive in the liquid or not depends on the comparison between the local enthalpy, associated with each building block, and the entropy associated with the short range order these building blocks establish. When the liquid is cooled, the constraints set by the need to densely occupy the available space increase as the density increases. For essentially steric reasons, many tetrahedral units are formed. These are not solid clusters; rather, they *fluctuate* spatially and temporally.

Whether chemical short range order will exist or not, thus, strongly depends on the symmetry associated with it. Any deviation from the (poly)tetrahedral structure implies there is an entropic contribution. Conversely, the realisation of chemical short range order with tetrahedral symmetry leads to an increase in local enthalpy when chemical bonds are formed without any contribution to configurational entropy.

The frustration that occurs while a liquid solidifies, when a lot of tetrahedral building blocks have to rearrange and coordinate themselves with *other* polyhedra, indicates that there is an energy barrier due to the new kind of chemical short range order, and that this barrier is higher the stronger the chemical bonds are.

The height of the barrier is insufficient to allow vitrification in most of the pure metals, under reasonable supercooling conditions. Elements that stabilise the amorphous phase have to be added to the system; they form a dilute alloy with the metal. The greater the cohesion energy of the alloy is, the greater is the efficiency of these impurities, which operate as nucleation centres of the glassy phase. This corresponds to the formation of stable chemical bonds that can associate atoms together to form structural units, e.g. tetrahedra, still in the liquid state. During the cooling phase, and starting from these localised units, a well defined short-range chemical order is developed which, in the end, freezes in the amorphous state.

Among the experimental results to prove the existence of tetrahedrally coordinated units in the glassy state, on the one hand we may mention the great similarity between the intrinsic magnetic properties of the transition metal-metalloid ferromagnetic alloys in the crystalline and in the amorphous states and, on the other, the tiny density reduction, about 2%, in the glassy alloys compared to the corresponding crystals. Such value is incompatible with the strong density reduction, about 15%, in the amorphous materials, as provided for in a model that does not include local structural organisation, such as the dense random packing of hard spheres.

However, the most direct proof of structural organisation in metal glasses comes from X-ray and neutron diffraction, which often give rise to a pre-peak and to a sharp, narrow first peak.

The pre-peak is found in alloys with two constituents, such as $Ni_{40}Ti_{60}$, $Ni_{35}Zr_{65}$, $Ni_xNb_{(1-x)}$. The reason for this pre-peak has already been discussed for NiNb (see Sect. 4.3). This pre-peak is an index of the considerable degree of chemical short range order, thereby hetero-coordination is preferred to a random atomic arrangement. A tetrahedral packing model for the atoms of the minority species often gives the best agreement between the calculated distances between first neighbour atoms of the various constituents and the measured interatomic distances.

A dense packing of tetrahedra best describes the structure of the rare earth-aluminium glasses with a composition ratio of four to three. The local arrangement of the structural units is very similar here to the crystalline alloys with a strongly packed tetrahedral structure, as exemplified by the prototypical alloy $Zr_{57}Al_{43}$. The rare earth atoms are arranged in Frank–Kasper normal coordination polyhedra and the aluminium atoms are icosahedrally coordinated.

Even in those alloys where the chemical factors are less efficient in determining short range order, this order is often tetrahedral. For example, when we compare the structure of amorphous $Gd_{57}Al_{43}$ with that of crystalline $Gd_{57}Cu_{43}$, we see that the need to fill the space as efficiently as possible generates topological short range order that is extremely similar to tetrahedral order in the Frank–Kasper phases. Yet, we do not observe a distribution in the distances between first neighbours, which clearly indicates the presence of chemical short range order. This same order, which is determined by the dense packing of tetrahedral units, has been observed, among others, in equiatomic amorphous GdY. The most probable hypothesis about the nature of such an ordering is that it is originated by the spherical symmetry typical of the pair potential which governs interatomic interaction in metals.

The bond directionality in *covalent* materials greatly simplifies the study of short range order. This can be defined by referring to a specific *local* coordination polyhedron. As shown in Fig. 4.47, the description of the local topological order in a binary compound XY only requires us to know the number of first neighbours for the atom x, n_y, the subtended bond angle for atom x, φ_x, the bond length X_{xy} and the corresponding quantities when we consider atom y as the origin, namely φ_y and n_x.

The number of faces, edges and vertices that each coordination polyhedron shares with the other surrounding polyhedra, namely the *connectivity*, is more important when studying medium range order than short range order.

When atoms of different species are coordinated about the central atom of a coordination polyhedron, the chemical short range order takes on prime importance. If the composition is not stoichiometric, and we assume that both the valence and the atomic coordination are unchanged, then the only possible way to arrange the excess atoms is through homopolar, thus "wrong" bonds. Consequently, the chemical order associated with stoichiometric composition is at least in part broken. The significant order parameter in this case is given by the relative abundance of wrong bonds.

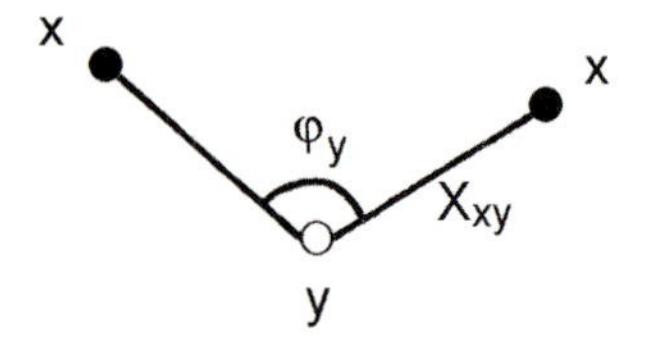

Fig. 4.47. Schematic illustration of the elements defining short range order in amorphous covalent materials: number of first neighbours, n_x (two in the figure), bond angle, φ_y and bond length, X_{xy}

In some cases we notice that atoms of a given chemical species have *different* bond charge and connectivity states, even when the composition is stoichiometric. The consequence is that, for a given polyhedron, these atoms, which are coordinated about a central atom, will realise a degree of chemical short range order which is less than the CSRO in an identical coordination polyhedron in which all the atoms have equal valence and connectivity. The trend can continue until the limiting case is reached where chemical order can be completely discarded.

The description of chemical order in covalent systems, is particularly simplified in the case of the binary compounds $X_{(1-c)}\,Y_c$ where, if elements X and Y belong to column x and y of the Periodic Table, the local coordinations $N_x = 8 - x$ and $N_y = 8 - y$ are achieved, which is an agreement with the normal valence of the two elements and with the "$8 - N$" rule.

In a compound with arbitrary stoichiometry, bonds X–X, X–Y and Y–Y can coexist. The possible distribution of these bonds can be derived from two alternative models. In the first model, the Chemically Ordered Network (CON), the X–Y bonds are more likely to form. At the stoichiometric composition, $c_s = N_y/(N_x + N_y)$, the phase is perfectly ordered from the chemical point of view. If the compound is enriched in element X, namely for $0 < c < c_s$, then apart from the X–Y bonds there are only X–X bonds. Conversely, if the compound is over-stoichiometric in element Y ($1 > c > c_s$), X–Y and Y–Y bonds only are observed. The bond statistics in the CON model, in the case of a XY_2 system (like SiO_2) in the region enriched in $X(0 < c < 0.67)$, can be obtained if we take into consideration that each divalent Y atom coordinates through two bonds, so

$$N_{\mathrm{X\text{-}Y}} = 2c. \tag{4.88}$$

We can obtain the number $N_{\mathrm{X\text{-}X}}$ of X–X bonds by considering the number of bonds atoms X can enter, $4(1 - c)$; this is equal to

$$N_{\mathrm{X\text{-}X}} = \frac{1}{2}\,[4(1 - c) - 2c] = 2 - 3c. \tag{4.89}$$

The second term in the brackets has to be subtracted from the first to take into consideration the already formed X–Y bonds. The 1/2 factor is introduced to ensure we count each bond once only. Given the stoichiometry, then obviously

$$N_{\mathrm{Y\text{-}Y}} = 0. \tag{4.90}$$

In the region enriched in the divalent element Y, ($1 > c > 0.67$), the bond statistics becomes

$$N_{\mathrm{X\text{-}X}} = 0. \tag{4.91}$$

The number of Y–Y bonds is obtained from the difference between the total number of bonds, $(1/2)(2c)$, and the number of X–Y bonds, namely $(1/2)[4(1 - c)]$, so

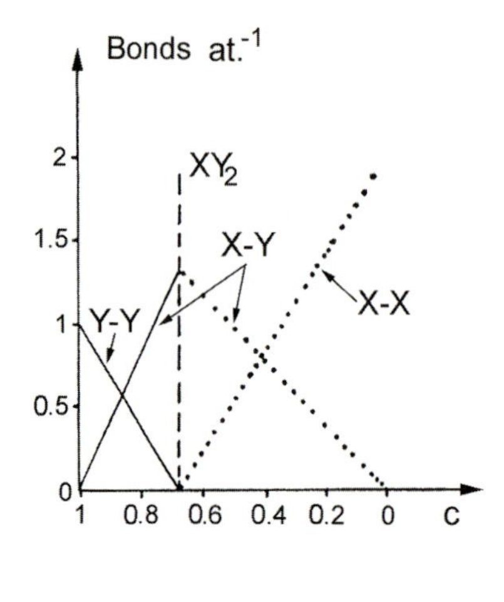

Fig. 4.48. Trend in the number of bonds per atom as a function of composition, for a binary XY compound, analysed using the chemically ordered network (CON) model. The compound with stoichiometric composition XY_2 separates the regions over-stoichiometric (—) and sub-stoichiometric (...) in Y (adapted from [4.9])

$$(1/2)[2c - 4(1 - c)] = 3c - 2 \tag{4.92}$$

where the 1/2 factor is necessary again to ensure we count each bond once only.

In Fig. 4.48 the curves represent the number of bonds per atom as a function of composition c for the Chemically Ordered Network model.

The second model, the *Random Covalent Network* (RCN), deals purely with the statistical bond distribution and is only determined by the local coordinations N_X and N_Y and by the composition c. Any effect due to preferential ordering, that arises from the possible differences in bond energy, is disregarded; thus, for any composition, except for the extremes $c = 0$ and $c = 1$, the X–X, X–Y, and Y–Y bonds are allowed. Once again, referring to a compound with a ratio of 2 : 4 between the constituent coordinations, we obtain the statistics for the various bonds from the total number of bonds taken singularly,

$$\frac{1}{2}[4(1 - c) + 2c] = 2 - c. \tag{4.93}$$

The first term in the brackets in equation (4.93) takes into account the valence four of the X atoms, and the second term the valence two of the Y atoms.

The number of X–X bonds is proportional to $4(1 - c)^2$, the number of X–Y bonds is proportional to $2[2c(1 - c)]$ and the number of Y–Y bonds is proportional to $(1 \cdot c^2)$. Factors 4, 2 and 1 refer to the various possibilities that atoms with different valences have to combine together. Thus, in a purely statistical model, a pair of next neighbour tetravalent X atoms is more likely to bond together since each one has four available atomic orbitals, whereas two divalent Y atoms each have two atomic orbitals. Thus we obtain the following system:

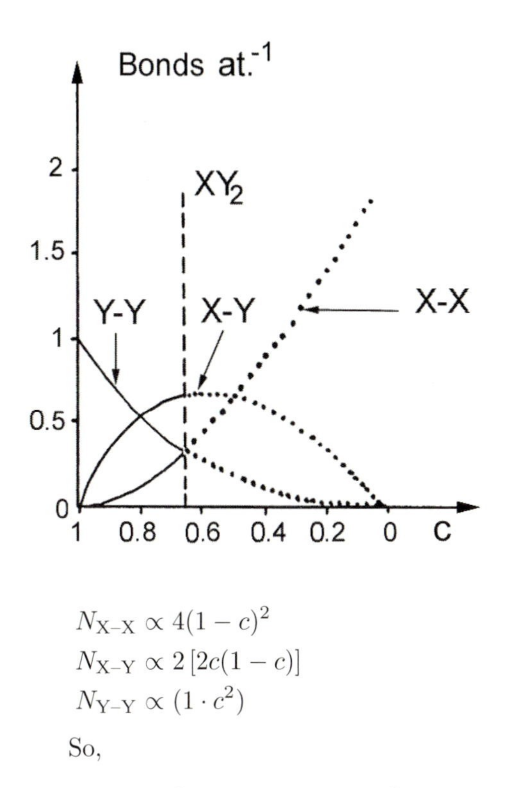

Fig. 4.49. Trend in the number of bonds per atom as a function of composition, for a binary XY compound, analysed using the random covalent network (RCN) model. The compound with stoichiometric composition XY_2 separates the regions over-stoichiometric (—) and sub-stoichiometric (...) in Y (adapted from [4.9])

$$\begin{aligned} N_{\text{X-X}} &\propto 4(1-c)^2 \\ N_{\text{X-Y}} &\propto 2\left[2c(1-c)\right] \\ N_{\text{Y-Y}} &\propto (1 \cdot c^2) \end{aligned} \tag{4.94}$$

So,

$$4a(1-c)^2 + 4ac(1-c) + ac^2 = 2 - c \tag{4.95}$$

from which we obtain $a = 1/(2-c)$.

When we insert this value into (4.94) we obtain

$$\begin{aligned} N_{\text{X-X}} &= 4(1-c)^2/(2-c) \\ N_{\text{X-Y}} &= 4c(1-c)/(2-c) \\ N_{\text{Y-Y}} &= c^2/(2-c). \end{aligned} \tag{4.96}$$

The curves in Fig. 4.49 represent the dependence of the number of bonds per atom on stoichiometry in the random covalent network model.

Both models are simplified pictures and can only be applied to covalent systems with well defined directional bonds. Of the two models, CON provides a deeper physical insight, and gives a good description of most chalcogenide glasses.

4.8 Medium Range Order

On the basis of the remarkable amount of short range order in all the kinds of non-crystalline systems, it is quite normal for us to wonder if there is any structural correlation on a longer scale, typically in the interval 0.5–2 nm.

We certainly understand less about medium range order (MRO) than we do about local order. This is immediately confirmed by the lack of a single definition of medium range order. The simplest definition accounts for the elements accepted when the defining short range order, namely the two and three particle correlations that determine, respectively, the bond lengths and angles as well as the local symmetry of the site under examination. Four to ten particle correlations regard ordering over intermediate distances. On the other hand, we can claim that an amorphous solid exhibits medium range order when it is structurally *non*-random, even beyond the first neighbour shell of an atom taken as the origin.

Just like local order, the development of medium range order must be considered separately in covalent and metallic systems. Furthermore, in the former, where order on an intermediate level is undoubtedly clearer, we can identify various *levels* of structural organisation on scales of increasing length, yet within the already specified interval. The short range order is often described by well defined coordination polyhedra; in turn the first level of intermediate order depends on the kind of connections between those polyhedra and on their relative orientation.

Even if we start from a single polyhedron, which constitutes the most elementary structural unit, then the obtained structure when a number of units are interconnected in such a way as to share *vertices* or *edges* or *faces* is highly differentiated, already at the level of local order.

Medium range order is more evident when the relative orientation of adjacent polyhedra is locked by the interconnection procedure. We expect the medium range order associated to the sharing of edges to be more marked than the MRO determined by the sharing of vertices, which in principle would allow the various structural units to freely rotate around common bonds. In this case, however, we are more likely to obtain certain orientations than others. In the covalent systems, these orientational correlations are measured using the *dihedral* or torsion angle Φ (Fig. 4.50). When we consider two interconnected structural units, and having chosen a reference bond, then Φ is the angle the homologous projections of the other bonds in the structural unit under examination must be rotated about in the plane normal to the bonds to be brought into coincidence. Deviations from a uniform distribution for Φ values, $P(\Phi)$ are indices of medium range order.

The difference in SRO for chemically similar systems with the same local structural organisation (thus with the same coordination polyhedra) gives rise to different *dimensionality* of the networks of disordered bonds, which we can obtain when we force adjacent polyhedra to share different elements (vertices, edges or faces).

One characteristic example is given by the X(Ge; Si) Y_2 (S; Se) glass family. In these cases the characteristics of the short range structure are due to the prevalence of XY_4 tetrahedra; however, the kind of connection between polyhedra changes from material to material. While the adjacent tetrahedra

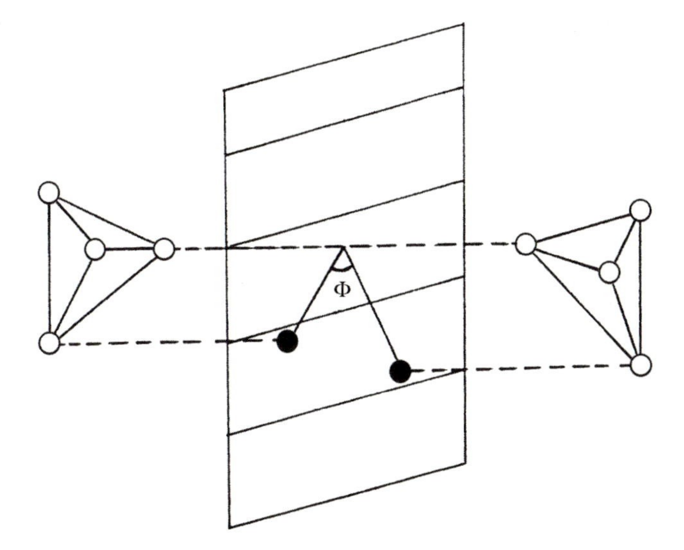

Fig. 4.50. Schematic illustration to define the dihedral angle, Φ

of SiO_2, both crystalline and glassy, are connected through the vertices and give rise to a three-dimensional giant molecule structure, $GeSe_2$ and GeS_2 exhibit a fraction of structural units interconnected at the edges, both in the amorphous and in the crystalline states, and these edges produce local two-dimensional structures. Lastly, the polymorphous crystalline modifications in $SiSe_2$ and in SiS_2 exhibit edge sharing only between tetrahedra. This type of connection predominates in the glassy state, too, and gives rise to one-dimensional local structures.

If we consider a greater scale than 0.5 nm, then the medium range order is characterised by *extended* structural units. These units consist of, for example, rings of atoms with a particular shape and size, and are found in the system with significantly greater frequency than the purely statistical one. The local structure within each ring is determined by short range order.

So far our examination of medium range order has developed by considering that the structural order over an intermediate distance is a consequence of a specific kind of short range order. This is largely proven in covalent systems where the short range order is well defined, and is strictly dependent on the constraints of chemical bonding. Conversely, in the metallic amorphous systems, where non-directional bonding prevails, these kinds of constraints are absent.

However, when a considerable degree of short-range order is found, like in the transition metal-metalloid alloys, where short range order has mainly topological origin, we may assume that such a SRO is the consequence of a well defined medium range order, unlike in the covalent systems.

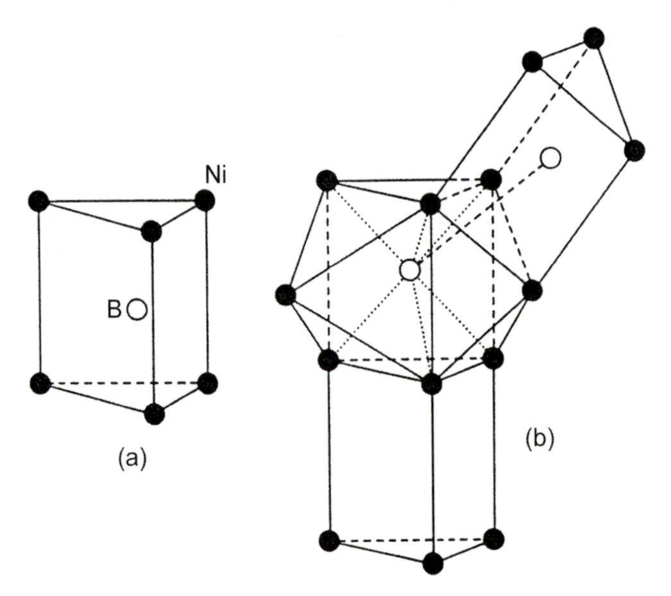

Fig. 4.51. (**a**) Trigonal prismatic elemental building block of cementite $Ni_{75}B_{25}$; (**b**) construction of the disordered structure for amorphous $Ni_{80}B_{20}$, using the same kind of structural element

As such, diffraction studies on amorphous $Ni_{80}B_{20}$ show that the average coordination number around the metalloid atoms is 9.3 (in crystalline $Ni_{75}B_{25}$, with a cementite structure, for which $Fe_{75}C_{25}$ is a prototype, the same coordination number is 9) and around the transition metal atoms it ranges between 12 and 14. It is noteworthy that there are no metalloid-metalloid first neighbours. We require that the same kind of elementary structural unit, namely trigonal prismatic, capped, already adopted for crystalline $Ni_{75}B_{25}$ is the base element for the amorphous $Ni_{80}B_{20}$ structure. This choice has the specific objective to highlight the considerable degree of medium range order in the system.

Each prism contains a metalloid atom surrounded by nine transition metal atoms. These units are efficiently connected together by sharing vertices and edges, and are highly constrained both in terms of position and orientation. Layers of densely packed trigonal prisms (Fig. 4.51) make up the core of a computer simulated model for the $Ni_{81}B_{19}$ alloy. This system exhibits extra planes of densely packed nickel atoms which separate the various layers; three connecting rules govern the reciprocal arrangements of structural sub-units within each domain:

1) the nickel atoms can only occupy the vertices, or they must cap prisms. In this way we ensure that the trigonal prismatic coordination extends throughout the structure;

2) the connection between adjacent prisms can only occur by way of shared vertices, or edges. The structure is thus similar to the crystalline structure;
3) if an atom does not occupy a prism vertex, then it occupies a (common) vertex of two semi-octahedra which cap adjacent prisms.

With great care not to create grain boundaries, the model produces various *positionally* ordered domains similar to, though not identical to, some of the crystalline phases of the system. The medium range order obtained may extend to distances of 2 nm, as shown in Fig. 4.52. A further structural relaxation in a Lennard–Jones potential with 6 – 12 exponents (Fig. 4.44) does not bring about any significant topological modifications. This means that the system in its initial configuration already lies in the neighbourhood of a local minimum of the potential energy.

It is noteworthy that the structural model for this class of metallic glasses does not include, by construction, any icosahedral local order. Glass formation is caused by the freezing of configurational defects rather than by geometrical frustration.

Whether or not we need local icosahedral order in the formation of amorphous metallic structures is a problem that greatly influences the study of medium range order for these systems. So far only a few model alloys have been analysed, using Molecular Dynamics simulations, and the orientational order of the bonds in the liquid phase has been measured. Attention has been placed particularly on those glass forming alloys involving aluminium; these can be divided into two groups: the first group includes those systems which easily turn amorphous, that do not produce quasicrystals and have a rather low re-crystallisation temperature. Among these systems we shall consider Al–Ni as our prototype. The second group covers the alloys that are more difficult to vitrify, that do form quasicrystals and re-crystallise at high temperature. We shall consider Al–Mn as the prototype.

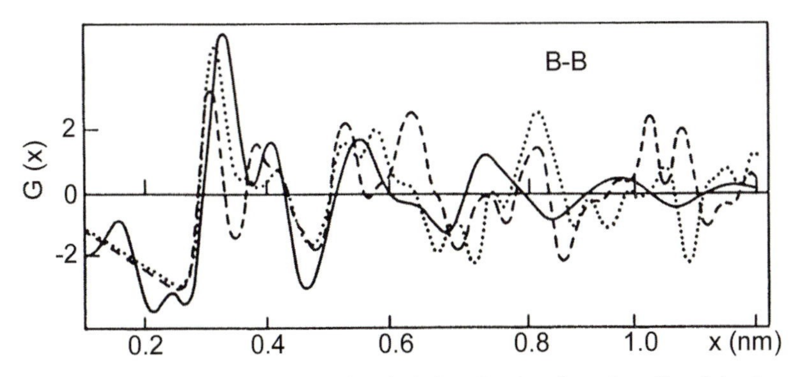

Fig. 4.52. Trend in the reduced radial distribution function $G_{BB}(x)$ of amorphous $Ni_{80}B_{20}$. —: experimental; – – –, ..., obtained using the structural model in part (**b**) of Fig. 4.51 which presents non-crystalline domains with medium range order extending typically up to 2 nm and 1.2 nm, respectively (adapted from [4.10])

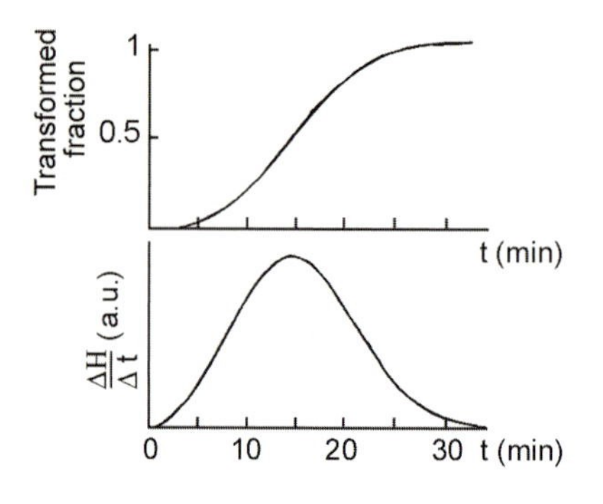

Fig. 4.53. Crystallisation via isothermal nucleation and growth; the top curve shows the fraction of transformed material, the bottom curve gives the corresponding enthalpy change (adapted from [4.11])

The way crystallisation occurs in the two groups is clearly different since in the first group crystallisation occurs through a nucleation and growth mechanism, whereas for the second group no nucleation stage exists. As such, we are given to think that the "amorphous solids" in the second group are in actual fact quasicrystals whose crystallites are no larger than 1.5 nm, and, as such, cannot produce narrow diffraction peaks. This is not surprising; "amorphous" films have been deposited, e.g. $Ag_{52}Cu_{48}$, whose experimental structure factor, with apparent non-crystalline features, coincides perfectly with that of a fcc cluster of the very same alloy composed of a small number of atoms, 125 for the specific case.

Microcalorimetry shows that the crystallisation kinetics of a system that undergoes a nucleation and growth process is characterised by sigmoidal trend as a function of time of the fraction of the crystallised material, as shown in Fig. 4.53. Conversely, the observed trend, where there is no nucleation stage (Fig. 4.54), for example in the "crystallisation" of $Al_{82.6}Mn_{17.4}$ films supposed to be amorphous, involves monotonic growth in the size of the *crystalline* grains already present in the alloy.

When we consider the partial correlation functions obtained by neutron scattering from Al–Ni and Al–Mn in the liquid state, they seem very simi-

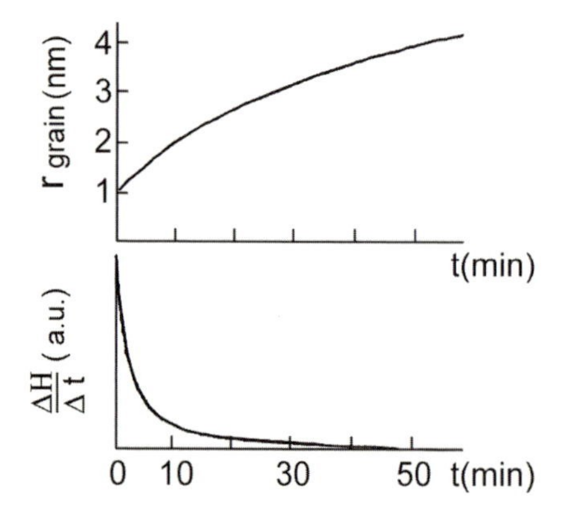

Fig. 4.54. Transformation of a nanocrystalline material via isothermal grain growth; the top curve shows the evolution in the average grain size; the bottom curve gives the corresponding enthalpy change (adapted from [4.11])

lar, except for the first peak in the Al–Ni distribution which is consistently broader than the Al–Mn peak. The calculated pair interaction potentials are very different from each other and from the Lennard–Jones potential. These potentials have been used in Molecular Dynamics simulations to analyse the orientational order of atom clusters with initial random configuration corresponding to the liquid state. Each bond connecting the first neighbour atoms together in a cluster is given a spherical harmonic

$$\boldsymbol{Q}_{lm}(\boldsymbol{x}) = Y_{lm}[\theta(\boldsymbol{x}), \phi(\boldsymbol{x})]. \tag{4.97}$$

In this equation $\theta(\boldsymbol{x})$ and $\phi(\boldsymbol{x})$ are the polar angles for the bond being considered, measured in a reference coordinate system, and where $(\boldsymbol{x})$ is the coordinate of the mid-point in the bond. We calculate averages on suitable groups of bonds,

$$\boldsymbol{Q}_{lm} = \langle \boldsymbol{Q}_{lm}(\boldsymbol{x}) \rangle .$$

The analysis of the orientational order parameters $\boldsymbol{Q}_{lm}(\boldsymbol{x})$ allows us to determine the characteristic intervals orientational order extends on. In an isotropic system, only $\boldsymbol{Q}_{00}$ is not null, after having calculated an average on the volume of the sample.

We then consider the combination of $\boldsymbol{Q}_{lm}$ which are invariant under rotation, for given values of l,

$$\boldsymbol{Q}_l = \left[\frac{4\pi}{2l+1} \sum_{m=-1}^{+1} |\boldsymbol{Q}_{lm}|^2 \right]^{1/2} . \tag{4.98}$$

$\boldsymbol{Q}_l$ is called a second order invariant.

Analysis of $\boldsymbol{Q}_l$, as shown in Fig. 4.55, indicates that only liquid Al–Mn consistently deviates from a random atomic packing, and exhibits a kind of order very similar to the order in the α–AlMnSi phase, which is a crystalline approximant of the icosahedral quasicrystalline structure (see Chap. 6).

Very similar results have also been obtained for the Al–Pd–Mn liquid alloy. We may be given to think that local orientational order of icosahedral kind is not dominant in all alloys in the liquid state; easy glass forming alloys, such as Al–Ni, that do not exhibit icosahedral organisation in the disordered phase, often exhibit a crystalline phase with cementite structure. This has also been observed in $Ni_{75}B_{25}$ and in $Pd_{75}Si_{25}$, whose local structural units are also found in the glassy structure. In all these cases, vitrification seems to be caused not so much by the topological frustration, but by frustration of the connectivity scheme for the structural units, as already discussed.

Only very recently have structures caused by medium range order been directly observed on an atomic scale, using high resolution electron microscopy. One very clear example of this kind of structure is given in Fig. 4.56, and refers to two amorphous alloys with composition $Pd_{82}Si_{18}$ prepared at cooling rates of 10^5 Ks^{-1} and 6×10^5 Ks^{-1}. Domains with an inter-fringe distance

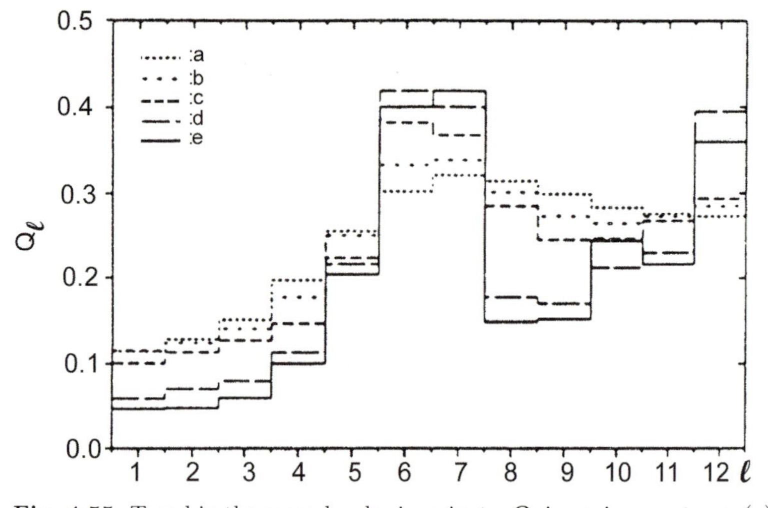

Fig. 4.55. Trend in the second order invariants, $\boldsymbol{Q}_l$ in various systems; (*a*) random packing of identical atoms; (*b*) liquid $Al_{80}Ni_{20}$ at $T = 1320$ K; (*c*) liquid $Al_{80}Mn_{20}$ at $T = 1320$ K; (*d*) quasicrystalline icosahedral AlMnSi; (*e*) α–AlMnSi, crystalline (adapted from [4.12])

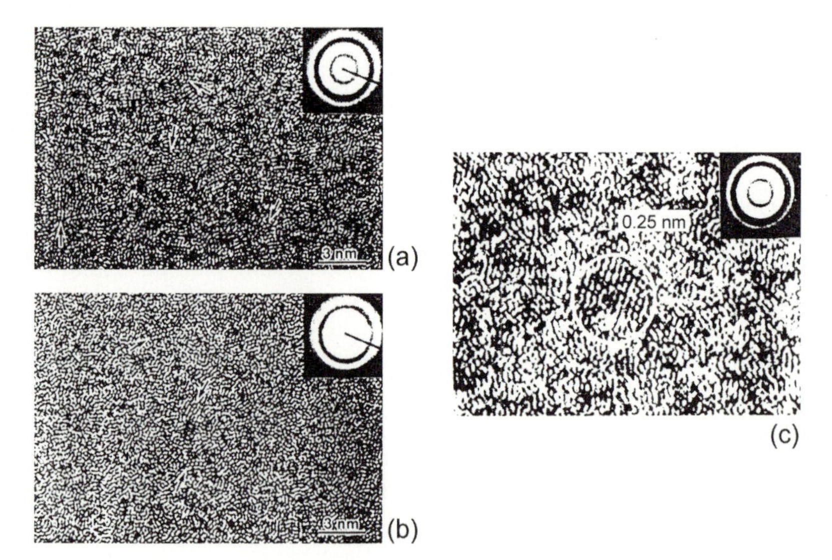

Fig. 4.56. High resolution electron microscopy images of glassy $Pd_{82}Si_{18}$ alloys prepared under different fast-quenching conditions; (**a**) $-\frac{dT}{dt} = 1 \times 10^5 \mathrm{Ks}^{-1}$; (**b**) $-\frac{dT}{dt} = 6 \times 10^5 \mathrm{Ks}^{-1}$; (**c**) shows the drastic structural modifications in the alloy with changed composition, $Pd_{75}Si_{25}$ (adapted from [4.13])

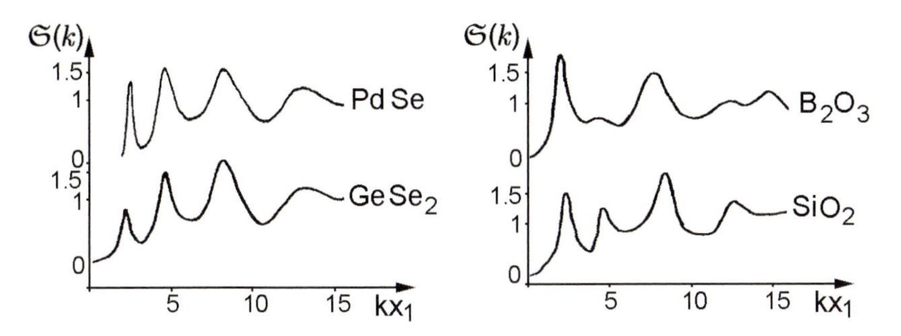

Fig. 4.57. Experimental evidence of medium range order in amorphous systems; the first sharp diffraction peak is invariably around $kx_1 = 2.5$

of 0.22 nm have been identified, whose mean size and volume fraction grow with the decrease in the quenching rate. These domains have a very similar structure to the fcc structure. The structure and inter-fringe distances change drastically on changing the stoichiometry of the material. The inter-fringe distance in $Pd_{75}Si_{25}$ grows up to 0.25 nm, whereas the super-structures due to the medium range order display very similar characteristics to $Pd_{75}Si_{25}$ crystalline cementite with Pd_6Si trigonal prisms. The above observations confirm that when the concentration of metalloid in the transition metal-metalloid alloys exceeds 20 at. %, medium range order with a chemical origin is developed.

The most immediate experimental proof of medium range order in *non-metallic* glasses is given by the presence in the diffraction patterns of the so-called First Sharp Diffraction Peak (FSDP), or pre-peak, in the reciprocal space. While this structure is very evident, as shown in Fig. 4.57, there is usually no specific well defined index for medium range order in any correlation functions in real space, typically in the radial distribution function.

The pre-peak has been observed in many chalcogenide glasses and in other covalent materials, such as $SiSe_2$, $GeSe_2$, GeS_2, B_2O_3, P_4Se_3, As_2Se_3, B, P, As, Sb, whereas it is *not* present in pure, amorphous Se and Ge. It has also been observed in liquid metallic alloys like KPb.

We must not confuse the FSDP with the pre-peak observed in amorphous metallic solids with two constituents which is caused by chemical short range ordering of the minority species (see Sects. 4.3 and 4.7). In this case, if k_{p} is the wavevector associated with the maximum of the pre-peak, and x_1 is the bond distance between first neighbours, then the reduced parameter $k_{\mathrm{p}}x_1$ has values between 4.3 and 5, very close to $kx_1 = 4.43$, which is characteristic of the tetrahedral packing in amorphous Ge.

The alternative name for FSDP, namely pre-peak, comes from the fact that when the structure factor $\mathfrak{S}(k)$, whether including the above feature or not, is Fourier transformed, the correlation functions in real space coincide among themselves. This indicates that the peak under examination *does not* contain structural information concerning short range order, but that it is associated with the realisation of structural arrangements that can be defined at the level of medium range order.

The $\mathfrak{S}(k)$ features that are to be attributed to medium range order can be estimated by considering the function $(\sin kx)/(kx)$, whose first maximum is located at $kx = 7.725$. This is a reasonable estimate of the FSDP position, which is correct within the limits of validity of the Debye equation (see (4.38)). The significant scattering vector is given by $k = 7.725/x$, where x is evaluated starting from the short range order information on the system under examination.

If we take a fairly common structural unit, the XY_4 tetrahedron, where the X–Y distance is d, then a medium range order generating element consists of three tetrahedra with one common vertex. We can define the maximum distance this cluster of tetrahedra extends along. This distance is the radius r_{max} of the sphere centred on atom X in one of the tetrahedra, and whose surface touches the centres of the other two tetrahedra. As such $r_{\mathrm{max}} \simeq 2d$, and the features of the $\mathfrak{S}(k)$ curve at k values not greater than $7.725/2d$, can be interpreted as being due to medium range order. As such we can set the lower threshold for the correlation lengths that give origin to medium range order as $x_{\mathrm{c}} = 2d = r_{\mathrm{MRO}}$.

The main features of the pre-peak in covalent systems are:

1) when we represent $\mathfrak{S}(k)$ as a function of reduced parameter kx_1, the peak is located at $kx_1 \simeq 2.5$ for *all* the systems it has been observed in. The full width at half maximum of the peak, Δk, is such that $\Delta k \cdot x_1 \simeq 0.6$; the correlation length x_{c} for the medium range order is $x_{\mathrm{c}} \simeq \frac{2\pi}{\Delta k} \simeq 10x_1$, a broad correlation interval for materials in which, by definition, long range order is absent. We must note that this estimate for x_{c} is the result of a "crystalline approach" to the amorphous structure where we combine the Bragg equation (see (4.25)), for the interplanar spacing d in a crystalline lattice, with the most general definition for the scattering vector k;
2) the peak intensity increases with the temperature (this has not been fully validated), whereas the height of all the other peaks of the structure factor decreases with the temperature, in agreement with the Debye–Waller factor. The peak is visible even in the liquid state;
3) when we apply a hydrostatic pressure the peak intensity falls and its position shifts towards higher k values.

Lastly, the pre-peak intensity increases as the atomic number of chalcogen X in a particular chalcogenide glass lowers.

The origin of the pre-peak is connected to the hypothesis that the system consists of atomic *clusters* that correspond to a single broad peak in the correlation functions in direct space, located at distance x_{c}, which characterises the medium range order in the structure. In the reciprocal space, this corresponds to a highly dampened sine curve whose first and most intense peak is the pre-peak. We assume that the measured total structure factor results from two terms

$$\mathfrak{S}(k) = f(k) + d(k) \tag{4.99}$$

where $d(k)$ is the structure factor that describes the *inter*-cluster interference due to X-ray or neutron scattering from structural units. Here, x_c is the average correlation distance and $f(k)$ is the structure factor for *intra*-cluster scattering. Now, the fluctuations in x_c are much larger than the fluctuations in the bond length within a single cluster. Thus $d(k)$, whose contribution is dominant at small k values, is strongly dampened. As $f(k)$ is prevalent in the region of high k values, it is much less dampened.

Given that we can almost always observe the pre-peak in systems bearing an atomic component, typically a chalcogen, with a low coordination number that inhibits dense atomic packing in its neighbourhood, we can understand why these systems are so sensitive to externally applied pressure. In fact, the van der Waals bonds between atoms with low coordination in the directions where covalent bonds are absent, can easily be compressed.

A geometric interpretation for FSDP has recently been put forward. This interpretation stems from observations, with almost general validity (B_2O_3 is the only exception), that the position of the pre-peak in various kinds of glasses almost coincides with the first, and most intense, peak observed in elastic diffraction from the crystalline phase with a composition that is the same as, or very near to, the amorphous material under examination. This means that the pre-peak may be generated by atomic density oscillations corresponding to the Bragg planes in crystals. Taking amorphous SiO_2 as our example, the X-ray scattering data at low k values are interpreted in terms of the medium range order that leads to definite quasi-Bragg planes which, in turn, exhibit certain characteristics similar to those in the $\{111\}$ Bragg planes of cristobalite. From a modelling point of view the consequence of this is that the same set of operations leads to the formation of the amorphous structure and, with a few additional constraints, to the crystal structure. Since the main difference between the two depends on whether there is periodicity or not, we first analyse the (known) crystal structure using those operations that can be applied to both periodic and aperiodic structures.

We then introduce distortions and disorder, inherent in the aperiodic systems, into the structure. In the case of SiO_2, the requirement for tetrahedral coordination with defined average values and standard deviation for the bond distance, bond angles and the vertex angles among inter-connected tetrahedra produces short range order. The interconnections of SiO_4 tetrahedra, such that rings are formed, generates medium range order. The ring statistics is constrained in the crystal whereas it is relatively unconstrained in the glass forming material. For example, cristobalite exhibits an atomic plane structure consisting in planar layers. Each layer is composed of six membered rings connected together so that each ring shows just the "chair" configuration typical of cristobalite. These atomic layers are connected by a significant number of oxygen atoms arranged on anti-planes. The atomic connectivity within each layer is greater than the atomic connectivity in the anti-planes. We can introduce disorder by varying the Si–O bond distance in the plane and the distance between atomic layers, where the bonds are weaker. The first kind of disorder, which should have an effect on the (111) peak, is limited and leads to a well defined pre-peak, whereas the second is more important and corresponds to a much more badly defined peak, in qualitative agreement with the X-ray diffraction results.

A pre-peak similar to what is observed in covalent glass forming materials, as explained above, has been observed in some amorphous metallic systems. One significant example of the analysis of medium range order in these systems, starting from the simulation of the characteristic features of the pre-peak, is given by the metallic glass $Ti_{61}Ni_{23}Cu_{16}$. Experimentally, the pre-peak position shifts abruptly from 16.5 nm^{-1} at room temperature, to 15 nm^{-1} at 723 K (where the metallic glass has not yet re-crystallised), while the intensity grows with the temperature.

It has been assumed that the diffraction profile could be simulated by a set of atomic clusters with icosahedral structure. Each icosahedron (Fig. 4.58) contains a central titanium atom with a weighted statistical distribution of

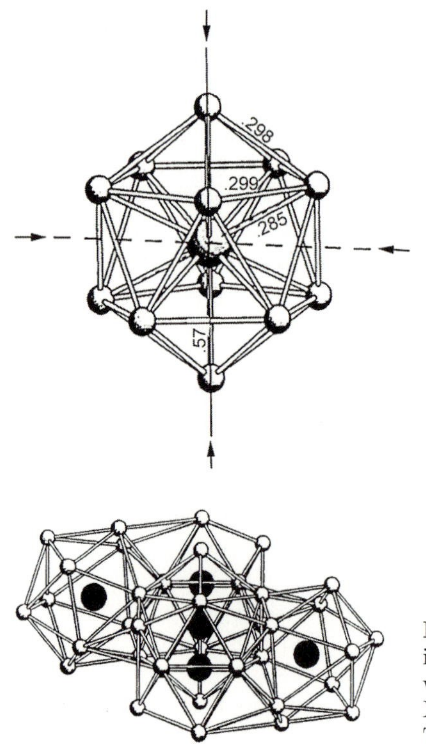

Fig. 4.58. Icosahedral structural unit in a model of amorphous $Ti_{61}Ni_{23}Cu_{16}$. The arrows indicate the directions the structure may be slightly deformed along (adapted from [4.14])

Fig. 4.59. Relative arrangement of four icosahedra to give qualitative agreement with medium range order features in the X-ray diffraction spectra of amorphous $Ti_{61}Ni_{23}Cu_{16}$ (adapted from [4.14])

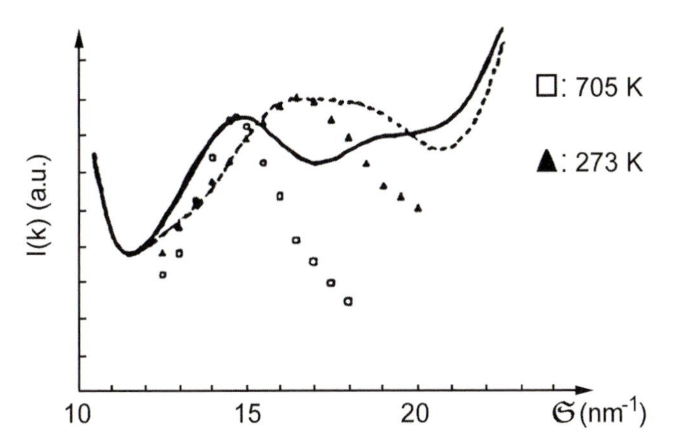

Fig. 4.60. Trend in scattering intensity in the region of the first sharp diffraction peak of amorphous $Ti_{61}Ni_{23}Cu_{16}$, as a function of the temperature (□, 705 K; ▲, 273 K). The simulated patterns were obtained using the structural elements in Fig. 4.59, with medium range order extending to 0.85 nm (...) and 0.925 nm (—) (adapted from [4.14])

titanium, copper and nickel atoms at the vertices. The size of the icosahedron can vary by contraction of both the pentagonal planes, normal to the icosahedral axis, and along the icosahedral axis. Four icosahedra connected to the central one through the faces, defined by the vertices of two adjacent pentagons, make up the reference extended cluster where 49 atoms strongly correlated together give rise to intermediate range order as shown in Fig. 4.59.

When we change the size of the icosahedra in order to obtain modest fluctuations, both in the centre-vertex distance and in the minimum vertex-vertex distance on the pentagon faces, we obtain excellent agreement between the experimental X-ray diffraction pattern and the simulated ones. Furthermore, when we translate a 49-atom cluster in the z direction over various distances, with the "double" cluster of 98 atoms, the pre-peak shift is well simulated without affecting the remaining diffraction pattern. Figure 4.60 qualitatively shows that the simulated patterns produce the same features as the experimental patterns, not only concerning the peak position, but also when the typical modification in the full width at half maximum is considered.

4.9 Structural Models

The inherent limitations to experiments on amorphous materials, especially concerning diffraction, prohibit us from obtaining a complete understanding of the structure, even in the more favourable case of covalent systems. From the analysis of the parameters that characterise the first two coordination shells we can calculate the distances and bond angles between the first neighbour atoms arranged in well defined structural units, but the information regarding the spatial organisation of the unit itself is becoming even more difficult to extract from experimental data obtained from spectroscopic techniques; it is in fact as yet impossible to exceed the picture emerging from the knowledge of the medium range order, which is a long way from giving us a global view of a disordered structure.

The information we have gathered from the various structural probes is complementary to the information obtained from the use of models, which are used to *simulate* the structure of a non-crystalline system. In order to realise an artificial amorphous material we have to merge two needs that often appear to be mutually not very compatible: the degree of randomness and the specific chemical-topological features in the structure, at least concerning short range order. Once again, this leads to a difference in the models used for systems where the bonding forces are non-directional, as the metals, and the non-metallic solids with covalent bonds.

The need to turn to models, both physical and computer simulated, is based on the lack of complete theories on the liquid state. When the model simulates a liquid the obtained atomic coordinates correspond to a snapshot of the atom positions in a real liquid. On the other hand, when the model simulates a glassy material, with frozen translational motions, we obtain a set of possible positions for the atom assembly.

Once the model has been built, we can then calculate the structural properties, in particular the radial distribution function and the density, in order to make a comparison with the experimental data for real system.

At the beginning of the 1960's, Bernal perceived that the structure of a simple liquid, made of spherical atoms or molecules, where the interatomic potential has no angular dependence, is given by the distribution of the volume that *is not* occupied when the spheres are brought into reciprocal contact. Since the density of a liquid is only a few percent lower than the density of its corresponding crystal, the constituent atoms must presumably maintain high coordination numbers in the liquid phase. In fact, the coordination numbers obtained from experimental data lie between 8 and 12.

The simplest structural model of an elemental, structurally disordered, condensed system, adopts hard spheres whose local spatial organisation depends on the constraints on space filling. A minimum distance between two atoms is required, which is equal to the diameter of the constituent spheres. The structure of the crystals made up of hard spherical atoms (inert gases and simple metals) is fcc or hcp. This is the best answer to the need for maximum density and for a crystalline lattice.

When this approach is extended to liquids, we come up against the problem of realising a dense packing of hard spheres without creating the slightest trace of crystallisation.

The first models were realised physically by filling a flexible container with thousands of steel balls (up to 7 934, [4.15]). The container was previously set on an irregular surface to avoid any formation of a crystalline surface. The

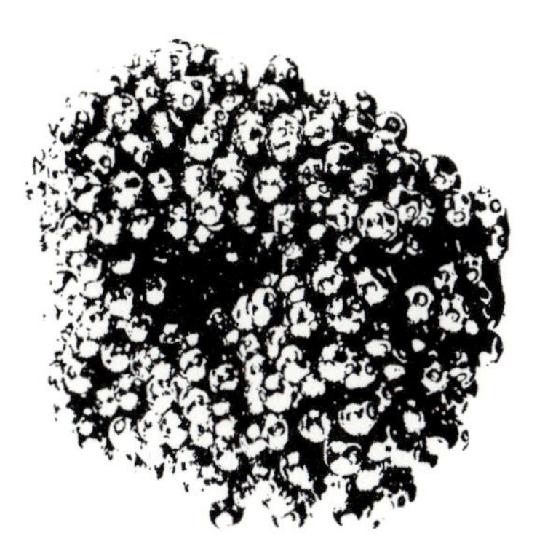

Fig. 4.61. A portion of the physical model of dense random packing of hard spheres (adapted from [4.15])

collection of spheres was then tightly wrapped in rubber strips to exert an eternal compressing force on the system. The spheres were kneaded to favour the spatial rearrangement and enhance the density to the most. In Fig. 4.61 is given a picture of the system after it was blocked by pouring a liquid glue into the structure and letting it harden. Lastly, the external "skin" was removed and the coordinates of each sphere were measured.

Experimental methods somewhat different from each other led to the very same results, namely that the maximum density obtainable in a dense randomly packed system of hard sphere is 0.6366, with an 86% packing efficiency compared to a close packed crystal. Further experiments to simulate the same structure using a computer resulted in radial distribution functions that were essentially undistinguishable from those obtained from the atomic coordinates in the physical models.

The calculations made on the distributions of interatomic distances and on the connectivities show that the dense random packing of hard sphere model (DRPHS) shows many of the structural characteristics of simple liquids. Since the model is static and the density is maximised (though models with lower packing densities can be built, namely the so-called loose random packing models), it is particularly suited as a model for an *ideal* glass with spherical atoms at 0 K.

However, as surprising as it may seem, a quantitative comparison between the experimental radial distribution function for amorphous $Ni_{76}P_{24}$ and that for the DRPHS model (Fig. 4.62) shows considerable agreement.

Among the available models, the DRPHS is the only one that reproduces the experimentally observed splitting into two sub-peaks of the second peak in the radial distribution function. This splitting is specific to the amorphous metals, and is absent in the liquid metals.

However, two serious difficulties are encountered when trying to compare the structure obtained using the DRPHS model with the experimental results.

The first problem is that the relative position and intensity of the two components of the second peak disagree with each other. This is essentially a quantitative problem and is overcome by adopting soft potentials in the

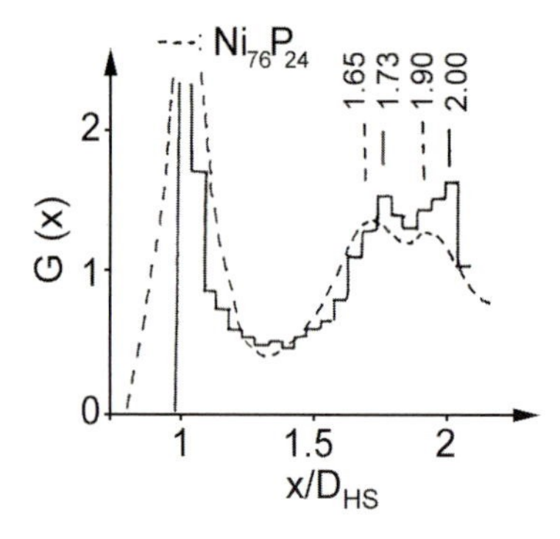

Fig. 4.62. Comparison between reduced radial distribution functions for the physical model of dense random packing of hard spheres (—) (adapted from J.L. Finney, 1970) and for amorphous $Ni_{76}P_{24}$ (– – –) (adapted from [4.16])

computer simulations. The second problem is a qualitative one. We have to identify a real two-component alloy with a single component model. Even if the nickel contribution is dominant in X-ray scattering, the phosphorus contribution, around 8%, to the scattered intensity cannot be overlooked. Furthermore, as there is a second component that promotes the vitrification process, namely the phosphorus, it undoubtedly has a significant influence on the structure of the metallic glass.

In view of this problem, the observed agreement does seem somewhat misleading in terms of understanding the DRPHS model, which may be more suited to representing the structure of an ideal, single-component glass than that of a particular amorphous alloy.

Since 1972 computer algorithms have been developed that generate numerical models of the dense random atomic packing. These models are realised about a pre-existing structural *core*, with defined geometry, whether a three-sphere triangle, or a tetrahedron with four spheres in mutual contact at its vertices. The spheres often have, though not necessarily, the same diameter, D. The computer adds to the structure one sphere at a time, keeping it in close contact with three spheres already present; then the added sphere is locked into its position. The process is then repeated up to a few thousand times to generate an irregular dense model that gradually grows and is completely characterised by the coordinates of each spherical atom. The disordered system can also be reproduced as a physical model by using, for example, hard plastic balls.

At each step of the process there are a number of stable positions where another sphere can be added. These so-called *tetrahedral pockets* are defined by three spheres, to which a fourth can be added that stays in contact with the three spheres already mentioned.

The structure of the model depends on the nature of the core and on the criteria used to judge the degree of acceptability for the positions of the additional spheres as they are added. The simplest criterion used to add the atoms is the *global* criterion where the pocket closest to the centre of the original core is chosen. The chosen site is, in fact, the point of minimum energy in a long range potential.

A second local criterion chooses the deepest pocket, thus favouring those sites that are supposed to be more strongly bound in a short range potential.

One example of a criterion for the acceptability of the new atomic positions, based on a modified version of the global criterion, follows these steps:

1) choose three spheres on the surface of the structure so that the distance between pairs of spheres is less than nD, where $n > 1$;
2) choose the position of another sphere that touches the other three spheres, without overlapping any other spheres present. This position is called the pocket;
3) label the position of all the pockets in the system;
4) place the new sphere into the pocket nearest to the centre of the structure and lock it into position;
5) calculate the new positions of the pockets in the "grown" structure, then repeat from step 4).

The whole calculation can be repeated for a set of values for parameter n. In Fig. 4.63 are compared the radial distribution functions for the DRPHS models, physical (part (a)) and computer generated, using the global criterion (part (b)).

The overall agreement between the two models appears quite acceptable even though the splitting of the second peak into two components is considerably less marked in the computer generated model.

A modification of the growth procedure for the global criterion model reduces the used pockets to a pre-specified set, defined on the basis of the tetrahedral perfection of each pocket, namely how close the four spheres, that

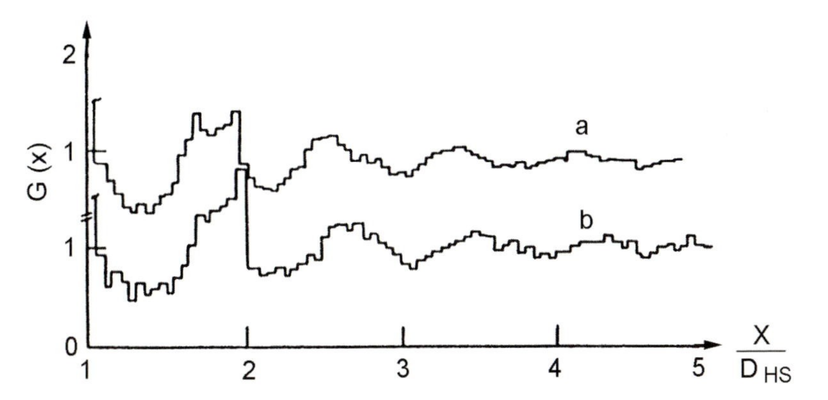

Fig. 4.63. Comparison between calculated reduced radial distribution functions; (**a**) physical model of dense random packing of hard spheres (adapted from [4.15]); (**b**) computer generated model according to the global criterion (adapted from [4.17])

make up the pocket, are to being an ideal tetrahedron. For a cluster of three spheres,

$$k_{123} = \mathrm{Max}_{\{12;13;23\}} \left[x_{ij}/(R_i + R_j) \right] \tag{4.100}$$

where x_{ij} is the distance between the centres of spheres i and j, and R_i and R_j are their radii. The ideal tetrahedral arrangement corresponds to $k_{\mathrm{Max}} = 1$. In this way models have been produced that contain up to 5 000 particles.

The sequential procedure of adding atoms raises a number of delicate aspects. The first regards the very sequentiality character in that once an atom has been added to the structure it cannot be removed. Due to similarity between the procedures the pockets are filled, the resulting packings look more like the experimental "loose" laboratory packings than the dense packings; as such, the model structure has to be made more dense. In fact, the extrapolated densities are rather low, less than 0.60.

To this low density we must associate the inability of these models to adequately reproduce the splitting in the second peak in the radial distribution function, as observed experimentally (Fig. 4.62). In other words, if we cannot reach an adequate density the system will not exhibit all the required structural properties. As such, the simulation methods using sequential build-up may give us misleading results if they are used singularly. Not surprisingly, the global procedure allows us to obtain the higher densities, whereas the density falls dramatically when we introduce progressively higher degrees of tetrahedral perfection. Models where $k = 1.2$ are highly *porous* and could be mechanically unstable.

Since the first component of the second peak in the RDF for $Ni_{84}P_{16}$ is found at about 1.65 hard-sphere diameters, a value that indicates that there is a large fraction of icosahedral structural units, an algorithm has been put forward that produces a model with a consistent number of these units. However, even though each icosahedral unit is rather dense ($\varrho = 0.67$), the global packing of the amorphous solid exhibits low density, $\varrho = 0.52$. This result can be explained by the existence of holes located at the interfaces between locally dense sub-structures, and is a further example of the topological frustration that inhibits effective extended packing of high density structural elements.

We should remember that, owing to the considerable polarisation towards local icosahedral order in the model, the position and relative intensity of the two components of the second peak for the radial distribution function reasonably agree with the experimental data. Yet, the unrealistically low global density demonstrates that this agreement has been obtained at the expense of the packing *constraints* that have to be present in a real material; as such the model is inappropriate.

The dense random packing models realised using equal-sized spheres are dense in that no holes are allowed that are big enough to contain normal-sized spheres. However, this does not mean no holes are present; on the contrary,

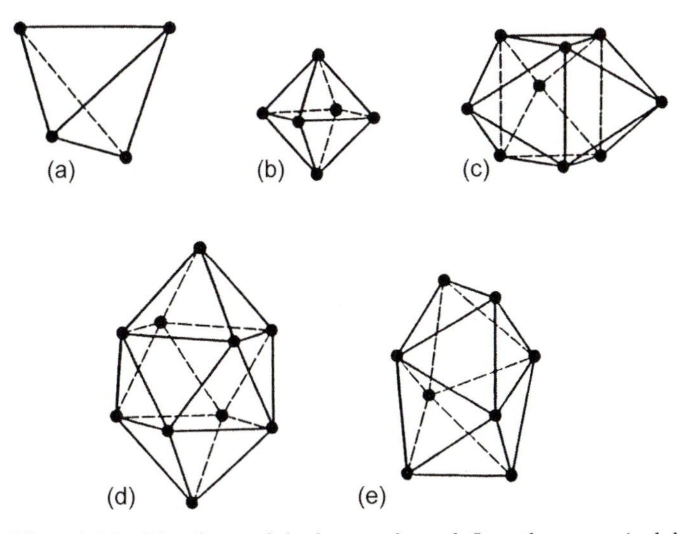

Fig. 4.64. The five polyhedra used to define the canonical holes: (**a**) tetrahedron; (**b**) octahedron; (**c**) trigonal prism, capped with three half-octahedra; (**d**) Archimedean anti-prism, capped with two half-octahedra; (**e**) tetragonal dodecahedron

the packing density difference, around 10% between these models and the densely packed crystalline structures, suggests that sizeable holes do indeed exist. These holes are classified with reference to five polyhedra, the so-called *canonical* holes (Fig. 4.64), whose vertices are defined by the centres of the involved atom spheres, provided we admit that the position of the polyhedra vertices can be shifted up to about 20% compared to the ideal ones.

The tetrahedron is the smallest polyhedron out of those under examination; it is also found with the highest frequency (73%, 2.9 per atom). This is followed by the octahedron (20%), the trigonal prism and the tetragonal dodecahedron (3%) and the Archimedean anti-prism (1%); the other polyhedra are even less frequent. The fact that there is such a high fraction of packing defects is reflected in the exaggeratedly high values of free energy calculated for many amorphous models.

The topology for the densely packed statistical structures can also be described with reference to the Voronoi polyhedra (see Sect. 4.7). On average each Voronoi polyhedron has 14.25 faces, compared to the 12 faces for the fcc and hcp cells and 14 faces for the bcc cell.

When we let the structure relax under a long range external potential, such as the Lennard–Jones $(6 - 12)$ one (Fig. 4.44), the total energy of the amorphous solid reduces by about 5%. Correspondingly, the arrangement of the Voronoi polyhedra is modified, resulting mainly in deformed tetrahedra (70%) and deformed octahedra (20%). These very same elements, undeformed, constitute the densely packed metallic crystals with coordination number twelve. Furthermore, the structural relaxation causes significant modifications in the radial distribution function. In particular, the height of the two components of the second peak is exchanged, which correctly reproduces the ratio between the experimentally observed intensities. Apart from this, the first peak shrinks and the first minimum is more pronounced. Lastly, we observe a striking increase in global density. The best results are obtained when we allow the system to relax, keeping its boundaries free. Unfortunately, this leads us to underestimate the importance of the packing constraints that exist in real structures.

To conclude, it is fairly simple to choose a model with either the experimentally observed density or a reasonable radial distribution function. However, we have to produce both at the very same time, and this certainly is not easy, probably impossible, to realise through the *infinite* packing of more or less hard spheres. Indeed, the above general considerations regarding the hard sphere models can also be applied to the more realistic dense packings of soft spheres, partially overlapping each other.

A dense random packing of (hard) spheres is at one extreme of the modelling procedures of disordered structures where the resulting short range order is given by the *geometry* of the arrangement of the spheres.

When we simulate systems with highly directional covalent bonds, the structural models must be significantly influenced by the valence of the system components, by the bond length and by the bond angles. The calculated correlation functions for the models, whether they concern particle pairs or triplets, must contain this information. We presume the number of first neighbours for a given atom must be small and must be determined by the chemical *affinity* of the compound constituents.

The *continuous random network* (CRN) of covalent bonds is the prototype model for covalent amorphous systems. It had a profound influence on the development of the representation of structurally disordered systems, including non-covalent ones.

We already examined some of the features of the CRN model when we discussed the criteria used to interpret glass-forming ease (see Sect. 3.3). The amorphous material is treated as a super-molecule in which connectivity is maintained throughout, the atoms keeping their normal valence, with the only exception of surface atoms. From the modelling point of view, in the case of SiO_2 we consider a tetrahedron with four oxygen atoms bound to a central tetravalent silicon atom as the structural unit. This same unit is found in the crystalline modifications of SiO_2. Let us now imagine that one oxygen atom occupies the common vertex between two adjacent tetrahedra. The oxygen atom is thus bound at the same time to two silicon atoms.

When we use the same connecting strategy with many structural units, by way of vertex sharing, we build an extended network without long range order. This occurs because we have introduced a randomness element into the system, namely a *distribution* around an average value of the bond angle O–Si–O. We can obtain this distribution by either stretching, or contracting, the bonds, or by rotating the tetrahedra adjacent to each other by an arbitrary angle, around a line that is given by the Si–O bond, or eventually, by using both methods together (see Figs. 4.47 and 4.50).

In the physical models for amorphous SiO_2, realised with small spheres connected together using rods and containing a few hundred tetrahedra, we observe that O–Si–O angle varies by $\pm$ 20% compared to the average value of 140°. Similarly, in a model of amorphous Si and Ge, 12% of the bonds are stretched by 10% of the average value.

These models can be traced to the traditions of both Zachariasen, given that short range order is determined by keeping the specific chemical bonds, and Bernal, given that we obtain essentially irregular extended packings.

We can measure the atomic coordinates, the density, the connectivity and the radial distribution function for the CRN models just as we did for the DRP model.

We can count one by one *closed* rings that are formed by bonds with four, five or n elements, giving us topological-chemical information that is useful for a theory of the structure of covalent glasses. If we consider, on the other hand, the inherent statistical imprecision due to the rather low number of atoms in the model and, on the other hand, the experimental errors and the limitations associated with converting data from the reciprocal space to the physical space, then the agreement between the radial distribution functions obtained from the models and from experiments is rather good. It is noteworthy that the models include well defined conditions for three and more atom coordinations, whereas the experimentally obtained radial distribution functions *do not* give us this information.

It is doubtful we shall ever realise complete agreement between models and experimental results; the realisation of continuous, statistical networks with unlimited sizes without having to insert defective bonds into the system, which is theoretically possible, corresponds to an *ideal* glass structure. Real materials, such as the prototype amorphous silica, come close to this ideal structure, but they do contain defects, just as the real crystals deviate from the perfect idealised crystalline structure.

Adoption of either the DRP or the CRN models does not mean that the CRN model includes chemical short range order, whereas this important structural property is absent in the DRP models and cannot be taken into consideration. One possible connection between the two kinds of models has been explored with the aim to discuss whether there is partial, or relevant, chemical short range order in a metallic alloy.

We obtain glassy structures with short and medium range order when first we choose a particular structural element, embodying the possible chemical order of the alloy, and second, we assemble these units using similar rules to those used for the continuous random network. The transition metal-metalloid alloys already discussed (see Sects 4.7 and 4.8) are examples of this approach.

5. Clusters

5.1 Definition of an Atomic Cluster

One of the foremost reasons for studying atomic and molecular clusters is to understand those mechanisms that relate the properties of the bulk material to the properties of the particles the material is made of. When we inquire into what stage of the "Aufbau" (in English, "construction") process single atoms collect together to form a cluster having the properties we know for an extended solid, we have to ask a number of questions which all have different answers depending on the specific property being considered. Indeed, for some properties, there is no answer because the properties change so slowly as the cluster grows that we cannot distinguish any clear dividing line between a cluster and a solid.

The term cluster means an assembly which, on the one hand, lacks the well defined composition, geometric structure and chemical bonds so typical of a molecule and, on the other, lacks any properties typical of bulk materials, where surfaces are ignored, unless we are specifically interested in studying the surface itself and the shallow layers immediately below it.

If we confine the field we are interested in, then these clusters of atoms, or molecules, contain any number of particles whatsoever ranging from a few atoms (a minimum of three) to some hundreds of thousands of atoms. A cluster is thus an example of a large finite system. A further classification is made for small, or molecular, medium and large clusters. The interval for cluster sizes does not depend on the nature of the cluster alone, but, to a great extent, also on the cluster forming technique. When the properties change so much with cluster shape and size that we cannot define any smooth functional dependence on the number of constituent particles, the cluster is considered small. It contains fewer than thirty atoms and has a very high ratio, f_{s}, between the number of surface atoms, n_{s}, and the total number of atoms in the cluster, N.

From the information in Table 5.1, concerning spherical clusters with increasing diameter d, in general f_{s} is proportional to $N^{-1/3}$. It is clear that the influence the surface atoms have on system properties increases rapidly as the size of the cluster gets smaller.

d(nm)	1	2	5	10
N	30	250	4000	30000
n_s/N (%)	99	80	40	20

Table 5.1. Variation in the number of atoms N and in the fraction n_s/N of surface atoms of spherical clusters with diameter d.

Even the structural, energetic and dynamic properties of the cluster micro-surfaces are of great interest. Owing to surface corrections, which scale with f_s, the various properties cannot be considered extensive.

The dividing line between small atomic clusters and small molecules is often blurred. A ring with eight atoms of sulphur, or a tetrahedron of phosphorus atoms, certainly cannot have all the requisites of a cluster. Such stable structural units exist in the vapour, liquid and solid phases and are traditionally called molecules. Conversely, the term cluster is adequate to define atomic aggregates that *are not* found in appreciable quantities in a vapour in equilibrium.

If the properties of a cluster change slowly enough as the cluster grows, but in the dependence there are still effects due to the low number of constituent particles, then the cluster is considered a medium-sized cluster. Lastly, the properties of a large cluster are very similar to the properties of a bulk material. While the radius of a large cluster may reach a few tens of nanometers, and thus it is classified as a nanocrystal, the average-sized cluster contains between thirty and one hundred atoms. Just to give an idea of their size, a cluster with one hundred atoms of gold has a radius of 1.1 nm.

The definition of a cluster, based on its size, is incomplete, and as such we have to add that an atomic cluster is

a) a finite collection of particles whose *composition* and *structure* can be changed by adding, or removing, units of the species that form it;

b) it is made up of a countable number of atoms which, in composition but not necessarily in structure, form an extremely small sample of the extended material.

An atomic cluster is thus different from molecules both in composition and in structure. In fact, usually the molecules consist of a small, definite number of atoms, they are completely specified in the stoichiometry and almost always exhibit a single structure.

Conversely, a cluster can contain *any* number of atoms, and both the cluster properties and the more stable structure it can take on are dependent on this number of atoms. Again, unlike molecules, the number of locally stable structures admitted for a cluster grows rapidly as the number of constituent atoms in the cluster grows; this is true of most clusters. Given a fixed composition, N, the number of chemical isomers also grows quickly. Even though we usually define an isomer as a distinct chemical species, in the case of the clusters with various structures but the same composition, these clusters are considered part of a single chemical species. However, there are cases where it is important to consider the isomers of clusters with fixed stoichiometry as being separate.

It is useful for us to make a distinction between *homogeneous* clusters, with a single kind of chemical constituent, both atomic and molecular, such as Ar_{14} and $(NaCl)_4$ and *heterogeneous* clusters made up of different kinds of atoms and/or molecules, such as $Os_5C(CO)_{16}$ or K_4Rb_7. Again, the distinction into two separate families is not completely clear cut; it is not easy to say whether the cluster with n molecules of RbI, given as $(RbI)_n$, is homogeneous or heterogeneous.

In normal conditions electrically neutral clusters are formed; however, in experimentally favourable conditions we can even obtain both positive and negative charged clusters. In the case of RbI, the clusters contain an extra atom of rubidium $(Rb_{n+1}I_n)^+$ or iodine $(Rb_nI_{n+1})^-$; it is not easy to define this kind of electrically charged cluster as being homogeneous instead of heterogeneous. Obviously, it is easier to explain the existence of charged clusters in alkaline halide, since bulk RbI can also be schematised as being composed of Rb^+ and I^- ions. However, many other kinds of electrically charged clusters have been studied.

Clusters have both ionic bonding forces, such as in the alkaline halide clusters, and bonding forces like those found in covalent bonding, such as in the elemental clusters of carbon, including the fullerenes, like C_{60} and C_{70}. In the clusters of the rare gases and the atoms with closed electron shells, such as magnesium, calcium and barium, the weakest form of electromagnetic forces are present, namely the van der Waals forces. Lastly, while large clusters of both simple and transition metals are subject to the forces that give rise to the metallic bond owing to delocalisation of the conduction electrons, in the small metallic clusters the forces are more like the forces observed in systems with covalent bonds.

This difference is based on the fact that in the extended matter, the quantum states are extremely close in energy to each other and give rise to bands of states whereas, in a small cluster, the energy difference between adjacent states cannot be overlooked. The latter depends on the size of the structure under examination and, as such, in small clusters it is similar to small, or medium-sized, molecules. In actual fact, quantum corrections are also observed in large clusters due to their reduced size.

When a cluster has grown to a critical large size it loses one of its distinctive features, namely it can no longer reconstruct its structure whenever an atom is added to it. At this point, a defined crystalline structure is frozen into the cluster and the cluster can then be classified as a nanocrystal.

Crystallisation occurs at highly variable cluster sizes, the bottom threshold being around 100 atoms. Obviously, in the case of particularly small nanocrystals, we observe deviation from the structure of the bulk solid, and this is adequately described as surface *relaxation*.

Up to the 80's atomic clusters were essentially looked upon as small molecules. In the specific case of metals, there was no reason to expect the properties of clusters, either different in size, or made of different materials, to be correlated to each other. Rather, each cluster was considered unique, the same way as a molecule.

At that time, extremely small clusters were synthesised containing a dozen atoms at the most, and it was not possible to identify any structural order. At the same time, much larger particles were considered as essentially similar to macroscopic solids. The breakthrough came in 1983 when it was possible to synthesise and unveil atomic clusters with up to one hundred atoms. On the one hand, these relatively large clusters exhibited a remarkable degree of structural order and, on the other, the electronic structure of the clusters exhibited the features of a potential well with spherical symmetry.

The study of clusters takes on many facets that regard, first of all, understanding the microscopic mechanisms by which a crystal grows. Whenever an atom, or a molecule, condenses on the surface of a cluster, the atoms in the cluster completely change their reciprocal arrangement, namely the cluster is *re-constructed* and, as such, in principle, we can follow the structural sequence the cluster undergoes step by step as it transforms from a molecule into a crystal. Then, based on the nature of the cluster and its size, we can recognise the dominant factors in the process. At the same time,

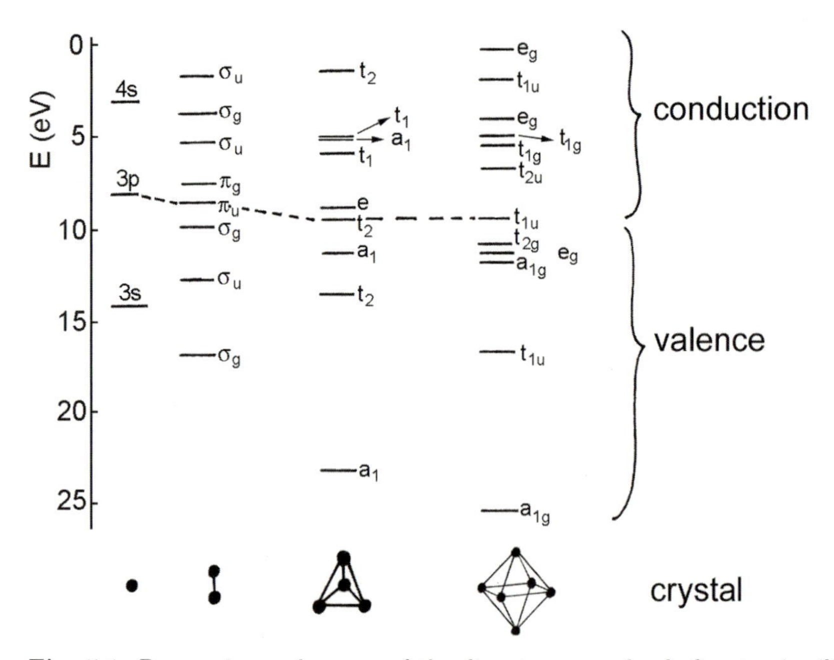

Fig. 5.1. Progressive coalescence of the discrete energy levels for atomic silicon into a band structure, as the cluster increases in size. The upper limit represented by the crystal bands is also reported

we can follow the development of the band structure of a solid through the progressive coalescense of the atomic levels to form these bands, and we observe the appearance of a gap between occupied and empty states. Figure 5.1 schematically shows this process for some small silicon clusters.

As it is possible to accurately select the mass of the produced clusters, we can study the trend in the significant chemical reactions in detail, for example in the catalysis, and determine the stability of the reaction products.

In the case of the metallic clusters, this study is of prime importance in understanding at what point of the molecule-crystal transition the characteristic features of a metal appear. These latter features determine the transport properties, which are so interesting from a technological point of view. In other words, we wonder when and how the electrons that form the bonds in the small clusters change their behaviour and give rise to the bands so characteristic of metallic solids. Photoelectron spectroscopy has, for example, allowed us to observe in detail how the occupation of $3d$ levels in Ti_N^- clusters changes as a function of their size, where N ranges between 3 and 65.

Bonding energy spectra are obtained when we subtract the measured kinetic energy distributions for the electrons emitted from the cluster surface from any given photon energy. Figure 5.2 shows all the photoelectron spectra gathered at the photon energy of 4.66 eV. In smaller clusters, up to Ti_7^-, we observe discrete spectra features. The spectra change greatly with N and reflect the essentially molecular nature of the clusters. Starting from Ti_8^-, a relatively narrow band appears in the spectra. This band progressively grows and broadens, and quickly converges into a single band, centred at about 2.7 eV, whose width tends to increase with N. The narrow low intensity peaks that appear in the higher bond energy region are spurious and are caused by statistical noise.

The only evident spectral feature in the "large" clusters is considerably similar to the broad band that appears around the Fermi energy E_F in the valence photoemission spectrum of titanium crystal. The full width at half maximum of this band is 2 eV and it is caused by the solid's $3d$ band. The broadening of the band in the measured cluster spectra is a function of N and exhibits a width of around 1 eV in the size interval between $N = 20$ and $N = 50$.

It is reasonable to assume that this band broadens monotonically as the cluster size increases until it finally converges onto the solid's valence photoemission band. At this point we are given to think that the solid's $3d$ band is exhibited already in small clusters, such as Ti_8^-. This kind of behaviour is rather different from the clusters of other transition metals, such as chromium and iron, where we observe narrow well resolved bands even when N is just greater than 20; this suggests that the titanium clusters presumably take on a structure similar to the solid even for small sizes.

The difference between the behaviour of the titanium clusters and the clusters of the transition metals with a greater number of $3d$ electrons, can

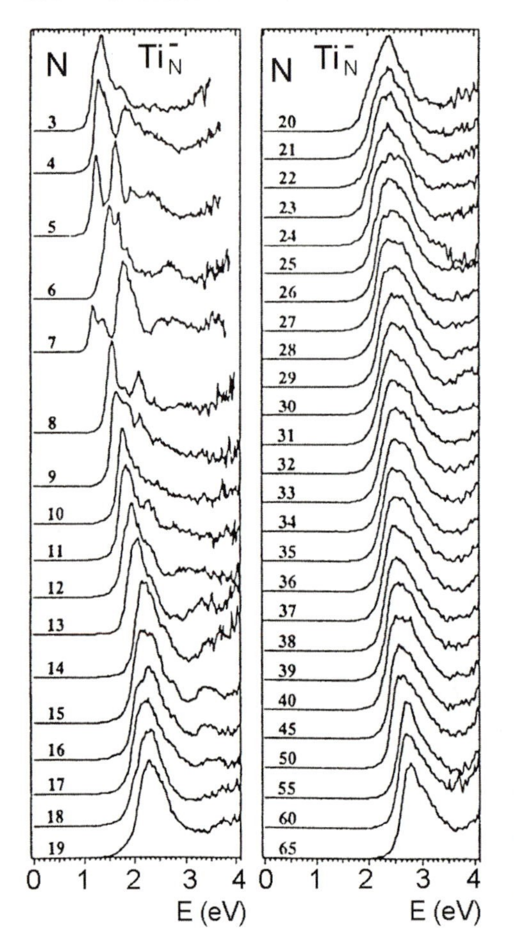

Fig. 5.2. Trend in binding energy E, deduced from photoelectron spectra, in ionised Ti_N atomic clusters, as a function of the number of atoms per cluster, N (adapted from [5.1])

be put down to the modifications in the $3d$ atomic orbitals along the first series of transition elements. The spatial extension of these orbitals falls with the atomic number Z owing to the increased nuclear charge. These orbitals in titanium are highly delocalised, overlap between first neighbour atoms in the cluster and exhibit a strong valence character.

Since there are two $3d$ electrons, the antiboding d orbitals are empty, thus allowing spatially compact clusters to form with strong bonds, as shown by the high dissociation energy in the small Ti_N^+ clusters. Such energies are very high, compared to those of elemental clusters in the first row of the transition metals. Also the icosahedral structures with high packing efficiency suggested for certain titanium clusters confirm their stability. In this respect, the photoemission spectrum for Ti_{55} in Fig. 5.2 is an exception to the general trend discussed above, and exhibits a single band which is much narrower than the bands of the surrounding clusters. Since $N = 55$ is a magic number for the clusters with icosahedral structure (see Sect. 5.3), this anomalous band width is an index of a structure that is highly symmetrical in this cluster.

It is thus possible to attribute the fast convergence of the electronic structure of titanium clusters to that of the solid both to the strong delocalisation of the 3*d* atomic orbitals and to the densely packed structures of these clusters.

In this chapter much attention is placed on the structure of clusters and on their evolution towards the macroscopic structure of the corresponding solid. The analysis is developed by examining elemental clusters of two families, namely van der Waals systems and neutral alkali metals, which stand as a valid model also for the study of polyvalent metal clusters. We shall also briefly examine the structural properties so specific to the covalent cluster of carbon C_{60}, the fullerene. Finally, we shall examine the structure of the recently discovered and technologically promising cluster-assembled solids.

5.2 Synthesis and Detection of Atomic Clusters

Cluster physics has progressed through the development of sources and detectors of clusters present in molecular beams, and thus in an environment that is essentially free from any interactions.

The size distribution and the mass analysis can be performed either on free clusters, namely clusters under vacuum, or in an inert gas. The structure of clusters is studied mainly using electron diffraction; the electron beam travels perpendicular to the direction of the cluster beam being examined in order to perform a scattering experiment. As an alternative, we can deposit a certain number of clusters onto a substrate, usually a film of amorphous carbon. Cluster beams are presently synthesised and analysed for just about all the elements, with sizes that vary from a few atoms to tens of thousands of atoms.

Each of the various kinds of sources has been developed to synthesise clusters of a certain class of materials with definite size distributions.

The *seeded supersonic* nozzle source is used for substances with a low boiling point. We obtain the highest intensities for the synthesised beams and fairly narrow cluster velocity distributions. Using this technique we can obtain clusters with up to a few hundred atoms per cluster.

The material is vaporised in an oven and mixed with an inert carrier gas. The vapour/gas mixture is then sprayed into the vacuum through a small nozzle. The supersonic molecular beam expands adiabatically in the vacuum and very quickly cools. The cooled vapour becomes supersaturated and condenses in the form of atomic, or molecular, clusters.

The large clusters further cool as the surface atoms evaporate, which demonstrates that they are near the evaporation temperature.

The relative abundance of clusters is determined by thermodynamic processes and is sensitive to the binding energies. For this reason the measured mass spectra clearly show structures.

Using the *gas aggregation* technique we synthesise large clusters, containing 10^4 atoms or more. The intensities are much lower than the intensities obtained from supersonic beam sources, and the size distribution of the clusters is quite broad. These sources are used for materials whose boiling point is below 2 000 K, such as the alkali and noble metals, germanium and tin. The temperatures of the synthesised clusters may be as low as 100 K.

The material is vaporised and fed into a cold inert gas flow (He). The vapour is then supersaturated, and produces a dense smoke containing mainly nanocrystals between 1 and 10 nm in diameter. Since the support gas is at a low temperature, the cluster grows by way of successive single-atom additions. As any re-evaporation is negligible, then the cluster abundances are independent of their thermodynamic stability and show a smooth dependence on their size.

The *pulsed* laser *vaporisation* method allows us to synthesise clusters of any material, both neutral and positively or negatively ionised, whose sizes range from a few atoms to some thousands of atoms. The laser pulse vaporises the atoms of the target material; the clusters then form as the vapour is transported and cools into a low temperature pulsed jet of helium. The gas/cluster mixture is then sprayed into the vacuum through a nozzle and is further adiabatically cooled to an estimated temperature of around 100 K. Using this source we can also synthesise atomic clusters of carbon.

The *sputtering* source produces continuous, intense and singly ionised beams of clusters for most of the metals and the alkaline halides. The clusters, around the boiling point, are synthesised by irradiating a solid surface with a high current (up to 10 mA) beam of inert ions (Xe^+; Kr^+), with medium-low energy ($5 - 20$ keV). The bombardment causes the emission from the surface of secondary ions and of ionised clusters. The cluster formation processes are not fully understood. The produced intensities fall exponentially with the size of the clusters, whose initial temperatures are extremely high. Cooling occurs by way of evaporation during the flight phase.

The mass spectra give us direct information about the thermodynamic stability of the clusters, as a function of their size, which is dependent on the cluster binding energy.

The mass analysis of the cluster beams is basically performed using Wien-filters, quadrupole mass filters, or, lastly, time-of-flight mass spectrometers. In the first case the mass separation is achieved with crossed homogeneous electrical, $\boldsymbol{E}$, and magnetic, $\boldsymbol{B}$, fields, perpendicular to the ionised cluster beam. If we define q as the charge for a cluster, with a mass of m, which is accelerated by a potential, V, up to speed v, then the cluster will reach the energy qV. The cluster undergoes a null force if $E = Bv$. As the clusters pass through the filter they are not deflected, and can be selected by suitably positioned collimators only when

$$m/q = 2V(B/E)^2. \tag{5.1}$$

The resolution $\delta m/m$ is around 10^{-2} and the mass interval that can be analysed is between 1 and 5 000 amu.

Using the quadrupole mass filter only ionised clusters with a charge-to-mass ratio that corresponds to trajectories compatible with applied ac and dc currents can pass through the filter. Typical resolution is around 10^{-3}, and the mass interval that can be analysed is normally 1 000 amu.

When we use a time-of-flight spectrometer, the ionised clusters are accelerated by a sequence of homogeneous electrical fields (ion gun) until they enter a free flight region and, in the end, fall onto an ion detector. The ion arrival times are recorded, which give us the mass, or, rather, the mass-to-charge ratio for the cluster. A typical resolution using this apparatus is 10^{-4}, but limits of 10^{-5} can be reached by reflecting the ions backwards using an electrostatic mirror (a uniform electric field directed in the opposite direction to the incident ions) located at the end of the flight tube. In the electrostatic mirror, the time required for an ion to revert its direction depends on its speed. For the same mass value, the slowest ions have enough time to cluster with the faster ions, and all ions with the same specific mass reach the detector simultaneously.

5.3 Structure of van der Waals Clusters

Many of the experimentally observed properties of condensed matter have been interpreted with structural models based on hard sphere packing. Such properties include the morphology of single crystals, the structure of liquids and certain classes of amorphous materials and the fivefold symmetry of small metallic particles.

The first micrographic observations of clusters with fivefold symmetry date back to 1966, when thin films of gold were evaporated onto substrates of sodium chloride. These anomalous crystallographic morphologies (see Chap. 1) were later observed in clusters of other metals, including silver, nickel, platinum and palladium, deposited onto substrates by physical, or electrochemical methods or, again, evaporated in a flowing inert gas. Structural models of small and medium-sized atomic clusters with N atoms are obtained by progressively clustering together atoms assumed to be spherical. The highest degree of coordination, which, in a real system, is equivalent to minimising the number of open surface orbitals, has unique geometric solutions up to $N = 5$. These structures correspond respectively to dimers, equilateral triangles, ideal tetrahedra and bi-tetrahedra, as shown in Fig. 5.3. At $N = 6$, we obtain two isomers with equal coordinations, that consist of the packing of four adjacent tetrahedra, or octahedron (Fig. 5.4). Of the two

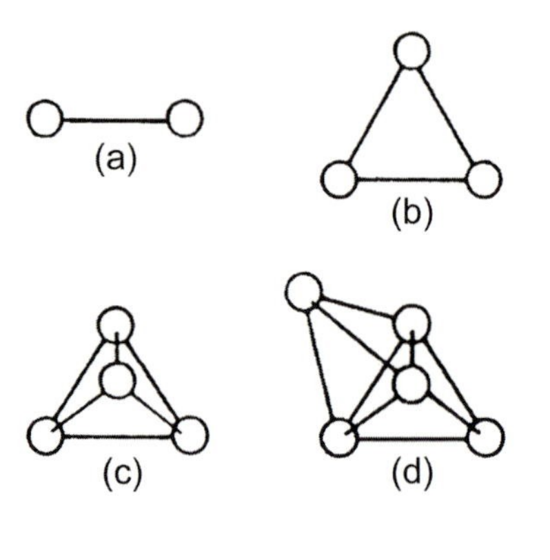

Fig. 5.3. Geometric arrangements of identical atoms with maximum coordination. Up to five atoms, we have a single solution (**a**–**d**)

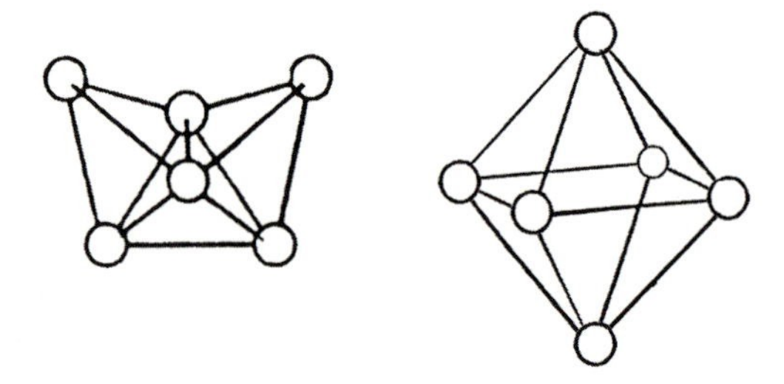

Fig. 5.4. Six identical atoms admit two alternative dispositions with maximum coordination

kinds of arrangement the octahedral packing is the most stable. When we add a seventh atom (Fig. 5.5), the first structure, this time with the packing of five tetrahedra, is favoured since the re-entrant angle is an energetically favoured site. If we require that the tetrahedra are composed of hard spheres, then they have to be ideal, which leads to packing frustration, namely to the opening of a gap in the structure of the five tetrahedra (see Sect. 4.7). In the case of real atoms, we can obtain a more stable configuration by slightly distorting the tetrahedra in order to close the gap, thus forming a regular pentagonal bi-pyramid, or decahedron. In so doing, we obtain a higher coordination, at the expense of modest elastic distortion. The decahedron exhibits one more bond than any other seven-atom arrangement, and is the most stable. It consists of a regular pentagon with five atoms centred on each of the vertices, and a further two atoms that cap the pentagon in the two opposite

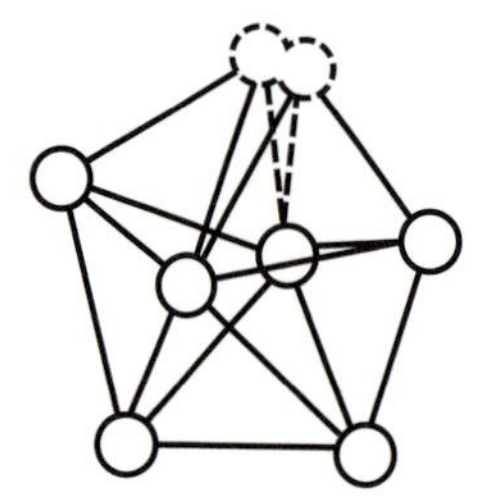

Fig. 5.5. Five adjacent tetrahedra constitute the energetically favoured packing of seven identical atoms. The dashed atom shows geometric frustration

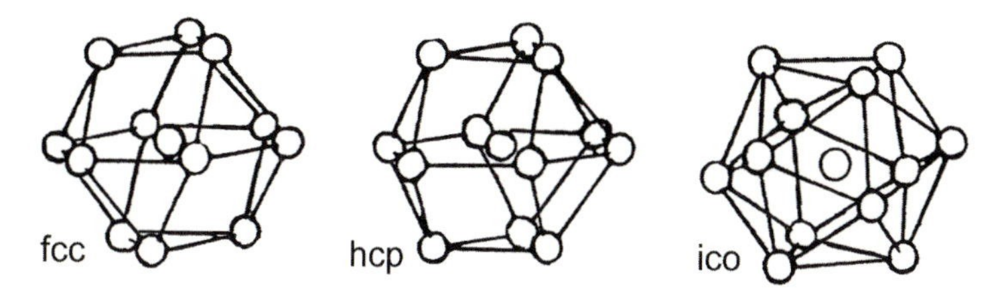

Fig. 5.6. Three possible solutions out of the 988 theoretically possible for thirteen atoms: respectively the cuboctahedral fcc, hcp and icosahedral clusters

directions. Alternatively, the structure may also be constructed by adding a fifth atom to the four atoms that make up the base of the octahedron.

As the number N of atoms increases, so does the possible number of isomers; at $N = 13$, and, if we adopt a Lennard–Jones interatomic potential, we can obtain 988 stable isomers. Among these isomers, the icosahedron, with its twelve surface atoms located at the vertices, exhibits the maximum number of surface orbitals (Fig. 5.6). In particular, the competing fcc and hcp clusters exhibit four bonds between each surface atom and its surface first neighbours. The icosahedron has five of these orbitals, provided we slightly distort it, by compressing the radial orbitals, as compared to the surface orbitals. Both the decahedron and the icosahedron exhibit *fivefold* symmetry axes, in numbers of one and six respectively (see Sect. 1.2). This kind of symmetry, which allows both structures to increase their bond energies at

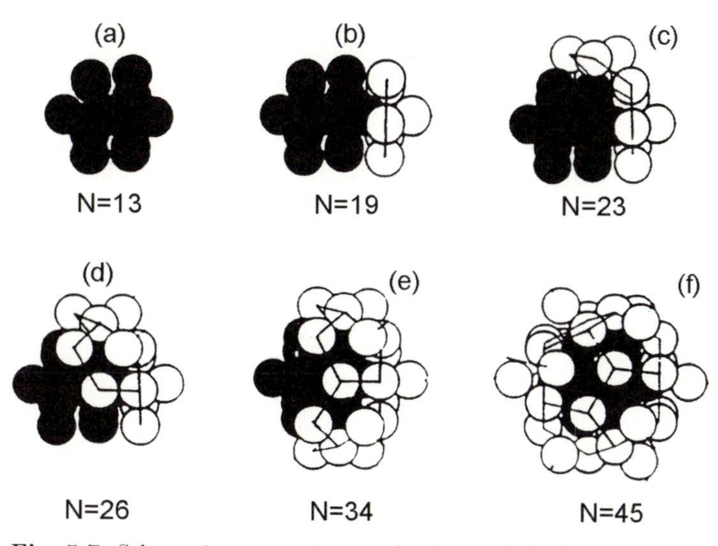

Fig. 5.7. Schematic representation of polyicosahedral packing: from (**b**) to (**f**). N is the number of atoms corresponding to the increasing fractions of covering of the parent icosahedron (**a**). Each covering is complete

the expense of the slight elastic distortion mentioned, is obtained by packing regular tetrahedra together and then distorting them.

If we proceed with this method we can increase the sequence to beyond $N = 13$ by adding pentagonal rings around each icosahedral vertex. The process does not give us any guarantee that the resulting isomers will have the greatest degree of stability; however, on hindsight, the method can be justified owing to the fact that it leads to the first sequence of magic numbers observed in the mass spectra for the atomic clusters of inert gases. When we add a pentagonal ring, centred about a fivefold axis, and a further atom in the "capping" position, we obtain a double icosahedron, with $N = 19$ (Fig. 5.7), made of two interpenetrating icosahedra which, in turn, have in common two central atoms and a pentagonal ring (part (a)). At $N = 23$, an isomer can be obtained by placing a second pentagonal ring about another axis, provided that this ring has the highest number of atoms in common with the first pentagonal ring (part (b)). In order to obtain this result we just have to add three atoms to the structure, keeping the external atom, which is the capping atom, in its position.

Likewise, by adding another two atoms, plus an external one, we can build an $N = 26$ isomer (part (d)), and so on to achieve $N = 29$, $N = 32$, $N = 34$, (part (e)). The structural model of this last isomer is characterised by having five axes with fivefold symmetry, and consists of seven interpenetrating icosahedra, which gives us 16 double icosahedra. Once all the possible pentagonal rings have been added to the parent icosahedron it will be completely enwrapped; now $N = 45$, (part (f)). A structure that consists of double interpenetrating icosahedra, or of icosahedra that are joined together by way of common faces, is called a *polyicosahedral* structure.

In the experimental mass spectra, the relative intensities for clusters containing N atoms, as N progressively increases, exhibit a slowly increasing trend. However, for specific N values it emerges that the intensity increases, giving us a local maximum; immediately after, the intensity falls abruptly.

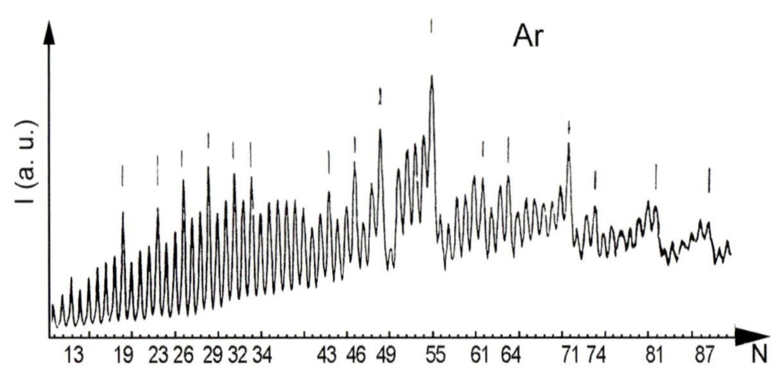

Fig. 5.8. Mass spectrum for atomic cluster beams of argon with magic numbers highlighted (adapted from [5.2])

Clusters with pathological intensity, namely those abnormally abundant clusters in the beam, as compared to the global trend, are characterised by particular structural stability. We expect these abnormally abundant clusters to exhibit the minimum number of open surface orbitals. The mass numbers N pertinent to these clusters are called magic numbers.

The sequence of the first seven values for N, for the construction based on double icosahedra, is $N = 13, 19, 23, 26, 29, 32, 34$. It is particularly interesting that such a sequence coincides exactly with the sequence numbers N at which the first intensity maxima are observed in the mass spectrum for argon clusters synthesised in the adiabatic expansion of a seeded supersonic beam (Fig. 5.8). With this technique the growth in cluster size is thermodynamic and the relative intensities are given by the relative stability.

Figure 5.9 shows a similar spectrum for xenon clusters; when we compare the two spectra we observe that the trends in the peaks for the mass abundance are highly regular.

The structure of argon clusters with just a few tens of atoms has been reproduced using Molecular Dynamics (MD) calculations. This computer simulation technique consists in examining the time evolution of motion, as governed by Newtonian mechanics, for a set of atoms, or molecules. The set usually includes one hundred to one thousand particles, interacting through a given potential (generally a Lennard–Jones potential, but phenomenological potentials are also used). Particle dynamics is a function of the change in

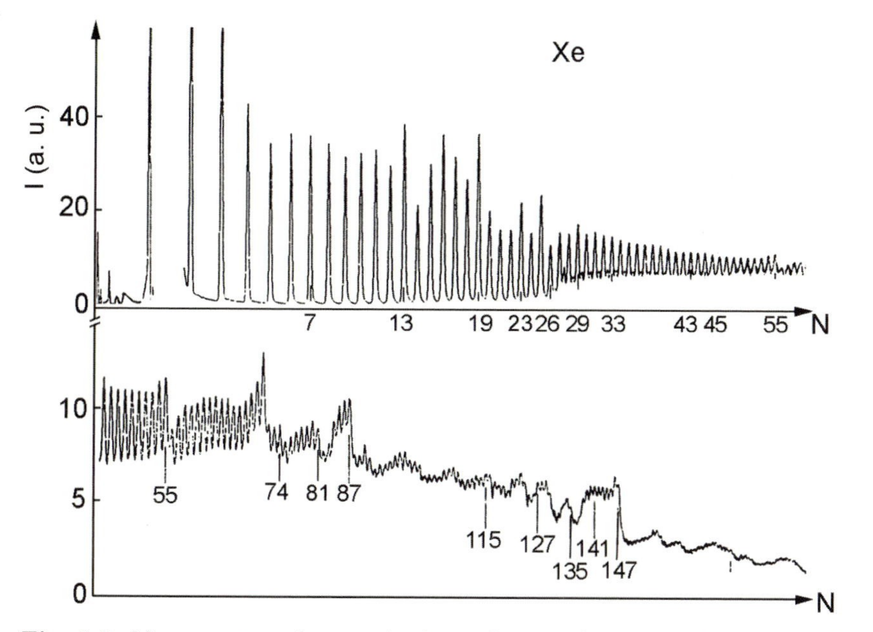

Fig. 5.9. Mass spectrum for atomic cluster beams of xenon with magic numbers highlighted (adapted from [5.3])

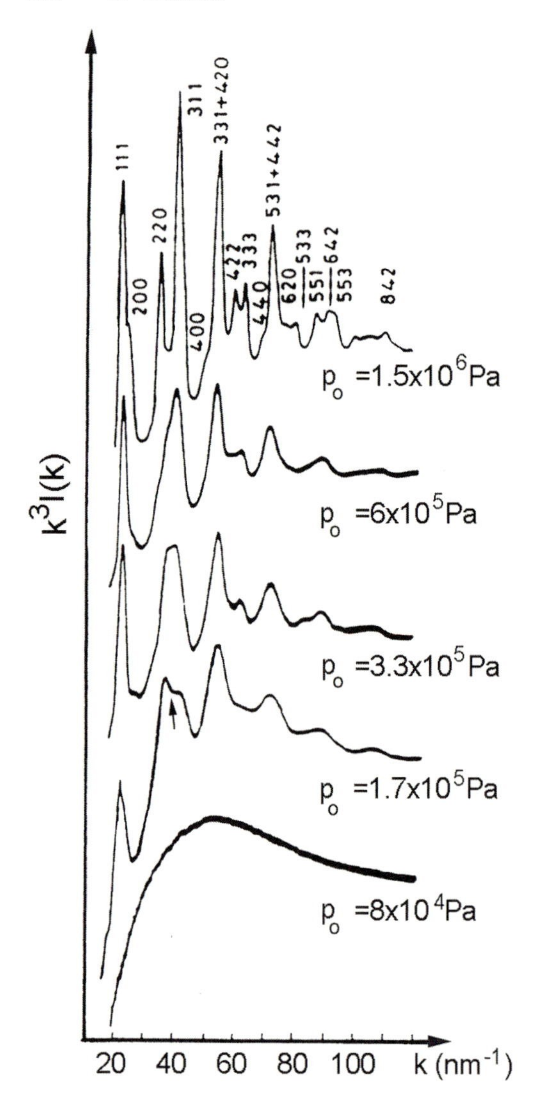

Fig. 5.10. Electron diffraction patterns from cluster beams of argon obtained at various gas inlet pressures, p_0 (adapted from [5.4])

temperature, or other thermodynamic variables. In practice, if we start from the liquid state of the system at given temperature, we observe the liquid as it solidifies. We obtain a complete simulated thermal history in terms of the trajectory and conjugated moments of all the cluster atoms.

Initially, if we want to minimise the surface effect, the atoms have to be confined to a cube, which is subject to periodic boundary conditions; the sizes of the cube are such that they reproduce the density of the material in the required aggregation state. Each atom is assigned an initial random position; a Boltzmann distribution of random velocities is then imposed which typically corresponds to a temperature of some thousands of Kelvin degrees.

With a set of successive time steps, each being typically around 10^{-15} s, we solve the motion equations. The system is cooled by reducing the temperature one finite step at a time; we then perform a Molecular Dynamics isothermal simulation, which requires around 10^{-10} s, to ensure that the system reaches thermal equilibrium.

The results of MD simulation on inert gas clusters are particularly interesting both because of the small number of atoms, even in real clusters, and because the structural properties of these materials are well described by a Lennard–Jones interatomic potential. As such, by using this potential the simulation is realistic.

As regards argon cluster beams, the patterns obtained by electron diffraction from the beam, at different gas inlet pressures, p_0, between 8×10^4 Pa and 1.5×10^6 Pa (Fig. 5.10), were compared with the diffraction patterns simulated for various possible structures of differently-sized clusters.

The experimental pattern, for the lowest inlet pressure, corresponds to a molecular beam and represents electron scattering from atomic argon. At the other end of the inlet pressures, namely where $p_0 = 1.5 \times 10^6$ Pa, the great number of large atomic clusters gives rise to a crystallographic diffraction pattern where the peaks are caused by an fcc structure, the very same structure taken on by argon when it solidifies under normal conditions.

The predicted structure for the heavy noble gases in the solid state is hcp; however, the observed structure is fcc. Details of the transition from the icosahedral structure to the fcc structure, at a critical cluster size of around $N = 1\,500$, could shed some light onto why this solid system prefers the fcc structure to the hcp structure. In actual fact, the detailed properties of the diffraction pattern obtained from the argon clusters synthesised using $p_0 = 1.5 \times 10^6$ Pa *do not* exactly correspond to the properties for an ideal fcc system, nor can they be reproduced using various distorted fcc structures involving multiple twinning and defects in the stacking of atomic planes.

The best simulation of the experimental pattern is obtained using a structure where the largest clusters consist of a non-crystalline "core" with an atomic arrangement that exhibits fivefold symmetry identical, or very similar, to the smaller clusters, as discussed below, completely surrounded by matter with an fcc single crystal structure. Figure 5.11 shows the electron diffraction patterns for five kinds of equal-sized argon atomic clusters ($N = 3\,000$).

The best simulation is obtained using the "mixed" structure, which gives us pattern (1). When we adopt this model we first have to solve two problems: how to embed local atomic arrangements with fivefold symmetry into a crystalline matrix, and how these arrangements stimulate that matrix to grow.

One possible solution can be traced to the experimental observation that, unlike in perfect macroscopic crystals, the four $\langle 111 \rangle$ directions are not equivalent in very thin fcc structures. For example, they produce three crystallographic reflections such as $\bar{1}11, 1\bar{1}1, 11\bar{1}$ and one diffuse reflection, 111. This is due to $\{111\}$ faceting, namely that thin triangular or hexagonal lamellae

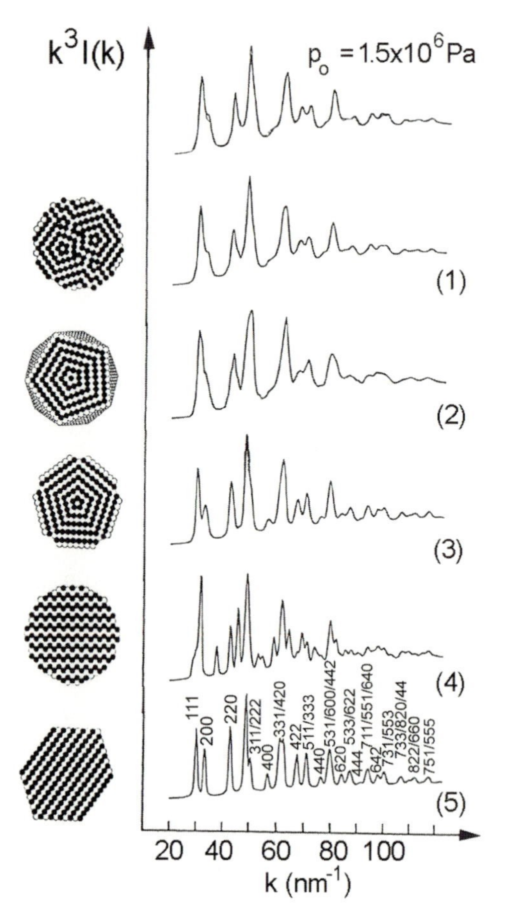

Fig. 5.11. Simulated electron diffraction patterns obtained from equal-sized argon clusters ($N = 3\,000$) with different geometric structures; (1) "mixed" fcc-icosahedral cluster; (2) nine-shell icosahedron; (3) decahedron; (4) spherical hcp; (5) nine-shell fcc cuboctahedron. The simulated pattern (1) (thin line) has been superimposed to the experimental diffraction pattern ($p_0 = 1.5 \times 10^6$ Pa), obtaining nearly perfect coincidence (adapted from [5.5])

give rise to anisotropy of the $\langle 111 \rangle$ growth rates in the crystal. The diffuse reflection corresponds to the direction of slow growth, perpendicular to the triangular, or hexagonal, faces. This is apparently the direction along which disorder in atomic plane stacking occurs; this very same disorder prevents stacking disorder developing in the other three $\langle 111 \rangle$ directions. The high growth rate along these directions suggests that faces $(\bar{1}11)$, $(1\bar{1}1)$ and $(11\bar{1})$ are modified by the stacking defects in the [111] direction in such a way that the nucleation of new layers on these faces is much more probable than on the faces with low growth rate, which are parallel to the stacking defects.

In order to obtain high growth rates without disorder in the stacking of atomic planes along all $\langle 111 \rangle$ directions, we have to introduce two twin lamellae into the crystal, and ensure they are not parallel to each other. Each lamella is defined by two twin planes very close to each other. Since the two lamellae must intersect each other, then the atomic arrangement in the intersection region deviates significantly from the ideal fcc arrangement. Such a local atomic arrangement represents the cluster structure on the size interval where a structural "transition" is expected from behaviour dominated mainly by pentagonal atomic dispositions to one where the fcc crystalline structure becomes progressively dominant.

Crystallography shows that the twin lamellae cannot intersect each other in a hcp crystal, whereas the defects in atomic plane stacking in one direction favours the fcc sequence growth in other directions. This may well be the reason why we observe fcc and not hcp structures in the solidified heavy noble gases.

If we examine Fig. 5.11 once again, we will notice that the hcp structure, which gives us pattern (4), is completely inadequate to reproduce the experimental diffraction pattern. Furthermore, if the structural transition exhibited a dependence on the cluster size, then the experimental result should be compared to a weighted average of simulations: (2) icosahedron, (3) dodecahedron and (5) fcc cuboctahedron. Even if we chose smaller or larger clusters to those used, with such an average we would not be able to adequately reproduce experimental results.

Unlike the diffraction pattern for the clusters synthesised with high inlet pressure, the experimental pattern for clusters synthesised using intermediate values of p_0 cannot be interpreted using crystalline models; we need non-crystalline structural models. The pattern for $p_0 = 1.7 \times 10^5$ Pa exhibits a number of oscillations, and in particular a shoulder in the second peak on the side of high values for wavevector $\boldsymbol{k}$. This shoulder is a characteristic feature of diffraction patterns for amorphous metals and is attributed to the kind of short range order they have. Simulations using Molecular Dynamics calculations on the solidification of droplets with fewer than 50 argon atoms are very similar to each other, and correctly reproduce the shoulder in the second peak. We obtain the best agreement between simulated and experimental diffraction patterns using a cluster with $N = 42$, as shown in Fig. 5.12. The structure of the computer simulated solid cluster has three icosahedra; of these, two have seven atoms in common and form a double icosahedron, whereas the third has four atoms in common with this double icosahedron, namely two faces. As such we are faced with a polyicosahedral structure.

The ability to reproduce the characteristic splitting of the second diffraction peak is peculiar to polyicosahedral structures. Figure 5.13 shows the

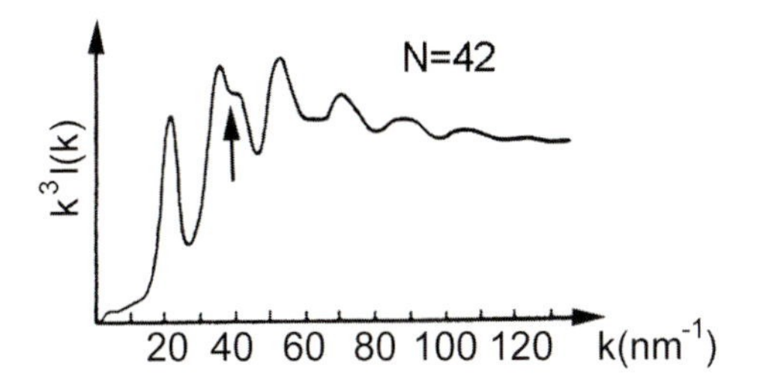

Fig. 5.12. Simulated electron diffraction pattern obtained from a 42 atom argon cluster with polyicosahedral structure (adapted from [5.4])

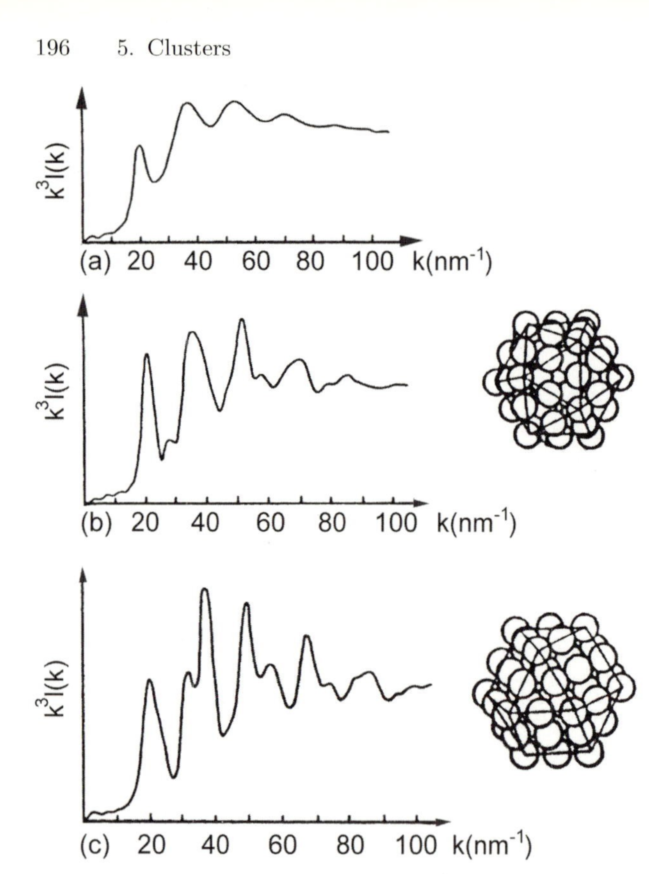

Fig. 5.13. Simulated electron diffraction patterns obtained from (**a**) a liquid argon drop containing 56 atoms, (**b**) a cuboctahedral fcc cluster with 55 argon atoms and (**c**) a 55-atom icosahedral cluster (adapted from [5.4])

diffraction intensity patterns for a liquid droplet containing 56 atoms, (part (a)), a cluster of 55 atoms with the cuboctahedral structure (fcc), (part (b)), and an analogous cluster with an icosahedral structure, (part (c)), all calculated with Molecular Dynamics. It is clear that the first two structures cannot correctly reproduce the typical trend in the second peak.

The ability of the polyicosahedral model to simulate experimental structural data is limited to the understanding of small clusters, whereas the correct sequence of magic numbers in the mass spectrum of argon, and other noble gases, for clusters with more than fifty atoms is not predicted. This sequence is correctly interpreted by the *multilayer icosahedral* structure.

This is an extended structural arrangement with, once again, the external shape of a regular icosahedron with twelve vertices and twenty faces, though it can be extended to infinity. Despite the perfect fivefold symmetry, the local order of this atomic arrangement is similar to that of a crystal. The internal structure is made up of a packing of twenty identical tetrahedra with one

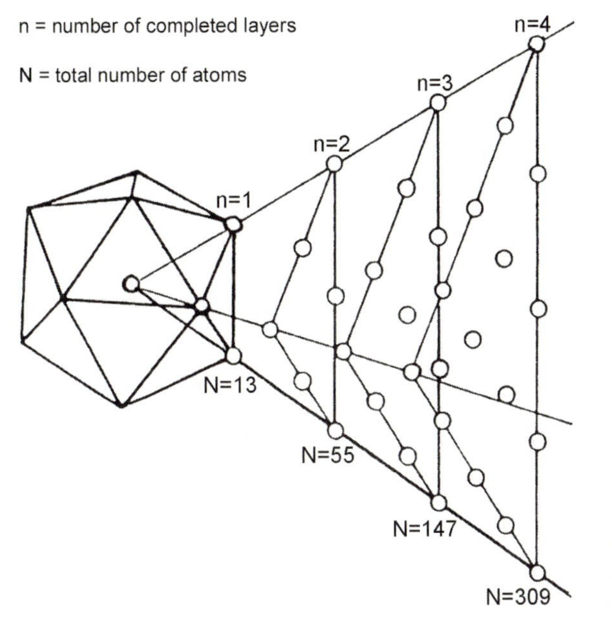

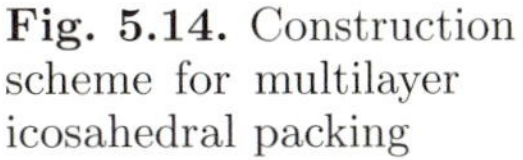
Fig. 5.14. Construction scheme for multilayer icosahedral packing

common vertex occupied by the atom at the centre of the structure. From Fig. 5.14 we see that the twenty tetrahedra are connected to each other by adjacent faces, and each face gives rise to twinning planes. The atoms in each tetrahedron are arranged on planes parallel to the surface, just like in the fcc structure.

However, the radial interatomic distances are 5% shorter than the tangent distances, as illustrated in Fig. 5.15. An icosahedron with n complete layers exhibits $n_e = (n + 1)$ spherical atoms located on each edge, just like the tetrahedra it is made of. The sequence starts with the icosahedron, at $n = 1$, which accommodates two atoms on each edge. Each face, whose side has $(n + 1)$ atoms, holds n_f atoms, n_f being equal to the summation of the first $(n + 1 - 3)$ integers, which, using the Gauss relation, can be written as

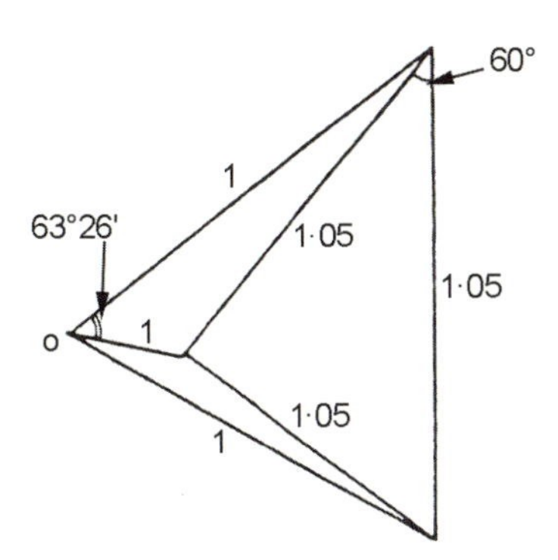

Fig. 5.15. Distortion of the tetrahedral building block in the multilayer icosahedral packing

$$n_{\mathrm{f}} = \frac{(n+1-3)(n+1-3+1)}{2} = \frac{(n-2)(n-1)}{2} = \frac{n^2-3n+2}{2}. \quad (5.2)$$

The icosahedron displays twenty faces and thirty edges, each of which contains $(n+1)$ atoms. Considering that each vertex (12) is common to five edges, we have to subtract (12×4) atoms from the total. In the end, for the total number of atoms contained in the layer with index n, n_{l}, we obtain,

$$n_{\mathrm{l}} = 20\frac{n^2-3n+2}{2} + 30(n+1) - (12 \times 4) = 10n^2 + 2. \quad (5.3)$$

The total number N of atoms in a multilayer icosahedral structure with complete external layers can thus be deduced quite easily. This structure, whose construction is shown in Fig. 5.14, is extremely stable since it allows the cluster to take on an almost spherical shape and has high atomic surface density due to the (111) faces. When we complete the successive layers we obtain a second sequence of numbers of atoms per cluster which corresponds to maximum structural stability; the sequence includes, in order 13, $55[13 + 10 \times 2^2 + 2], 147, 309, 561, 923, ...$ (Fig. 5.16).

That the multilayer icosahedral arrangement is stable is confirmed by the fact that an fcc crystallographic structure, such as the cuboctahedron, can be transformed into a regular icosahedron by simply uniformly shortening the distances from the vertices to the centre by 5%. This process will transform each external square face of the cuboctahedron into two equilateral triangles. Molecular Dynamics calculations on clusters with 55 atoms relaxed in a Lennard–Jones potential suggest that these clusters easily transform into a two-layer icosahedron.

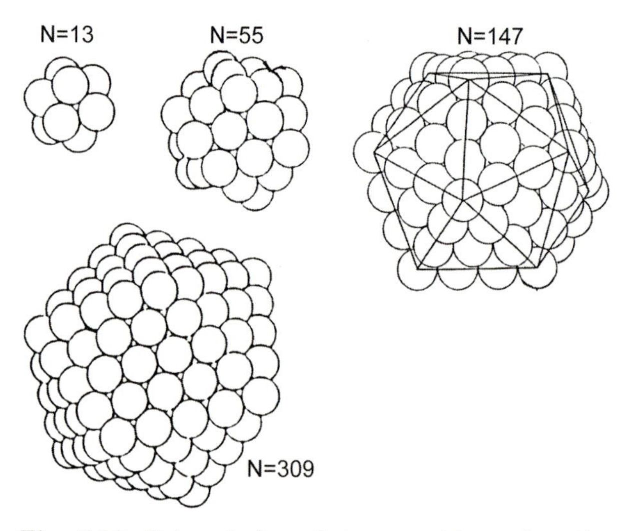

Fig. 5.16. External view of clusters with number of atoms that correspond to filling the first four layers in the multilayer icosahedral packing

When we use multilayer icosahedral structural models for clusters with $N = 147$, three-layer icosahedron, and with $N = 420$, 4.5-layer icosahedron, the obtained diffraction patterns show excellent agreement with the experimental results for argon clusters synthesised with $p_0 = 3.3 \times 10^5$ Pa and $p_0 = 6 \times 10^5$ Pa respectively, as shown in Fig. 5.17.

When we compare experimental diffraction patterns with the calculated ones for clusters at various temperatures, we obtain the best agreement for clusters whose simulated temperature is 32 K.

For an even higher beam inlet pressure, p_0, a number of fine structural details of the experimental diffraction patterns are not reproduced by the diffraction functions calculated using purely multilayer icosahedral models. For example, the pattern for $p_0 = 9 \times 10^5$ Pa agrees with the superposition of two weighted functions, for multilayer icosahedral model (70%) and fcc (30%), respectively, as shown in Fig. 5.17, curve (3).

This is an indication that the clusters of the noble gases undergo a structural phase transition when they contain around 1 000 – 1 500 atoms. The elastic deformations in the multilayer icosahedral structure are no longer efficiently compensated for by the surface energy when the ratio of the number of surface atoms N_s to the number of cluster atoms, N, falls to below a critical value.

Apart from the principal magic numbers, which correspond to complete icosahedral layers closing, other less evident magic numbers correspond to

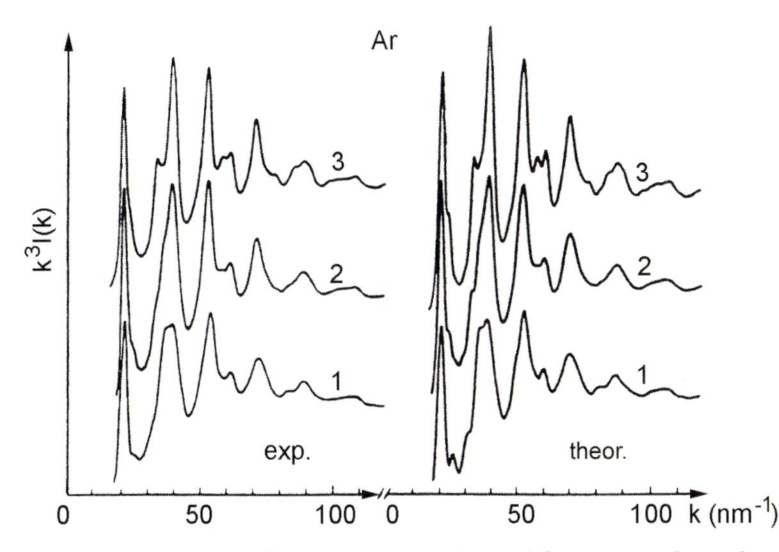

Fig. 5.17. Electron diffraction patterns obtained from argon cluster beams of various sizes, increasing with gas inlet pressure, p_0, in the order (1), (2), (3) and simulated electron diffraction patterns obtained from argon clusters packed (patterns (1) and (2)) according to the multilayer icosahedral scheme (MI), and with contribution from both MI packing, and from fcc packing, (3) (adapted from [5.4])

the closing of sub-layers. The very existence of sub-layer closing confirms that icosahedral packing occurs, since, with fcc packing, we can obtain stable clusters with closed layers giving the *same* sequence of the main magic numbers observed with icosahedral packing.

Apart from the inert gases, 13 has experimentally been observed as a magic number for some metals: for example barium, when it is in the form of a cation in clusters with various charge states; magnesium, which exhibits clusters with an icosahedral structure in the size interval $147 < N < 2869$, and for which there is proof that the smaller clusters, in turn, have this structure during cooling; aluminium, gallium and indium, whose clusters exhibit particular stability at $N = 13$, or $N = 14$, subject to their charge state. Clusters of Pb_{13}, Sm_{13} and Yb_{13} correspond to considerably intense peaks in the mass spectra of the atomic clusters of those elements. When a silicon cation cluster with 13 atoms reacts with various molecules it is chemically inert. Lastly, recent reactions with H_2 and NH_3 molecules indicate that even the Ni_{13} and Co_{13} clusters have icosahedral structure.

Rarely have sequences of magic numbers been observed in the mass spectra for transition metal clusters because laser vaporisation is usually used to generate cluster beams for those elements, and as such cluster growth is an off-thermodynamic equilibrium process. Furthermore, one feature of the chemistry of transition metals is their ability to adopt various oxidation states, which makes counting the cluster electrons difficult, unlike what occurs for alkali metal clusters. On the other hand, interesting results have emerged recently regarding nickel and cobalt cluster structures from laser vaporisation beams, over the size interval $50 < N < 1000$. Magic mass numbers are observed for both the closing of the main shells $(147, 309, 561, 923)$, and for the sub-shells expected in icosahedral packing.

Unfortunately, the laser vaporisation sources are, as yet, not efficient enough to produce beams of clusters from these elements where N is greater than 1 200, which is a useful size range in analysing the transition to crystalline structures and the relative mechanisms. These mechanisms may be quite different in these two metals since nickel crystallises in the fcc structure, whereas in cobalt the fcc and hcp structures compete, even though they exhibit the same coordination numbers (see Fig. 1.20).

The small clusters of alkali and noble metals do not follow the same strategies as the inert gases to achieve maximum structural stability. Indeed we observe magic numbers in these metals at $N = 8, 20, 34, 40, 58, 92, \ldots$ These magic numbers indicate that the maximum stability coincides with *electronic* shell closing. The electronic structure of alkali atoms immediately leads to a structural stabilisation scheme in which electronic factors play a major role. On the other hand, the electronic configuration of the noble metals is $(n+1)s^1 nd^{10}$. Some characteristic mechanisms of electronic shell closing are often found in the clusters of these elements; this is because the nd shells are complete and spatially contracted, so the bonds mainly involve orbitals

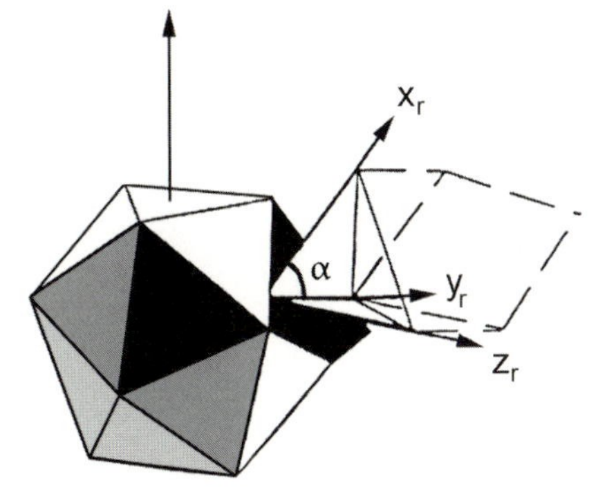

Fig. 5.18. Crystallographic relation between tetrahedral and rhombohedral (r) cells in multiply twinned icosahedral particles; angle α is $63°44'$

$(n+1)s^1$. If we take collections of large gold and silver clusters with diameters between 4 nm and 8 nm, with a few thousand atoms per cluster, and subject the clusters to annealing treatment, then the fraction of clusters with fivefold symmetry, caused by multiple twinning in the smaller particles with an fcc structure increases from 70% to around 100%. The features of the diffraction patterns of these large clusters, obtained using high resolution electron microscopy, change sharply from point to point, and this suggests that the two extreme fcc and icosahedral structures compete with each other, and that there are configurations with intermediate structure. In the multiply twinned icosahedral particles where the relation between tetrahedral cell and rhombohedric cell, given by r, is shown in Fig. 5.18 (see Chap. 4), the single units are twinned along the $\{100\}_r$ faces. Such particles share $< 100 >_r$ edges, and the $[111]_r$ directions, which correspond to c axis in the hexagonal reference, coincide with the threefold axes through the centre of the icosahedron and the centres of its triangular faces (see Chap. 1).

The single crystallites exhibit a rhombohedric unit cell where $\alpha = 63°44'$, and this corresponds to a (c/a) ratio $= 2.267$ in the hexagonal reference. The remarkable difference from the ideal close-packed cubic arrangement, where $\alpha = 60°$ and $(c/a) = 2.45$, explains why the multiply twinned icosahedral particles are observed in metals with a close-packed cubic structure only as nanoparticles; such icosahedrally coordinated particles are found in larger grains when the latter are grown under non-equilibrium conditions.

5.4 Structure of Alkali-Metal Clusters

The alkali metals are the prototypes for such systems that can be described using the free electron model. By analogy, the clusters of these elements are considered the prototypes of metal clusters. It is, however, necessary to question if those methods, that are recognised in solid state physics, can be applied for small clusters. In other words, starting from what cluster size will the free electron model correctly describe the properties of alkali metal

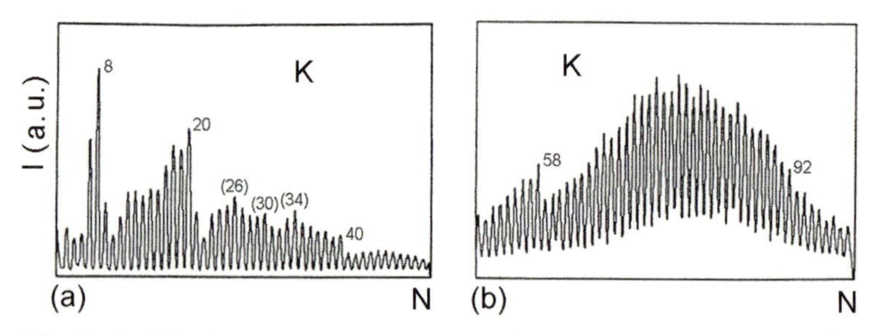

Fig. 5.19. Mass spectrum for potassium clusters: (**a**) $N = 3$ to $N = 51$; (**b**) $N = 50$ to $N = 100$. Magic numbers (N) indicate closing of principal electronic shells and sub-shells (adapted from [5.6])

clusters? How will the reduced size influence the electronic structure of the system? From an experimental point of view, we observe that the properties of alkali metal clusters are determined by their size and shape, and with increasing size show a regular trend towards the properties of the bulk material itself. The study of alkali clusters has been concentrated mainly on the size interval corresponding to a number of atoms, N, between two and one hundred.

In Fig. 5.19 is reported the experimental mass spectrum for potassium where the magic numbers coincide with those for closing of the main electronic shells and sub-shells.

One efficient model used to represent the structural properties of the alkali metal atomic clusters (thus a model that is suited to those cases where there are no d electrons) is given by a "jelly" of ions, thus deliberately *lacking* structure, embedded in a sea of indistinguishable itinerant electrons. Such a model system is called jellium.

At a first approximation the conduction electrons, whose density is uniformly distributed throughout the volume of the system, are totally delocalised. In actual fact, the electrons shield the positive charge in the ion core and the positive charge, in turn, is represented as being delocalised and uniformly distributed within volume V.

The picture the jellium gives us of two equal, opposite and congruent charge distributions does seem somewhat unrealistic, especially when we compare it to the usual scheme where the atoms, which form a crystalline structure, are bonded together by localised orbitals. Once we eliminate the crystalline structure the quantum energy levels for the valence electrons are determined solely by the *shape* and the *symmetry* of the region they are confined in. In the case of a solid, this may be represented by a macroscopic box, whereas for atomic clusters we often refer to a microscopic spheroid.

The volume V_c of a spherical cluster with n atoms is

$$V_c = 4\pi N R_{WS}^3/3 \tag{5.4}$$

where R_{WS} is the radius of the Wigner–Seitz sphere for the material. Radius R_c for the cluster of N atoms is

$$R_c = R_{WS}N^{1/3}. \tag{5.5}$$

Independently from the geometry of the volume that contains the electrons, the model represents an extreme situation where the electrons are confined within a well under the action of a *uniform* attractive potential. The valence electrons experience the effects from the positively charged background, the average potential due to the other electrons and the constraints imposed by the confining potential well.

When we introduce the mean field approximation the electrons undergo the same potential at all points and the calculation for the wave functions and energies depends on the *shape* of the potential well. As such the geometric structure of the well is dominant with respect to the geometry of the atomic packing.

The simplest shape for the potential well is a sphere, and the simplest shape for the deformed sphere is the ellipsoid. The axes of this ellipsoid are obtained by minimising the total electronic energy with respect to the axial ratio, keeping the volume constant.

Two of the results from studying jellium have important consequences in modelling small atomic clusters. The first is that, on an atomic scale, there is a sharp edge to the positively charged jellium sphere, with radius R_c, whereas the electron gas is not terminated at the jelliun edge but extends beyond the edge by a distance d, which is typically around 0.1 nm. In the case of small clusters, d is a significant fraction of R_c; for example, in a cluster with twenty atoms of sodium, $R_c = 0.58$ nm.

The second result is that the *electronic* energy is the dominant fraction of total energy of the system, and as such the equilibrium geometric structure for a cluster is that corresponding to the minimum electronic energy of the system. The energy and shape of the cluster are given in terms of the axial ratio for the ellipsoid.

Though we normally schematically consider a rigid jellium container in which the electrons are forced, in this case we observe the opposite. The shape of the jellium background adapts to the shape of the electron potential well.

It is experimentally found that in a first approximation, the details of the ionic core structure are not relevant to determine the structural stability of simple metal clusters. Moreover, the electrons may be treated as if they were free and confined within a potential well. While these results allow us to adopt the simplified jellium model, we are left with a complicated many-body problem in which the electrons are to be treated self-consistently.

We can adopt a highly simplified approach by using an effective *single* particle potential with simple geometric structure, such as a three-dimensional square well, possibly with a rounded background.

With reference to Fig. 5.19 we expect that for potassium clusters (but also for sodium, rubidium, ..., clusters) the potential the electrons experience is spherically symmetric and it gives rise to a spherical shell structure where the electrons progressively fill the energy levels.

Having approximated the potential the electrons undergo by way of a uniform well, where each particle is subject to the same average field, the independent particle states show *degeneracies*, which depend on the shape of the potential well. The last electron to complete a degenerate state closes an electronic shell. This condition corresponds to a particular state of stability in the system compared to the immediate neighbouring configuration. The electronic magic numbers suggest that the principal electron shells are closed, which corresponds to discontinuities in the system stability as well as in other properties such as the first ionisation potential.

The demand for clusters to take on spherical symmetry does not always coincide with a configuration of minimum energy; in fact, infinite clusters with electronic shells that are allowed to deform, keeping the volume constant, take on an ellipsoidal shape where values of semiaxes x_0, y_0, z_0, correspond to the minimum energy. The degree of deformation with respect to the spherical configuration is measured by the deformation parameter η.

The problem is to determine the geometric structure of the cluster where N atoms arrange themselves, provided that the total energy in the system is minimum. As a first approximation, the energy levels occupied by the cluster atoms are calculated by solving the single particle Hamiltonian for the harmonic three-dimensional oscillator

$$\frac{E}{\hbar} = \left(n_x + \frac{1}{2}\right)\omega_x + \left(n_y + \frac{1}{2}\right)\omega_y + \left(n_z + \frac{1}{2}\right)\omega_z \tag{5.6}$$

where n_x, n_y, n_z are the quantum numbers for the three axes and ω_x, ω_y, ω_z are the corresponding oscillation frequencies. Each quantum state can host a maximum of two atoms at most.

The total energy of the system is given by the sum of the energies of all atoms. These energies are determined as functions both of the occupations of the stationary levels (each being identified by a set of three quantum numbers) and of the cluster's geometric structure (the oscillation frequencies depend on the of x_0, y_0, z_0 values for the ellipsoid axes), provided that the potential energy is constant over the entire ellipsoid surface,

$$2E_{\mathrm{p}}/\left(m\omega_0^2 R_{\mathrm{c}}^2\right) = (x/x_0)^2 + (y/y_0)^2 + (z/z_0)^2 \,. \tag{5.7}$$

Given that, as the cluster deforms from a spherical geometric structure to an ellipsoidal geometric structure, we require the total volume to remain constant, then the normalisation condition must be fulfilled,

$$x_0 y_0 z_0 = R_{\mathrm{c}}^3 \tag{5.8}$$

where R_{c}, as given in (5.5), is the radius of the spherical configuration.

The energy state of the cluster is obtained from the set of equations

$$\begin{cases} E = E(\omega_x, \omega_y, \omega_z) \\ \omega_x = \omega_x(x_0, y_0, z_0) \\ \omega_y = \omega_y(x_0, y_0, z_0) \\ \omega_z = \omega_z(x_0, y_0, z_0) \\ x_0 y_0 z_0 = R_c^3 \end{cases} \tag{5.9}$$

Once the stationary levels occupied by the atoms are known, when we substitute the equations for ω_x, ω_y, ω_z into the equation for energy, we obtain the explicit dependence $E = E(\omega_x, \omega_y, \omega_z)$.

Before we proceed to calculate the values for the semiaxes x_0, y_0, z_0, which correspond to the minimum of the function $E(x_0, y_0, z_0)$, we should notice that the normalisation condition (5.8) reduces the problem by one degree of freedom because a relation among the ellipsoid semiaxes is introduced.

We can further simplify the problem by assuming the particular geometric structure where, for example, is $x_0 = y_0$, so that only two axes of the ellipsoid are different to each other. In this case the energy only depends on x_0. If we impose the variation $\mathrm{d}E/\mathrm{d}x_0$ to be null, we obtain the value for parameter x_0 where the energy is minimum. The semiaxis z_0 is given by the equation

$$z_0 = R_c^3/(x_0 y_0).$$

Having just outlined the general solution procedure, we have to express in explicit form the way the ellipsoid geometric structure depends on angular frequency.

Given the condition of equipotentiality of the cluster surface, again provided $x_0 = y_0$, is

$$\frac{1}{2} k_z z_0^2 = \frac{1}{2} k_x x_0^2 \tag{5.10}$$

hence

$$\frac{k_z}{k_x} = \frac{x_0^2}{z_0^2}. \tag{5.11}$$

On the other hand, $k_x = m\omega_x^2$ so, substituting,

$$\frac{x_0^2}{z_0^2} = \frac{\omega_z^2}{\omega_x^2} \tag{5.12}$$

When $\omega_x = \omega_x(x_0)$ and $\omega_z = \omega_z(z_0)$, from (5.12) we obtain

$$\omega_x \propto x_0^{-1}. \tag{5.13}$$

Let us now return to the calculation for the energy eigenstates in the harmonic oscillator approximation. The energy levels may be represented as a function of deformation parameter η, in the so-called Nilsson diagram, (Fig. 5.20).

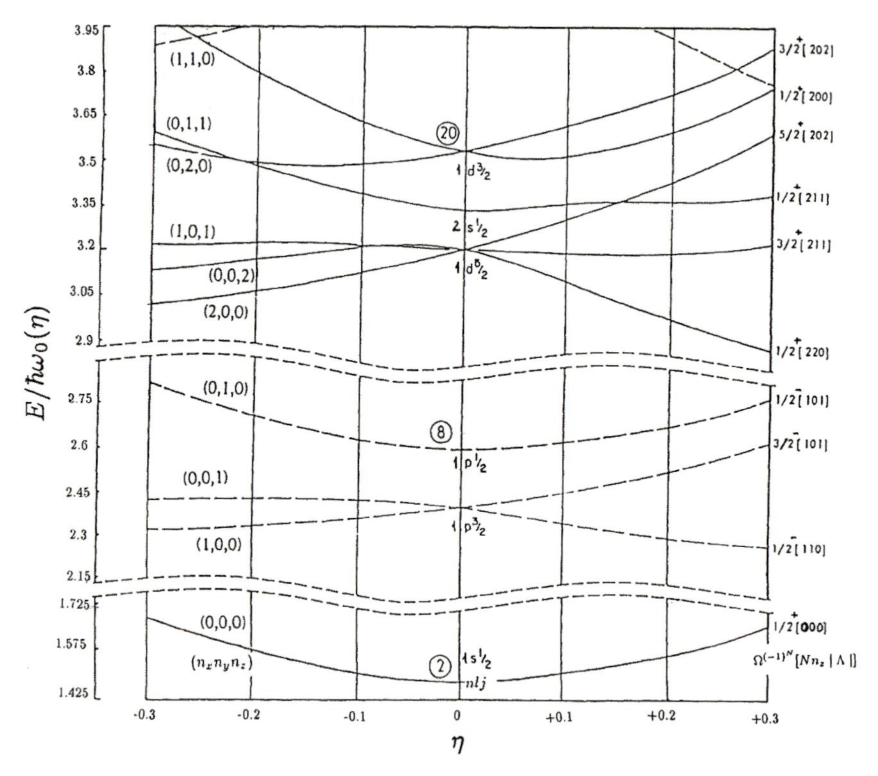

Fig. 5.20. Nilsson diagram for $N \leq 20$

The Nilsson model develops as an extension of the three-dimensional isotropic oscillator problem where the deformation parameter η is null. In this case, in order to consider spin-orbit interaction, the Hamiltonian operator is

$$\hat{\mathrm{H}} = \hat{\mathrm{H}}_0 + C(\boldsymbol{l} \cdot \boldsymbol{s}) \tag{5.14}$$

where $\hat{\mathrm{H}}_0 = -\frac{\hbar^2}{2m}\nabla^2 + \frac{1}{2}m\omega_0^2 r^2$ refers to the harmonic oscillator and $C(\boldsymbol{l} \cdot \boldsymbol{s})$ includes the spin-orbit interaction. $\boldsymbol{l}$ and $\boldsymbol{s}$ are respectively the orbital and spin momenta.

When we solve the Schrödinger equation for the energy eigenstates we obtain the eigenvalues

$$E_{N,l,j} = (N + 3/2)\hbar\omega_0 - a_{jl} m\omega_0^2 \tag{5.15}$$

The first addendum is the harmonic oscillator solution. The quantum number N is called the principal quantum number and coincides with the sum of the three quantum numbers n_x, n_y, n_z needed to solve the three-dimensional case projected onto the orthogonal reference x, y, z. The zero point energy ($3/2\ \hbar\omega_0$) is the sum of the zero point energy ($1/2\ \hbar\omega_0$) in each of the projection directions.

When we have analytically solved the equation for the energy eigenstates we observe that the principal quantum number may take on a set of infinite values

$$N = \{0, 1, 2...\}\,.$$

For the harmonic oscillator $\hat{\mathrm{H}}=\hat{\mathrm{H}}_0$, and each level E_N corresponds to a number of states with different values for l; l varies between 0 and N. The eigenfunction for the considered eigenvalue must have the same parity as N. Since this parity is $(-1)^l$ then the values for l associated to a specific N must have the same parity as N; thus not all the l values between 0 and N are acceptable. When we arrange the possible energy eigenvalues we obtain

N	allowed l	spectroscopic notation	available states
0	0	$1s$	2
1	1	$1p$	6
2	0 2	$2s$ $1d$	12
3	1 3	$2p$ $1f$	20
4	0 2 4	$3s$ $2d$ $1g$	30
...	...	...	...

In the spectroscopic notation l is the index for the type of orbital, in agreement with the atomic physics convention, whereas n means that the following l value occurs for the nth time in the sequence.

It is remarkable that this very same sequence of energy levels is found in the spherical jellium model.

Having set l, $(2l+1)$ degenerate quantum states are available, each of which can be occupied, following the Pauli principle, by two particles. The degeneracy degree may be verified by observing that

$N = \sum n_k$	compatible triplets (n_x, n_y, n_z)	number of orbitals	available states
0	$(0,0,0)$	1	2
1	$(1,0,0)$ $(0,1,0)$ $(0,0,1)$	3	6
2	$(2,0,0)$ $(0,2,0)$ $(0,0,2)$	6	12
	$(1,1,0)$ $(1,0,1)$ $(0,1,1)$		
...	....	...	...

Having solved the complete equation for the three-dimensional isotropic oscillator, the eigenvalue (5.15) contains the term

$$-a_{jl} m\omega_0^2$$

which takes into consideration the spin-orbit interaction. The total angular quantum number is given as j, where

$$j = l + s,\ l + s - 1, ... \left| l - s \right| . \tag{5.16}$$

When we consider particles with a spin angular momentum of $s = \pm\,(1/2)$, the possible values for j are $(l + 1/2)$ and $(l - 1/2)$, so

$$a_{jl} \begin{cases} +l & \text{if} \quad j = (l + 1/2) \\ -(l+1) & \text{if} \quad j = (l - 1/2). \end{cases}$$

The spin-orbit interaction lifts the spin degeneracy, splitting the generic nl level into the two levels $(nl)_{l+1/2}$ and $(nl)_{l-1/2}$. For the state ns, where $l = 0$, only $j = (l + 1/2) = \frac{1}{2}$ is allowed by (5.16).

The sequence of these energy levels is given in the Nilsson diagram in correspondence to $\eta = 0$.

The non-spherical shape of the potential for the clusters with any kind of geometric structure is considered in the energy eigenstate problem defined by the operator

$$\hat{\mathrm{H}} = -\frac{\hbar^2}{2m}\nabla^2 + \frac{1}{2}m\left(\omega_x^2 x^2 + \omega_y^2 y^2 + \omega_z^2 z^2\right) + C(\boldsymbol{l} \cdot\ \boldsymbol{s}) + D\boldsymbol{l}^2 \tag{5.17}$$

where $\frac{1}{2}m\left(\omega_x^2 x^2 + \omega_y^2 y^2 + \omega_z^2 z^2\right)$ is the anisotropic oscillator potential and $D\boldsymbol{l}^2$ is a correction to the eigenstate energy for high values of angular momentum $\boldsymbol{l}$.

The simplest non-spherical configuration, to which often we can reduce the geometric structure of a generic cluster, is a rotation ellipsoid. As such, we assume first that the field is characterised by axial symmetry and second that $\omega_x = \omega_y \neq \omega_z$. If we take that the cluster volume is constant, we can introduce a single deformation parameter η.

The equation for $\hat{\mathrm{H}}$ reduces to

$$\hat{\mathrm{H}} = \hat{\mathrm{H}}_0 + C(\boldsymbol{l} \cdot\ \boldsymbol{s}) + D\boldsymbol{l}^2 + \hat{\mathrm{H}}_\eta . \tag{5.18}$$

$\hat{\mathrm{H}}_\eta$ includes all the effects due to the potential distortion and is given as

$$\hat{\mathrm{H}}_\eta = -B\eta\ m\omega_0^2 r^2 Y_{20}$$

where Y_{20} is the spherical harmonic.

Since the operator associated with $j_z = (l_z + s_z)$, where z is the symmetry axis for the potential, commutes with the operator $\hat{\mathrm{H}}$, the field eigenstates (see (5.17)) are defined by definite values for the projection of the total angular momentum, Ω

$$\Omega = \left| j_z \right|, \text{where} \quad j_z = \pm\ 1/2, \pm\ 3/2 ... \pm\ j.$$

When $\eta \neq 0$, no analogous property holds for l^2 and j^2, which are no longer constants of the motion; as such, j and l are no longer good quantum numbers. Conversely, Ω is still meaningful.

The principal quantum number N remains a good quantum number even when $\eta \neq 0$. The eigenvalues of the Schrödinger equation for the deformed potential are

$$E_\alpha^{N\Omega}(\eta) = \left(N + \frac{3}{2}\right) \hbar\omega_0(\eta) - \left(\frac{C}{2}\right) r_\alpha^{N\Omega}(\eta). \tag{5.19}$$

In (5.19), index α numbers these very eigenvalues. The equation shows that the energy is directly dependent on the deformation parameter η, for every quantum state (N, Ω).

If η is not null, then a given isotropic oscillator eigenstate $(nl)_j$ splits into $(2j+1)/2$ eigenstates, and each eigenstate has a different Ω value. If the projections of the orbital and the spin angular momenta on the z symmetry axis of the potential are respectively Λ and Σ, then $\Omega = \Sigma + \Lambda$.

Since quantum numbers l and j lose physical meaning as deformation increases, then when η tends towards infinity the single particle states are described by the quantum numbers N, Ω, Λ and n_z. If Ω and Λ are known, Σ is univocally determined, whereas n_z is the number of nodal planes for the eigenfunction in direction z.

In the Nilsson diagram the quantum numbers are conventionally represented as

$$\Omega^{(-1)^N}[N, n_z, |\Lambda|]$$

Since these quantum numbers are meaningful only for large deformations (when $\eta \to 0$, N, l, j remain good quantum numbers) they are called asymptotic quantum numbers. When we introduce the deformation parameter η the degeneracy of states (nl) is lifted, so we put the eigenstates $\Omega^{(-1)^N}[N, n_z, |\Lambda|]$ into bi-univocal correspondence with the triplets $(n_x\ n_y\ n_z)$, which fulfil the condition $\sum_k n_k = N$, where $k = x, y, z$.

If we use the Nilsson diagram we can define the geometric structure for a cluster of N atoms.

For example, at $N = 9$, the occupied states are:

eigenstate	asymptotic quantum numbers	(n_x, n_y, n_z)	number of particles per state
$1s$	$1/2^+\ [0,0,0]$	$(0,0,0)$	2
$1p$	$1/2^-\ [1,1,0]$	$(0,0,1)$	2
	$3/2^-\ [1,0,1]$	$(1,0,0)$	2
	$1/2^-\ [1,0,1]$	$(0,1,0)$	2
$1d$	$5/2^+\ [2,0,2]$	$(2,0,0)$	1

From (5.6) we obtain

$$
\begin{aligned}
\frac{E}{\hbar} = \quad & 2\left(\frac{1}{2}\omega_x + \frac{1}{2}\omega_y + \frac{1}{2}\omega_z\right) \\
& \underbrace{+2\left(\frac{1}{2}\omega_x + \frac{1}{2}\omega_y + \frac{3}{2}\omega_z\right)}_{001} \quad \underbrace{+2\left(\frac{3}{2}\omega_x + \frac{1}{2}\omega_y + \frac{1}{2}\omega_z\right)}_{100} \\
& \underbrace{+2\left(\frac{1}{2}\omega_x + \frac{3}{2}\omega_y + \frac{1}{2}\omega_z\right)}_{010} \quad \underbrace{+1\left(\frac{5}{2}\omega_x + \frac{1}{2}\omega_y + \frac{1}{2}\omega_z\right)}_{200} \\
& = \frac{17}{2}\omega_x + \frac{13}{2}\omega_y + \frac{13}{2}\omega_z .
\end{aligned}
$$

When we consider that z is the symmetry axis for potential, then

$$\omega_x = \omega_y \neq \omega_z \ ,$$

and thus

$$\frac{E}{\hbar} = 15\omega_x + \frac{13}{2}\omega_z \ .$$

Using (5.12) where $x_0/z_0 = \omega_z/\omega_x$, and substituting, we obtain

$$\frac{E}{\hbar\omega_x} = 15 + \frac{13}{2}\frac{x_0}{z_0} \ .$$

If we normalise to the unit

$$z_0 = (1/x_0^2)$$

so that

$$\frac{E}{\hbar\omega_x} = 15 + \frac{13}{2}x_0^3 \ .$$

Now, given $\omega_x \propto (1/x_0)$ (see (5.13)) we obtain

$$\frac{E}{\hbar} \propto \frac{15}{x_0} + \frac{13}{2}x_0^2 \ .$$

If we now impose the minimum energy conditions

$$\frac{\mathrm{d}(E/\hbar)}{\mathrm{d}x_0} = 0, \qquad \text{we obtain}$$

$$x_0 = 1.0488 = y_0; \qquad z_0 = 1/x_0^2 = 0.909.$$

Since the deformation parameter η is given by the relation

$$\eta = 2\frac{z_0 - x_0}{z_0 + x_0}$$

then, in this case, we obtain $\eta = -0.1428$.

Negative values for η correspond to oblate ellipsoidal geometric structure, whereas positive values correspond to prolate geometric structure.

When we determine the structure of a cluster with 8 atoms by the same procedure we obtain

$$\begin{aligned} x_0 &= y_0 = 1 \\ z_0 &= 1 \\ \eta &= 0. \end{aligned}$$

As such, a cluster with 8 atoms has, in correspondence to the minimum energy, a spherical configuration where shell closing occurs.

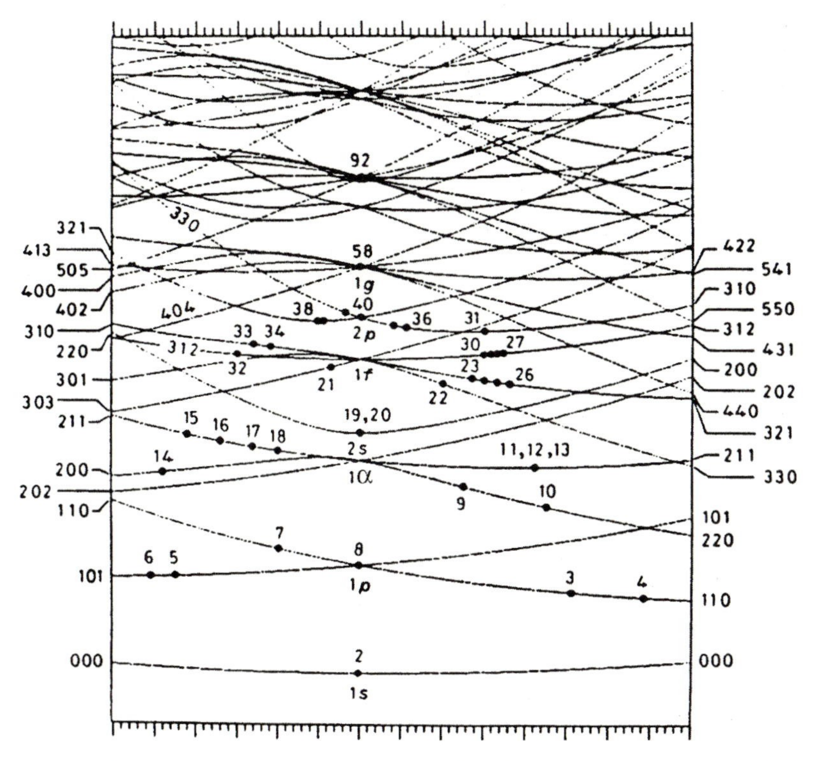

Fig. 5.21. Complete Nilsson diagram: besides closing of principal shells, with spherical symmetry, sub-shell closing is visible (adapted from [5.7])

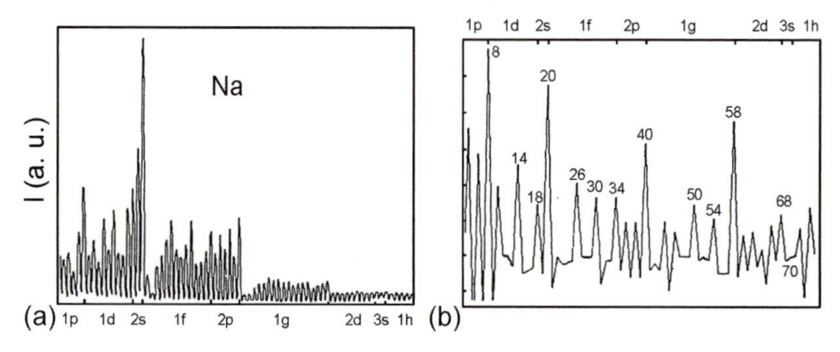

Fig. 5.22. Experimental mass spectrum for sodium clusters (**a**); prediction from Nilsson model (**b**). Notice the evident agreement in magic number sequence (adapted from [5.7])

In all the clusters where $\eta = 0$ we observe closing of electronic shells with spherical symmetry. This occurs at $N = 2, 8, 20, 40, 58, 92$, as shown in Fig. 5.21.

The deformation of deformed spherical clusters gives rise to *sub-shell* closing, which can be recognised in Fig. 5.21 in the fourfold sequences $15 - 18$, $23 - 26$, $27 - 30$, $41 - 44$.

The structure of clusters with $N = 18$ and $N = 34$ should consist of closed shells, according to the theory based on spherical shells, but when we examine the possible deformations, the latter exhibit a reduction in the corresponding energy gaps. These kinds of cluster correspond to *spheroidal* sub-shell closing and show excellent agreement with the experimental data deducible from the mass spectra. Figure 5.22 refers e.g. for sodium.

The geometric structures assumed by small and medium-sized clusters often raise delicate problems related to the equilibrium between binding energy and surface energy, to the kind of bonding orbitals and to their symmetry and, lastly, to the number of open orbitals.

One interesting example from this standpoint is given by germanium and silicon, for which we may well expect the same kind of cluster structural evolution as the size increases. When we perform Raman spectroscopy measurements on small silicon clusters, and then compare spectral features with the Raman vibrational frequencies for various test structures, we can assign its structure to each of these clusters. For a set of eleven well defined vibrational lines the maximum absolute deviation between measured frequencies and predicted frequencies is only around 10 cm^{-1}, as shown in Fig. 5.23. On these grounds it is concluded that Si_4 has a unique structure, which is planar, with a rhombus geometric structure; Si_6 takes on the structure of a slightly distorted octahedron, and, lastly, Si_7 is a pentagonal bi-pyramid. It is remarkable how these compact structures of clusters with very few atoms are so different to known microcrystalline structures.

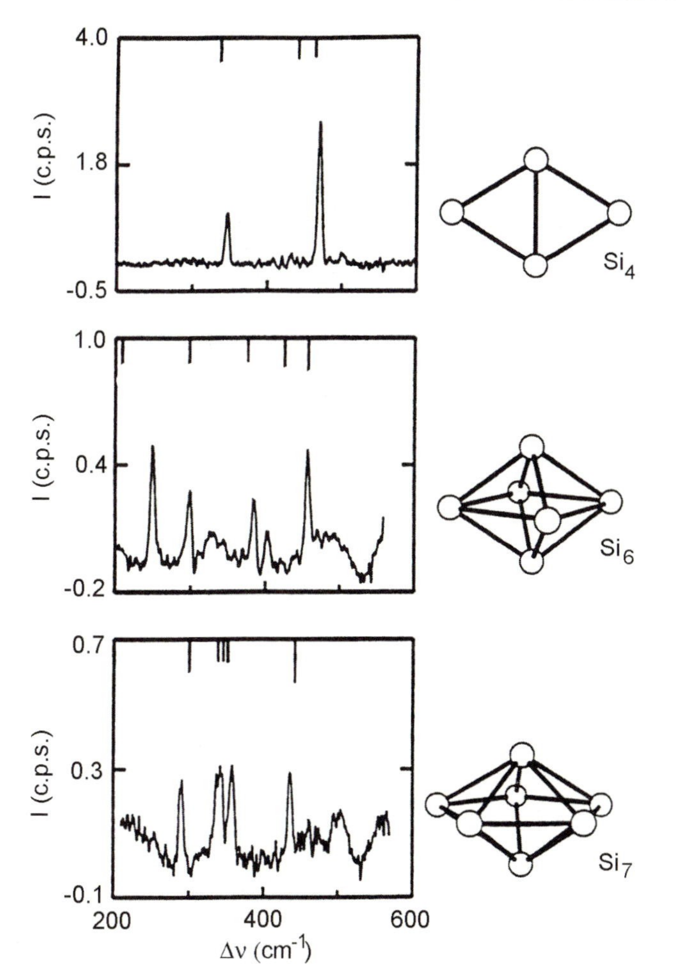

Fig. 5.23. Raman spectra for small silicon clusters and model structures to obtain the best agreement between simulated spectra and experimental data (adapted from [5.8])

If we consider larger, singly ionised silicon clusters, we observe a smooth decreasing trend in their mobility, down to $N = 27$; this is an index of prolate distorted spheroidal symmetry, which is associated with high surface energy. Then, on a short interval for cluster size we observe a sharp increase in mobility, followed by a further slowly decreasing monotonic trend. This kind of behaviour can be attributed to structural transition of the clusters that take on a more spherical geometric structure, with higher mobility. For prolate clusters where N is between 20 and 34, the most compatible structural models consist in sequences of capped trigonal prisms.

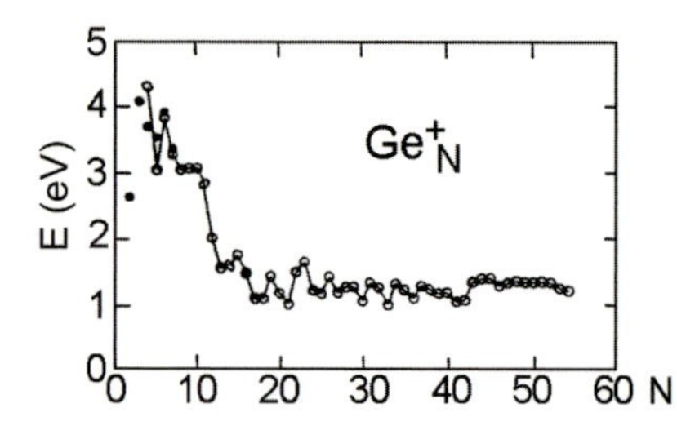

Fig. 5.24. Dissociation energy trend in Ge^+_N clusters (adapted from [5.9])

We even observe prolate geometric structure in ionised germanium clusters when the size of these clusters is between $N = 10$ and $N = 40$, which corresponds (Fig. 5.24) to dissociation energy values of around 1.2 eV. These values are much less than the binding energy in the bulk material, 3.85 eV atom^{-1}. Since the latter value is almost equal to the dissociation energy values for small clusters we can reasonably assume that, unlike silicon clusters, the larger germanium atomic clusters are made of weakly bonded packings of small stable clusters, such as Ge_7 and Ge_{10}. Atomic clusters with between 40 and 70 particles seem to be characterised by having a single geometric structure, still a prolate spheroid. Lastly, at N greater than 70, the clusters undergo sharp reconstruction which gives rise to essentially spherical geometric structure where the atomic disposition is similar to the bulk material, and differs greatly from the geometric structure of silicon clusters with similar size. The reason for this qualitative difference in the structural evolution of the clusters of these two materials is currently unknown.

Just as for isolated atoms, the trend in the first ionisation potential is an index of atomic cluster stability. The clusters of alkali metals with particular stability are also marked by relatively high ionisation potential. The ionisation potentials for small clusters of alkali metals were measured in the 70's: more recent measurements have been made with high energy resolution on clusters of sodium, potassium, lithium and silver, whose electronic structure is similar to that of the alkali metals.

The first ionisation potential is the minimum energy required for the reaction

$$X_n + h\nu = X_n^+ + e^- . \tag{5.20}$$

The atomic clusters are bombarded with low energy photons, so that they are ionised in one-photon processes; then the photo-ions are selected in mass. We study the intensity trend in the signal produced by a particular ionic mass as the ionising photon energy changes. As such, for each mass in the beam, we construct the photoionisation efficiency curve. The potential that corresponds to the extrapolated energy value at which the ion signal is null, is defined as the ionisation potential.

In general, given that the ionisation potential for single atoms is always higher than the extraction potential for the corresponding solid, we expect,

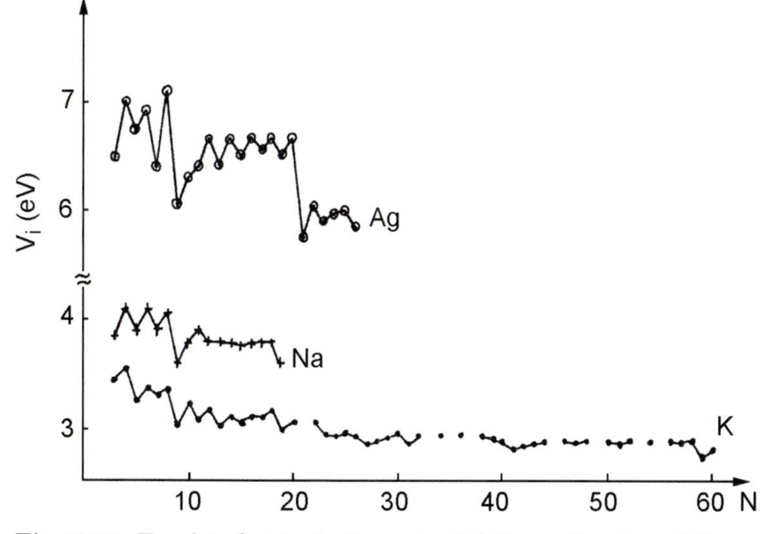

Fig. 5.25. Trend in first ionisation potential V_i as a function of the number N of particles in the cluster, for silver (○), sodium (+) and potassium (•) clusters

and indeed observe, a decrease in the first ionisation potential for atomic clusters as they become larger. The decrease law depends on the specific nature of the atoms under examination; when we compare equal-sized clusters of sodium and potassium we observe that the ratio between ionisation potentials, as reported in Fig. 5.25, is in fact independent from the size of the clusters and, starting with the isolated atom up to the crystal, is around 1.2.

In medium and large alkali metal clusters the measured values for the first ionisation potentials are in good agreement with the work function W estimated for an isolated spherical drop of metal with x radius and with Z charge. This function is the energy required to displace an electron from the drop to an infinite distance away, namely

$$W(x, Z) = W_\infty + \frac{e^2}{x}\left(Z + \frac{3}{8}\right). \tag{5.21}$$

In general, when medium and large-size clusters exhibit anomalies in the trend in the ionisation potential as compared to the trend in the metallic drop model, these anomalies coincide with strong changes in physical properties. A typical example is given by mercury and antimony clusters; in both cases the significant deviations compared to the smooth, regular fall in ionisation potential, that are observed when the clusters grow in size, are indices of metal–non-metal transition.

For small alkali-metal clusters we have to consider quantum effects. The description of a quantum metallic drop, with spherical symmetry, predicts a structure of electronic shells where we can observe closing in correspondence to a total number of electrons: $2(1s)$, $8(1p)$, $18(1d)$, $20(2s)$, $24(1f)$, $40(2p)$, $58(1g)$, $68(2d)$, $70(3s)$, $92(1h)$, $106(2f)$, $112(3p)$, $138(1i)$, $156(2g)$...; for small masses we observe most of the shells, whereas for the large masses we only observe the shells with quantum number $N = 1$.

Figure 5.25 is a representation of the ionisation potentials for clusters of sodium, potassium and silver with growing size as a function of the number of constituent atoms. We should notice, in the case of K, for example, that the falls in ionisation potential are observed at $N = 8, 18, 20, 40, 58, 92$ and correspond to spherical shell closing. Even if the absolute error in determining ionisation potentials is comparable to the observed fall in ionisation potential, the variation in potential from cluster to cluster may be determined to an accuracy of 10^{-2} eV, which allows us to recognise the falls in ionisation potential even in larger clusters, at $N = 40, 58, 92$. It is also possible to identify a "fine structure" in the trend in the variation of ionisation potential, which shows good agreement with the closing scheme for *spheroidal* sub-shells.

In general, the ionisation potential for a cluster with closed electronic shells is significantly greater than for larger clusters with open electronic structures. Lastly, the shell structure is more evident for silver than for sodium or potassium since the potential well in silver is deeper.

5.5 The Fullerene C_{60}

C_{60}, in the form of a free atomic cluster, was only clearly recognised for the first time in 1984 as an irregularity in the mass spectrum for atomic clusters of carbon, as shown in Fig. 5.26. These clusters are obtained when cooling the plasma resulting from the laser vaporisation of a graphite target at an initial temperature between 5 000 K and 10 000 K.

Despite being a relatively recent discovery, C_{60} is probably the most intensively studied molecule among those molecular compounds whose structure is based on a three-dimensional cage of trivalent carbon atoms, generally known as *fullerenes*; our present knowledge of the properties of fullerenes is comparable to our knowledge of methane and benzene. This is also due to the fact that C_{60} is the easiest fullerene to be synthesised, it has the most stable structure and the highest degree of symmetry, the same pertinent to the icosahedral point group I_h; thus C_{60} is an anomaly with respect to most fullerenes that adopt a relatively low point group symmetry.

Various experimental methods, including infrared absorption, Raman and photoelectron spectroscopies, X-ray diffraction and nuclear magnetic resonance, suggest that the sixty carbon atoms in each molecule, or cluster, are located at the vertices of a regular *truncated* icosahedron. This is obtained by substituting the twelve icosahedron vertices with twelve regular pentagons; the cluster surface is made of alternating regular pentagonal faces and regular hexagonal faces (twenty). The atomic sites coincide with the polyhedron vertices; every site is equivalent to every other site, and each site hosts one

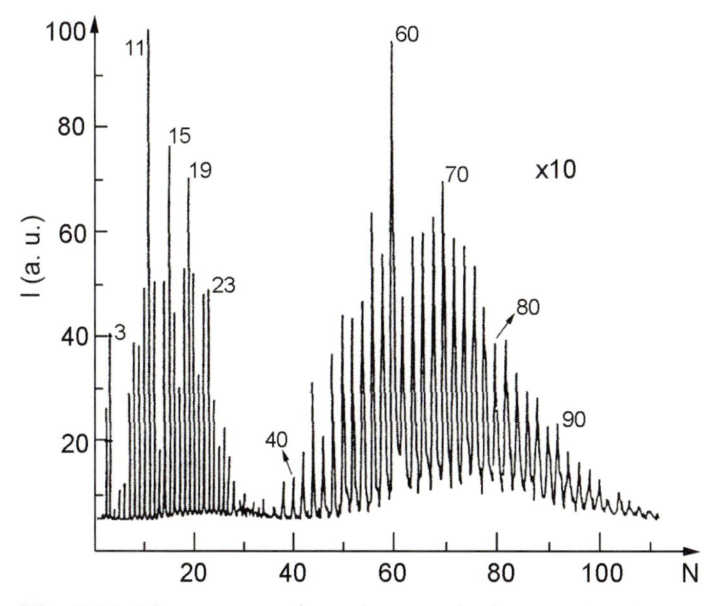

Fig. 5.26. Mass spectrum for carbon atomic clusters; abundance peaks corresponding to fullerenes are highlighted; in spectral region at $N > 38$ the mass abundance peaks are magnified (x10)compared to region at $N < 38$ (adapted from [5.10])

carbon atom. This polyhedron forms a closed cage and is shaped like a soccer ball with a radius of 1.02 nm (Fig. 5.27); given this shape the fullerene has been jokingly called a "buckyball", a contracted form of "buckminsterfullerene", because of its resemblance to the geodetic domes with a polyhedral structure designed and built by R. Buckminster Fuller.

All closed cage fullerene structures have hexagonal or pentagonal faces only. There are always twelve pentagonal faces whereas the number of hexagonal faces, in principle, in keeping with Eulero's theorem on polyhedra (see Sect. 1.1) is

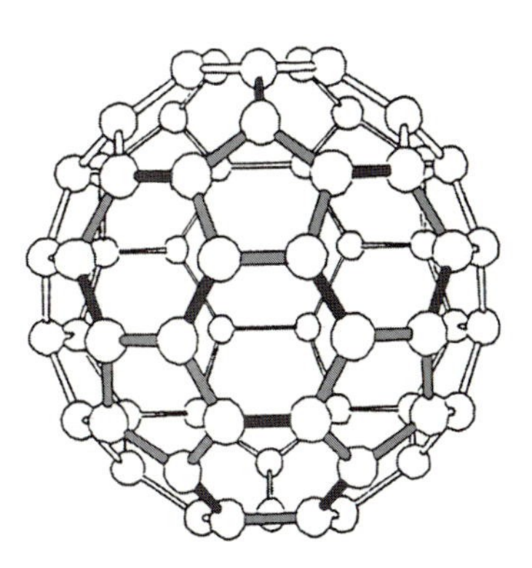

Fig. 5.27. Schematic representation of a C_{60} cluster; the 60 carbon atoms are distributed on the surface of the closed cage and are seen as pentagonal rings (12) alternating with hexagonal rings (20). Single bonds are grey, double bonds are black

$$N_{\mathrm{f}} + N_{\mathrm{v}} - N_{\mathrm{e}} = 2$$

where N_{f}, N_{v} and N_{e} are respectively the number of faces, vertices (carbon atoms) and edges (covalent orbitals) for the polyhedron. When we distinguish f_{h} hexagonal faces from f_{p} pentagonal faces, and bear in mind that two meeting faces make up one edge and that three adjacent faces have one carbon atom in common, then

$$\begin{cases} N_{\mathrm{f}} = f_{\mathrm{p}} + f_{\mathrm{h}} \\ 2N_{\mathrm{e}} = 5f_{\mathrm{p}} + 6f_{\mathrm{h}} \\ 3N_{\mathrm{v}} = 5f_{\mathrm{p}} + 6f_{\mathrm{h}} \end{cases} \tag{5.22}$$

so that,

$$6(N_{\mathrm{f}} + N_{\mathrm{v}} - N_{\mathrm{e}}) = f_{\mathrm{p}} = 12. \tag{5.23}$$

The smallest fullerene is thus C_{20}, a regular dodecahedron with twelve pentagonal faces. The fullerene size grows with the number of hexagonal rings, $(N-20)/2$, where N is the (even) number of carbon atoms. The structural stability depends on the ratio between the number of hexagonal and pentagonal faces. Indeed, C_{20}, which has not been experimentally observed, is energetically unfavoured since pairs of adjacent pentagons lead to excessive local curvature of the fullerene surface, which would require an unreasonable degree of elastic deformation. This is why, as epitomised in the "isolated pentagon" rule, fullerenes with much fewer than sixty carbon atoms are relatively unstable.

As N increases, all the clusters, except for $N = 22$, may form closed cage structures with pentagonal and hexagonal faces. Furthermore, the number of different ways the pentagons and hexagons are arranged increases rapidly. C_{60} will allow 1 812 cage isomers. However, only from $N = 60$ can at least one isomer exhibit non-adjacent pentagonal rings. This unique (for $N = 60$) fullerene structure is still today the only bare (that is made up of carbon atoms only) cage cluster that is stable at room temperature. Where N is greater than 60, apart from $N = 62, 64, 66$, and 68, we obtain isomers with non-adjacent pentagonal faces. These stable arrangements are unique for $N = 70, 72$ and 74, whereas the fullerene at $N = 76$ has two isomers with isolated pentagons, and that at $N = 78$ has five; with increasing N the number of isolated pentagon isomers increases rapidly.

Several examples in Nature illustrate the role pentagons play to round a flat surface; a simple one is the patchwork giraffe mantle. This mantle is light and spotted with dark spots that completely tile, without overlapping each other, the neck, back, sides and part of the legs. Such spots have the shape of polygons, often nearly regular, that include squares, triangles, hexagons and pentagons; it is the very presence of pentagons that allows the "tiling" to fit to both the convex curvature of the animal's body and to perfectly match adjacent spots.

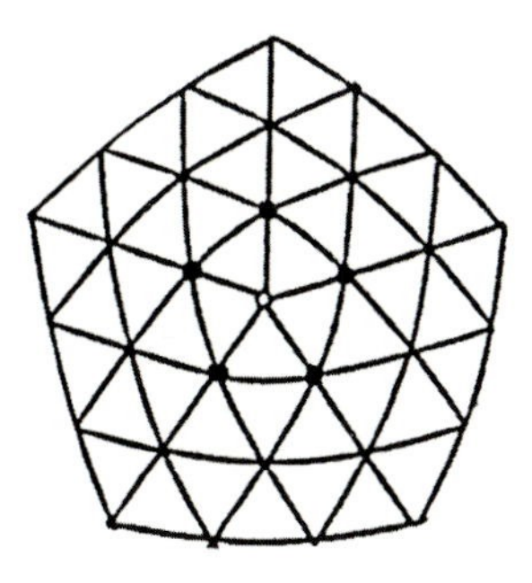

Fig. 5.28. Representation of a 60° positive cone disclination in a hexagonal plane lattice; the open point in the lattice is fivefold coordinated

Coming back again to microscopic structures, apart from the fullerene structure, other more complicated carbon structures have recently been synthesised under controlled conditions, and examined in detail. These structures are based on cages elongated in one direction: nanotubes, multilayer tubes and cone-shaped tubes. The local curvature of these geometric structures often requires heptagons, which suggests a greater number of pentagonal faces than $f_{\text{p}} = 12$.

In a curved structure, the corrugation of the flat hexagonal surface lattice of carbon atoms, typical of a graphene sheet, is due to energetic reasons. The sp^2 hybridised, open orbitals on the hexagon edges have a large dangling bond energy. Provided the temperature is sufficiently high and enough time is available to allow for annealing, the structure spontaneously re-arranges itself by folding and incorporating pentagonal defects until it closes (4π total disclinations, that is twelve pentagons); this way dangling bonds are eliminated. Each pentagon corresponds to a $+60°$ disclination, schematised in Fig. 5.28; thus a cone-shaped portion of matter with a vertex angle of 60° is removed from the lattice. If we weld the two remaining flaps of the lattice together, then the atom where the disclination is located is coordinated with five first neighbours to form the vertex of a cone.

In the very same hexagonal lattice the presence of a heptagon, namely a $-60°$ disclination (Fig. 5.29) coincides with the insertion of a cone of matter, with a 60° vertex angle. The coordination of the atom where the disclination is found is seven and the atomic plane corrugates with negative curvature.

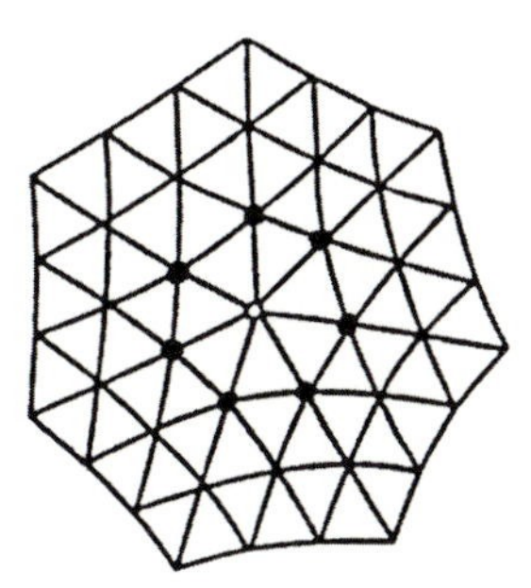

Fig. 5.29. Representation of a 60° negative cone disclination in a hexagonal plane lattice; the open point in the lattice is sevenfold coordinated

This kind of transformation from a cone to a cylinder favours the growth of nanotubes.

Even though the mechanisms for nucleation and growth of a cage structure are still hypothetical, recent experiments have allowed us to correlate the number of +60° disclinations we find in the nucleus from which the cones develop with the observed structure and topology. Seven kinds of apex geometric structure have been observed; they diverge from the initial nucleation centre and correspond to the number p of pentagons in the apex. If p equals zero, we observe two-dimensional discs, whereas when p lies between one and five we observe cones with a vertex angle of φ, such that

$$(\varphi/2) = \arcsin\left[1 - (p/6)\right] \tag{5.24}$$

φ is thus $113°, 84°, 60°, 39°$ and $19°20'$, respectively.

If p is six, we synthesise open cylindrical tubes, whereas whenever p is greater than six the structure falls into the closed cage structure for C_{60} ($p = 12$) with total disclination of 4π.

Apart from the geometric modelling, the structures of fullerenes deviate from the regular polyhedral structures due to the different lengths of the C–C bonds, which are determined by local deformation and by the electronic structure. This is why the symmetry of real fullerenes may be quite low; so far the observed point symmetries of isolated clusters have been $D_{5h}(C_{70})$, $D_2(C_{76}; C_{80})$, $C_{2v}(C_{78})$, $D_3(C_{78})$, $C_2(C_{82})$.

The particular stability exhibited by isomers that fulfil the "isolated pentagon" rule is further supported by the fact that when we have been able to determine the structures of experimentally synthesised fullerenes, these structures have always coincided with the structures of isomers with non-adjacent pentagonal faces.

Given the electronic configuration $1s^2 2s^2 2p^2$ for atomic carbon, the highest coordination number is four. When the bonds are $\sigma - \pi - \pi$, the orbitals are linear, *sp* hybridised. In the case of equivalent trivalent $\sigma - \pi$ bonds, the orbitals arrange themselves into sp^2 hybridised planar rings. Lastly, when the bonds are purely σ, tetravalent and equivalent, then the orbitals are three-dimensional, sp^3 hybridised.

This kind of structure has a surface with open orbitals; these orbitals are structure destabilising, just like graphene sheets.

No tetravalent bonds have been observed in carbon clusters; the average hybridisation is sp^n, where n is between 1 and 3. This hybridisation depends on the size of the cluster and its isomerisation. Experiments suggest that where N is greater than 32 the clusters arrange themselves in three-dimensional cage structures, whereas smaller sizes give rise to linear chains, or monocyclic rings.

Only C_{60} and C_{70+2n} are stable clusters, since their electronic shells are closed, whereas all other fullerenes are reactive due to their incomplete outer orbitals. It is the very existence of pentagonal rings in the graphitic layers in

C_{60} that force the sp^2 hybridised structure to bend up to the point where a closed cage of carbon atoms results, and the open orbitals are removed. The structure is further stabilised by overlapping adjacent π electrons.

We obtain sp^2 hybridisation when π states correspond to pure p_z orbitals, namely if the bond angle $\gamma_{\sigma-\pi}$ between σ and π bonds is 90°, like in the honeycomb graphite lattice.

When we introduce pentagonal rings we cause local curvature in the lattice and the disappearance of the pure p_z orbitals. As such, we obtain a different kind of hybridisation between the s and π states. The number of pentagons governs the degree of sp^3 hybridisation. In C_{20}, this is pure sp^3, where $\gamma_{\sigma-\pi}$ is 110°, a value that is very close to the observed value in the diamond (109°28′). As N increases, the average hybridisation tends towards sp^2 rather quickly: in C_{60} we obtain $sp^{2.3}$.

Hence, given that the number of pentagonal rings is unchanged, then the electronic structure of the carbon cages tends towards the graphite structure as the number of hexagonal rings increases, which corresponds to a reduction in the cluster curvature radius. This explains some of the similarities between C_{60} and graphite. On the one hand, the overwhelming number of hexagonal faces allows us to schematically depict C_{60} as a single layer of crystalline graphite where the twelve pentagonal defects generate the cluster curvature and, on the other, the bond coordination for each atom is *trigonal* with another three carbon atoms.

However, we observe two distinct kinds of orbital: each carbon atom is bound to three carbon first neighbours by two *long* bonds (about 0.145 nm) and one *short* bond (about 0.14 nm). The C_{60} bonds are neither pure simple bonds (about 0.154 nm), nor pure double bonds (about 0.132 nm), even though, often, they are respectively called *double* bond, or *simple* bond.

The *double* bonds are *interpentagonal* and realise a sequence where the *single* and *double* bonds alternate along the perimeter of each hexagonal face. This feature gives the hexagons threefold symmetry, thus maintaining one of the symmetry typologies of the icosahedral group (see Chap. 1), with its ten threefold axes (and not sixfold), that bisect the centres of the hexagonal faces, besides six fivefold axes that bisect the centres of the pentagonal faces, and fifteen twofold axes that bisect the centres of the edges that connect two hexagons together. The localisation of the bonds along the perimeter of the hexagonal faces is a feature peculiar to fullerenes; the C_{60} hexagons have a different chemical behaviour from both the benzene rings and the hexagonal arrangement taken up by carbon atoms in the graphite layers, even though the C_{60} surface is almost non-reactive, just like a sheet of graphite. The chemical inertness is given by the very low number of open surface orbitals, which is characteristic of a cluster with a magic mass number.

Several carbon-based structures, synthesised starting from fullerenes, have been studied both theoretically and experimentally. Among such structures a prominent one is obtained by assembling together ideal C_{60} units. The resulting solid, called a *fullerite*, is characterised by a very weak chemical bond between adjacent clusters, as realised by van der Waals forces. One of the problems we encounter when we try to interpret the crystalline structure of the fullerite is how the molecular fivefold symmetry can be compatible with the realisation of a crystalline lattice, given that perfect *fivefold* symmetry can only be realised in the quasicrystals (see Chap. 6), and the fullerite is not quasicrystalline.

Just as for all molecules with high point symmetry, C_{60} tends to crystallise in a structure where the molecule centres of gravity are arranged in a long-range periodic fashion. At room temperature, C_{60} crystals and thin films are characterised, just like inert gases, by fcc packing of hard spheres with a lattice constant of 1.42 nm (Fig. 5.30), as obtained directly by scanning tunneling microscopy. The hcp structure has also been observed, though it is less stable than the fcc structure. We observe that the various C_{60} clusters in the fullerite are disordered from the orientational point of view, since the molecules rotate quickly about preferential axes, and they are all oriented differently to each other. The frustration in the packing of icosahedral molecular structural units (see Sect. 4.7) is removed by rapid molecular re-orientation, which occurs at a frequency of 1.11×10^{11} Hz. The corresponding re-orientation times (about 9 ps), typical of free rotator in the gas phase, are extremely short for a cluster with a considerable moment of inertia. The calculated characteristic time of rotation of a free C_{60} cluster at room temperature is 3 ps. It is remarkable that if C_{60} is dissolved in solution, the re-orientation times become 50% longer than those observed in the fullerite.

Still at room temperature, the cluster surroundings in the disordered fullerite, analysed using nuclear magnetic resonance, suggest that there is a re-orientation mechanism with very small activation energy, around 690 K. This means that even those orientations corresponding to the maximum of

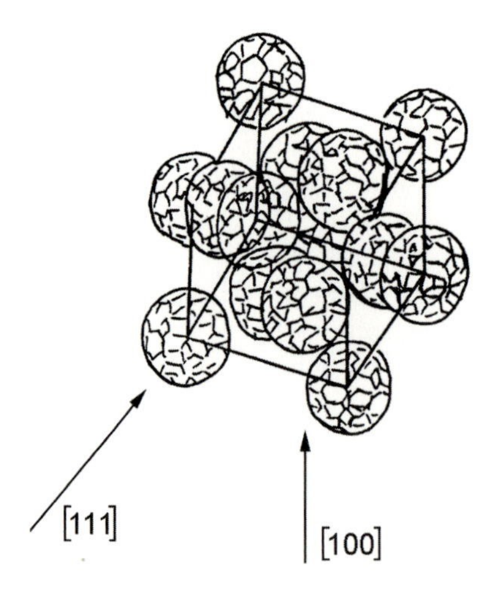

Fig. 5.30. The structure of the fullerite,at room temperature, is fcc; looking along [001] the fourfold symmetry is highlighted; looking along [111] the sixfold symmetry is highlighted

the activation barrier have a high probability, as high as 10% of the probability for lowest energy configurations.

At progressively lower temperatures, the fullerite exhibits various phase transitions to cubic structures, with lower symmetry than the icosahedral structure. In general we can interpret them by taking into account the progressively increasing weight of the properties of *each* cluster. First neighbour clusters may be reciprocally *locked* to specific relative positions, which gives rise to variations in the cubic lattice constant as the temperature changes.

At 260 K the activation barrier for C_{60} cluster free rotation is very high, 3 000 K. The molecules can only ratchet between various preferred orientations. The centres of the C_{60} clusters define a fcc lattice; the unit cell contains four clusters arranged in the space so as to form a tetrahedron; the orientation of each cluster is maintained within each tetrahedron. The tetrahedral structural units constitute a simple cubic crystal lattice; the relation between the fcc and simple cubic structures may be illustrated by construction by considering that the fcc structure is the result of the interpenetration of four cubic lattices. The first lattice contains the clusters located at the vertices of the cubes in the fcc lattice, whereas the other three are obtained from the first by translating the clusters along two of the x, y, z axes by a quantity $(a/2)$, where a is the lattice constant.

Each of the oriented clusters, which go to make up any tetrahedron, comes from a different cubic lattice, so the lattice formed by the tetrahedral structural units is simple cubic. Bearing in mind that the lattice sites occupied by clusters with different preferential orientation are not equivalent, then in order to represent the structure the simple cubic cell will be twice as large as the cell in the fcc structure. The interpretation of the data from nuclear magnetic resonance experiments highlights that there is a specific fast molecular re-orientation mechanism in the temperature interval between 260 K

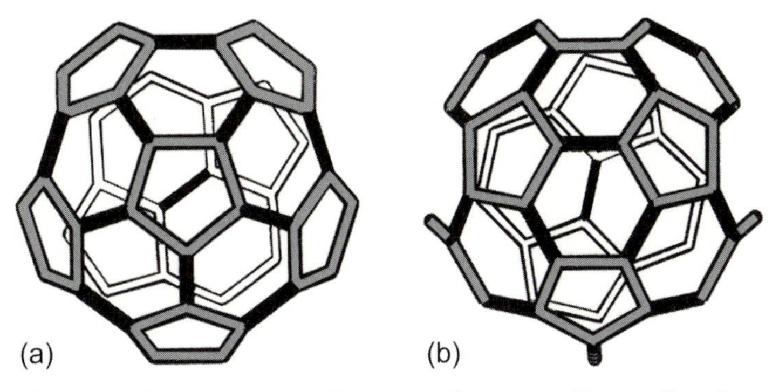

Fig. 5.31. Inter-atomic configurations between adjacent C_{60} clusters, as observed along [110]; (**a**) a pentagon with single bonds facing a hexagon with double bonds in the adjacent cluster; (**b**) a hexagon with mixed *single+double* bond facing a double bond of the adjacent cluster. Both configurations ((**a**) is energetically favoured) are observed in fullerite at low temperature, below 260 K

and 100 K. We assume that two distinct configurations with similar energies alternate. Figure 5.31 shows that when we observe the solid along the [110] bond direction in the first configuration we notice that a pentagonal ring of *single* bonds in a C_{60} cluster faces a *double* bond in a first neighbour cluster. The bond site 6 : 6 is exactly above the centre of a pentagonal face, and the torsion angle is $179°36'$. The consequence of this is that the pentagonal rings, with their low electronic density, are facing *double* bonds with high electronic density. In the second configuration, which is energetically less favoured, one hexagon faces the *double* bond of the adjacent cluster.

Lastly, at 5 K, the structure of the ordered fullerite is characterised by the coupling between facing pentagonal rings in adjacent C_{60} clusters.

The link between fullerene-based structures and diamond has been recently explored in detail. The problem consists in obtaining fully sp^3-coordinated lattices by assembling together small fullerenes. The possible stable structures of such crystals, called *clathrates*, are found after a topology-based analysis of periodic polyhedral networks. We use Eulero's theorem again (see Sect. 1.1) with the restrictions that the network can only be four-fold coordinated and only five- and six-membered rings are permitted. After total energy minimisation we find the ground state cohesive energy and elastic properties of each clathrate.

Using the four smallest fullerenes, C_{20}, C_{24}, C_{26} and C_{28} as building blocks, we construct the three most elementary three-dimensional periodic clathrate lattices. The simplest one is obtained from the coalescence of two C_{28} and four C_{20} per unit cell; the structure is fcc, with $[(2\times28)+(4\times20)]/4 = 34$ atoms per unit cell. We can view it as two interpenetrating diamond lattices with a relative displacement of $a/2$, a being the cube edge. The second lattice results from the combination of two C_{24}, two C_{26} and three C_{20} per unit cell; it is hexagonal, with 40 atoms per unit cell. We obtain the third lattice from the coalescence of two C_{20} and six C_{24} per unit cell; its structure is sc, with 46 atoms per unit cell.

Starting from clathrate structures infinite series of periodic lattices are obtained, provided we use carbon clusters with eclipsed configuration like the fullerenes. This is schematised in Fig. 5.32. We replace each atom in the

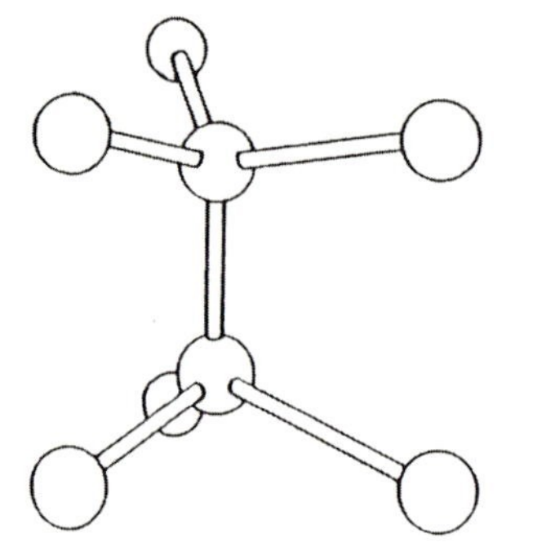

Fig. 5.32. Eclipsed tetrahedral bonding configuration

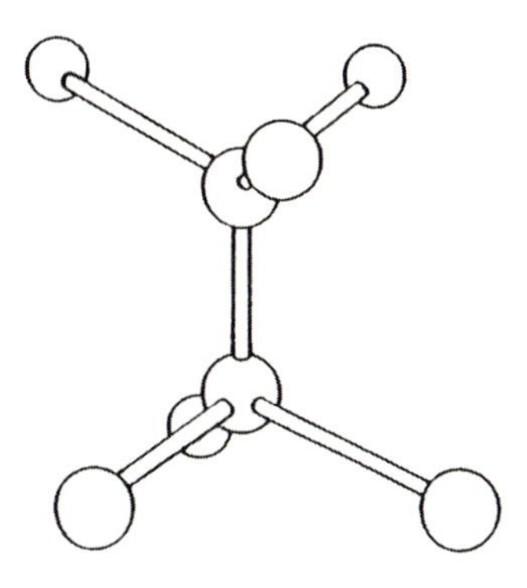

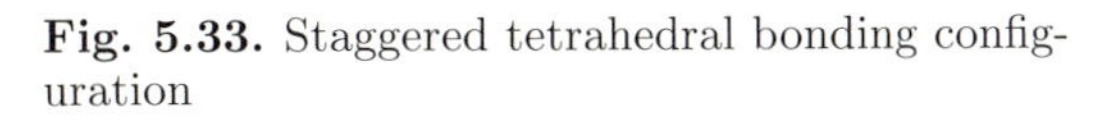

Fig. 5.33. Staggered tetrahedral bonding configuration

original structure with a tetrahedral diamond cluster of given size. A cluster that has n atoms along its edge contains $[n(n+1)(2n+1)/6]$ atoms per unit cell and from the clathrate with N atoms per unit cell ($N = 34, 40, 46, ...$) we can generate lattices with $N[n(n+1)(2n+1)/6]$ atoms per unit cell. Once equal-sized (i.e. with the same n) diamond clusters are bound together through triangular face sharing, every new lattice is a fourfold-coordinated structure in the eclipsed configuration.

In the smallest crystals ($n = 1$) small fullerenes coalesce, so that each atom belongs to four different fullerenic cages. In larger crystals ($n > 1$) fullerene cages are isolated and form a superlattice with increasing inter-cage spacing. The space between fullerenes is filled with a diamond structure of progressively larger size, with staggered configuration, as shown in Fig. 5.33. Such a family of carbon crystals has been named *hollow* diamonds. The fraction of diamond structure associated with the tetrahedral units that surround the hollows diverges with n. The limit structure is a diamond crystal. In this crystal the planes where tetrahedral clusters are joined (in an eclipsed way) form a regular network of stacking faults with the same periodicity and geometry as those of the starting clathrates. Structures of this kind, with n in the order of 2×10^4, have been recently synthesised; scanning electron microscopy pictures show that they are large trimmed tetrahedral clusters with regularly shaped large hollows at the cluster cusps.

5.6 Cluster-Assembled, Nanostructured Materials

The possibility to produce atomic and molecular clusters with well defined size and composition has recently opened the way to the synthesis of a new class of materials whose building blocks are clusters instead of atoms. By assembling together clusters with sizes in the nanometre range, typically containing hundreds to tens of thousands of atoms, bulk materials are obtained.

The properties of such cluster-assembled, nanophase materials, with grain size usually in the range between 5 and 25 nm, often differ from the properties of conventional materials. There are several reasons for the differences. A crystalline solid is characterised by a single length scale, the lattice parameter, by the specific nature of bonding force, whether ionic, covalent, metallic, or van der Waals, and by energy bands resulting from the overlap of atomic orbitals. A cluster-assembled solid presents *two* characteristic length scales, the intra- and the inter-cluster distances. Atom bonding in a cluster may differ from bonding between clusters (compare, e.g. the fullerene cluster to the cluster-assembled fullerite), and the energy bands result from the overlap of cluster orbitals that can significantly differ from each other, depending on cluster size, besides being different from atomic orbitals.

Cluster-assembled materials can be prepared using a variety of techniques; we limit our discussion to the structure of materials synthesised by the conceptually simple technique of modified gas condensation. This two-step procedure allows us to prepare samples with a large number of interfaces.

The precursor material is evaporated, usually by Joule effect, in an evacuated chamber (base pressure below 10^{-6} Pa) back filled with an inert gas, typically helium, at pressures around 1 kPa. Inert gases are used to prepare metal samples, while reactive gases, or gas mixtures are employed when synthesising ceramic materials. Evaporated atoms lose their kinetic energy through interatomic collisions with gas atoms in the supersaturated region near the vapour source and condense into clusters; these are continuously brought via convection to a cold finger kept at 77 K, where they land. Preparation of nanocrystalline systems under high vacuum conditions is crucial to obtain uncontaminated interfaces. Nanocrystals with contaminated surfaces usually do not have the properties discussed in the following. Particular atoms can be deliberately adsorbed at the crystallite surfaces before the compaction stage when we want to dope the grain boundaries in a controlled way.

As-deposited clusters form highly porous structures, they are easily scraped from the finger surface and collected in a piston-and-anvil device where they are compacted at pressures up to about 5 GPa. The resulting nanophase pellets are typically around 1 cm in diameter and 0.1 to 0.6 mm in thickness. The type and pressure of the gas and the evaporation rate mainly affect crystallite size distribution. The main impurities are light gases such as hydrogen, nitrogen and oxygen, up to a total concentration around 1.5 at. %.

The most impressive feature of a cluster-assembled material is the high number of incoherent interfaces it contains. In a polycrystalline sample with 5 nm grain size, typically a fraction of 30% of all atoms are found within one lattice spacing, or less from a grain boundary of average thickness between 0.5 and 1 nm. If the average grain size is 10 nm, the above fraction falls as low as about 15% and for 100 nm grain size it falls to about 1%. It is expected that, as the interaction energy between nearest neighbour atoms depends on the number of neighbours and their interatomic spacing, some physical properties of such nanostructured solids will strongly differ from the corresponding usual crystals. An example of this is the enhanced silver diffu-

Fig. 5.34. Schematic two-dimensional picture of a nanocrystalline solid. Full circles represent atoms in crystallites and empty circles represent atoms in interface regions

sion in cluster-assembled copper, which is twenty orders of magnitude larger than in single crystal copper over the same temperature interval, between about 300 and 400 K.

Wide- and small-angle X-ray and neutron scattering, as well as EXAFS, besides conventional and high-resolution transmission electron microscopy, have been used to investigate the structure of nanocrystalline materials. A simplified picture of the microstructure of a cluster-assembled solid (Fig. 5.34) shows that essentially it consists of small crystallites, separated by narrow grain boundaries, i.e. interface planes shared by two neighbouring crystallites. We are concerned with a twofold problem, the *crystal* structure and the *interface* structure.

The size distribution of isolated crystallites prepared by modified gas condensation is well described by a lognormal function, although abnormal grain growth in nanocrystalline metals results in a bimodal size distribution that deviates from the lognormal distribution. Presently in high purity elemental metals achievable grain sizes are in the range between 5 and 10 nm. The reduced crystal size often has the effect that the crystal structure of ultra-fine grained particles differs from the thermodynamic equilibrium structure in the corresponding coarse-grained polycrystalline, or single crystal material. For example nanoparticles of the bcc Cr, Mo and W metals crystallise in the $A15$ structure, which is non-equilibrium for coarse-grained samples. The structure of nanocrystalline Y_2O_3 is monoclinic and coincides with that of the high pressure phase. In freshly prepared metals and ceramics small-angle X-ray and neutron scattering give average density of the boundary regions around 60% and 80% of the crystal density, respectively. The estimated boundary thickness is between 0.5 and 1 nm. The density loss in the interface regions can be up to five times greater than the density reduction in an amorphous solid.

X-ray diffraction combined with EXAFS (see Sect. 4.4) allowed for the measurement of the nearest neighbour coordination number in the crystalline and in the interface regions of some elemental metals, including palladium, copper and tungsten, with crystal sizes between 5 and 10 nm. In the boundary regions coordination numbers are lowered by as much as 30%, with a reduction five times larger than in amorphous samples.

It is difficult to perform unambiguous quantitative measurements to define the atomic structure *in* the grain boundaries of a nanostructured solid. Using X-ray powder diffraction we obtain the average boundary atomic structure, provided we make a difficult separation of the contributions from crystallites. Imaging by high resolution transmission electron microscopy highlights those boundaries with special orientation relationships between the crystallites and could be not representative of all grain boundaries. The large static disorder emerging from EXAFS data can arise both from boundaries and from crystallites; again the separation of both contributions is questionable.

The initial picture of gas-like short range order in the boundary regions, thus a peculiar short range order, with no other analogues in condensed matter has been partly mitigated. If we refer to elemental metals, in freshly prepared samples about 10% of atoms are located in non-lattice sites, with a wide distribution of interatomic spacings across the grain boundaries, that is with no detectable short range order across the grain boundary planes. Associated to such high angle boundaries are large values of specific grain boundary area; this structural parameter is proportional to the grain boundary free energy. Thus the excess free energy ΔG_{nc} of a cluster-assembled nanocrystalline material scales with the inverse of the grain size; when this is very small, ΔG_{nc} becomes comparable to, or even higher than that of the corresponding amorphous phase. The essential difference is that an amorphous material is *metastable* with respect to the equilibrium phase, that is to the crystal, while a nanocrystal is usually *unstable* with respect to the approach to equilibrium via grain growth. Analysis of samples annealed at room temperature for some months indicates that during this time nearly all atoms relaxed to crystal lattice sites; this means that regions of perfect crystal lattice extend to the interface plane where the topological defect has become two dimensional. Pure nanocrystalline metals undergo spontaneous grain growth at temperatures that are a small fraction of the melting temperature. Stabilisation of the microstructure of cluster-assembled materials is presently a major problem with considerable technological implications. There is now evidence that the degree and type of grain boundary disorder are influenced by the nature of the material (metal, or ceramic, elemental, or compound, including the equilibrium structure of coarse-grained crystals) and by the sample history. In fact, in selected metals that have been studied in detail, it appears that nanocrystalline samples of the same material have similar properties, even if the structure of the boundaries is rather dissimilar.

It is worth mentioning that, apart from cluster-assembled crystalline solids, *nanoglasses* have been recently synthesised. These are obtained when amorphous spheroidal particles with nanometric size are produced and compacted together, with the same modified gas condensation technique used to synthesise cluster-assembled nanocrystalline solids. A nanoglass is a *bulk* non-crystalline solid; we can view it as an assembly of amorphous regions of nanometric size joined together by a network of narrow interfaces. The latter are formed when the initially free surfaces of the glassy spheres are welded together during the compaction stage. Similarly to what occurs when differently oriented crystallites are brought into contact, the boundary regions between adjacent amorphous spheres show reduced density and lowered nearest neighbour coordination.

6. Quasicrystals

6.1 Periodic and Aperiodic Crystals

We commonly refer to two atomic arrangements when describing the structure of an *ideal* solid, namely a homogeneous, chemically pure single-phase solid, the crystal, with its periodic structure (see Chap. 1) and the random arrangement typical of the amorphous state (see Chap. 4).

Given the periodicity of the microscopic structure, the crystal is considered a finite part of a structure with three-dimensional lattice periodicity. The surfaces, the atomic packing defects and the impurities can cause local deviation from the ideal atomic sites, as given by space group symmetry; however, long range order is not destroyed.

When a crystal undergoes X-ray, electron or neutron diffraction, we observe constructive interference between rays elastically scattered from atoms arranged on successive planes that belong to a given family $\{hkl\}$ (see Chap. 1), provided that the Bragg equation (see (4.25)) is fulfilled. Equation (4.25) may be rewritten as

$$2(d_{hkl}/n)\sin\theta = \lambda \qquad (6.1)$$

where n is the diffraction order, θ the scattering angle and λ the wavelength of the incident radiation; from this the nth order diffraction from $\{hkl\}$ planes is equivalent to the first order diffraction from the $(nh\ nk\ nl)$ plane.

The atomic periodicity makes the diffraction *coherent* along some privileged directions, as characterised by the geometric structure of the crystal. The diffraction pattern in these directions exhibits high intensity spots, or perfectly localised peaks, evenly spaced out.

The pattern's global symmetry (see Chap. 1) reveals the geometric shape of the elementary structural unit, and thus of the unit cell. The angular separation of the Bragg reflections is inversely proportional to the size of the unit cell. The intensity of the reflections allows us to deduce the crystal's spatial structure, by way of numeric simulation, namely the species and position of the least number of atoms required to build, by translation, the entire structure.

Since the lattice points are equally spaced out along each lattice line, then each lattice point is an inversion centre of the lattice, taken as a whole. The

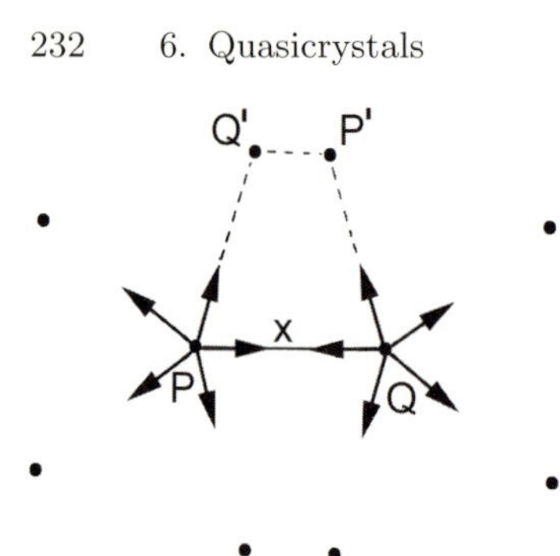

Fig. 6.1. No kind of two-dimensional or three-dimensional lattice will allow lattice points with fivefold rotational symmetry

same holds for each half-way point between any two lattice points. The lattices with the lowest symmetry do not exhibit any other symmetry operation. The lattices are then classified into families based on the symmetries that define their lattice points and on how the symmetry group operates on the lattice. The lattice symmetry is governed by geometry laws; an example of these laws is the Barlow theorem on crystallographic restrictions (see Sect. 1.4). Based on this theorem, fivefold rotations and higher than sixfold order rotations are incompatible with complete occupation of the three-dimensional space, without overlapping occurring; also, rotations of these same orders are incompatible with any periodic tiling of the plane. In this case, if we consider the pentagon, the 180° vertex angles are a fraction $(3.333...)^{-1}$ of 2π; so, if we juxtapose three regular pentagons so that they have one vertex in common, namely a lattice site, then a 36° angle of subtended space is left uncovered.

Alternatively, if we assume we have obtained a two-dimensional Bravais lattice with pentagonal cells, as schematised in Fig. 6.1, then the symmetry would require all the points about which fivefold rotations occur to be defined as a lattice, because of the translational symmetry. We shall call one of these points P. Since the lattice is discrete, there must be a minimum distance, x, between the points which, in turn, are translationally equivalent. Let Q be a lattice point at a minimum distance from P. Now, let us assume that P is surrounded by five of these points, Q', and that Q is surrounded by five points P'; however, this conflicts with the requirement that the distance between P and Q should be minimum.

We obtain the same result with three-dimensional lattices since the rotation about an axis (fivefold) affects rotations of the same order in the lattice planes perpendicular to that axis. As such, if a two- or three-dimensional crystal has periodic structure it cannot exhibit fivefold rotational symmetry. Nor can the outer shape of the crystal be either an icosahedron or a regular pentagonal dodecahedron, since the shape of a crystal cannot exhibit any kind of symmetry that is forbidden by its internal structure. In other words, a system that exhibits fivefold symmetry has a *non-periodic* structure.

Coming back to our two-dimensional case, and using the same reasoning as above, we cannot consider heptagonal tiles with a vertex angle of $128°57'$ a $(2.8)^{-1}$ fraction of 2π.

In general, the vertex angles for regular polygons with n sides is given by $\pi(n-2)/n$; the plane can be *periodically* covered if $p = 2n/(n-2)$ is an integer number (see Sect. 1.4). Apart from the heptagon ($n = 7$), where $p = 2.8$, we also observe that, for polygons with progressively increasing order, $p = 2.666$ ($n = 8$), 2.571 ($n = 9$), 2.5 ($n = 10$), 2.222 ($n = 20$), 2.143 ($n = 30$),..., and that p is always greater than 2.

Very few shapes with repeated juxtapositions can fill the space, even for three-dimensional crystals; these polyhedra do not include morphologies with fivefold rotational symmetry.

When we also consider translation, reflection and inversion operations, and when we define the 230 space symmetry groups, those space groups with fivefold, sevenfold and higher order rotations are excluded.

Since the set of possible symmetry operations is given by the *dimension* of the space the lattice is embedded in, the crystallographic restriction depends on that dimension. In particular, in four- or higher dimensional spaces, fivefold and twelvefold rotations are allowed, whereas sevenfold rotations are observed for the first time in six-dimensional space. On the other hand, atomic decorations characterised by *any* symmetry may be exhibited, and are observed, in many cases in the unit cells of a three-dimensional crystal, or on its faces. It is the symmetry of the unit cell that ensures the space be filled, regardless of the shape or symmetry of the decorations, since these decorations are in reciprocal relation through the symmetry of the underlying lattice.

The appearance of structural elements with "prohibited" symmetry is widespread both in chemistry and in biology. Figure 6.2 shows one among the simplest examples, namely the structure of the artificially synthesised dodecahedrane $(CH)_{20}$, where the carbon atoms are arranged on the face centres of a icosahedron, namely, in positions equivalent to the vertices of a regular dodecahedron (see Chap. 1). boron chemistry supplies us with a number of examples of molecules with icosahedral structure: $[B_{12}H_{12}]^2$, whose elemental structural unit is an icosahedron, the icosahedron B_{12}, as shown in Fig. 6.3 (part (a)), which is found in the allotropic forms of crystalline

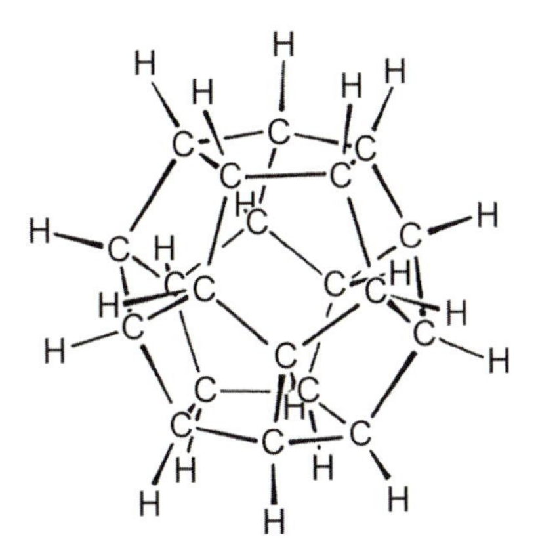

Fig. 6.2. The dodecahedrane $(CH)_{20}$ constitutes an ideal dodecahedron

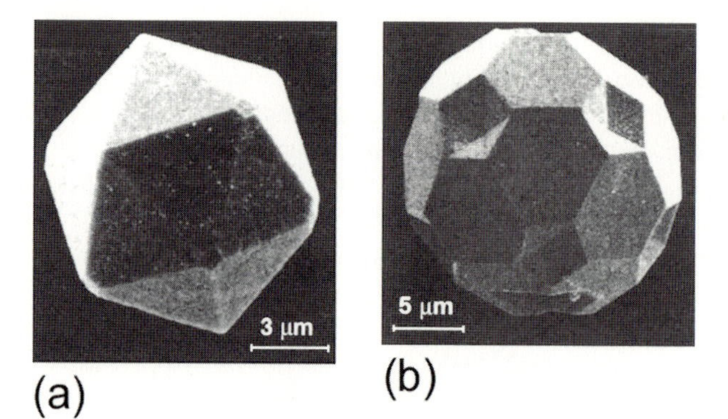

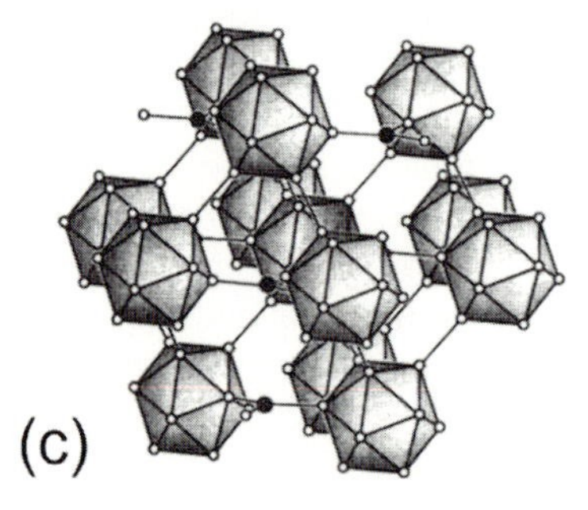

Fig. 6.3. (**a**) The icosahedral symmetry of B_6O is revealed by its morphology; (**b**) evidence of multiple twinning in B_6O; (**c**) structure of B_6O: full dots indicate oxygen atoms and open dots represent boron atoms; the centres of the B_{12} icosahedra occupy the vertices of a cuboctahedron

Boron, $B_{12}C_3$ (rhombohedric) and B_6O. From Fig. 6.3 multiply twinned particles (see Chap. 5) in this system have the same morphology (part (b)) as the C_{60} clusters (see Chap. 5) and are arranged in a compact cubic packing of B_{12} icosahedra (part (c)). The oxygen atoms are arranged on densely packed layers; each oxygen atom is coordinated with three boron atoms belonging to three B_{12} icosahedra. The B_6O particles are among the largest particles spontaneously arranged with icosahedral symmetry; as a comparison, a particle with a diameter of 20 μm contains some 10^{14} atoms, whereas a virus has around 10^7.

The fact that B_6O is an extremely hard material makes these particles suitable for such applications where high wear resistance is required. All the external faces have the same crystallographic orientation ($[1\ 1\ 1]_r$, see Chap. 5) and the same surface energy, minimum; this explains why they are all subject to the same, extremely low, wear rate.

It is remarkable that icosahedral SRO is observed even in amorphous boron, as revealed by a clear shoulder in the second peak of the scattering intensity curve $I(k)$, measured using high intensity synchrotron radiation. The distance between first neighbour boron atoms, as obtained from the well

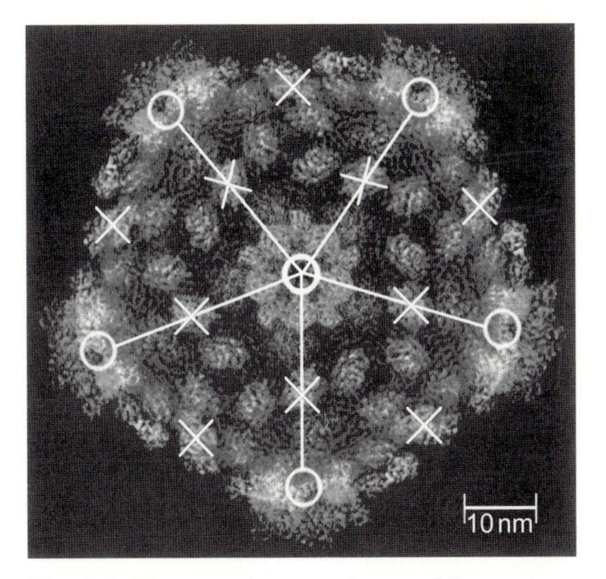

Fig. 6.4. Electron microscopy image of the reovirus core. Icosahedral symmetry is evident. The core is about 70 nm in diameter and contains five proteins out of the eight that make up a complete virion. The λ_1 protein defines the core symmetry and size; it is stabilised by the σ_2 protein (white crosses in the picture). A fivefold rotation symmetry axis lies normal to the central λ_2 structure. Other five equivalent λ_2 proteins are marked as open dots in the picture. (adapted from [6.1])

defined first peak of $I(k)$, on the one hand confirms a high degree of SRO and, on the other, coincides with the average distance between first neighbours observed in fragments of crystalline boron with icosahedral packing.

Boron ions are usually implanted into silicon to dope it p; after bombardment at doses higher than $10^{16}\,\mathrm{cm}^{-2}$, infrared spectra clearly suggest that boron arranges itself into B_{12} icosahedra; this corresponds to an absorption peak whose intensity linearly depends on the implanted dose.

The recurring icosahedral spatial arrangement is not an exclusive characteristic feature of boron; indeed, for example, icosahedral As_{12} units are encountered in Co_4As_{12}, a mineral (skutterudite) with a structure that is globally cubic. Even some primitive biological structures, such as the Radiolaria, a marine microfossil, have icosahedral skeletons, and many viruses exhibit icosahedral morphology (Fig. 6.4). The viruses are on the edge between living matter and non-living matter. Since perfect crystallisation is associated with non-living matter, then pentagonal symmetry, which we frequently observe in primitive living organisms, seems to constitute a defence strategy against crystallisation, as illustrated in Fig. 6.5.

At the end of 1984 it was reported that fast quenched thin ribbons of aluminium–manganese with composition $Al_{84}Mn_{16}$, obtained at a quenching rate of $-(\mathrm{d}T/\mathrm{d}t) = 10^6\,\mathrm{Ks}^{-1}$ and subject to electron diffraction, give rise to

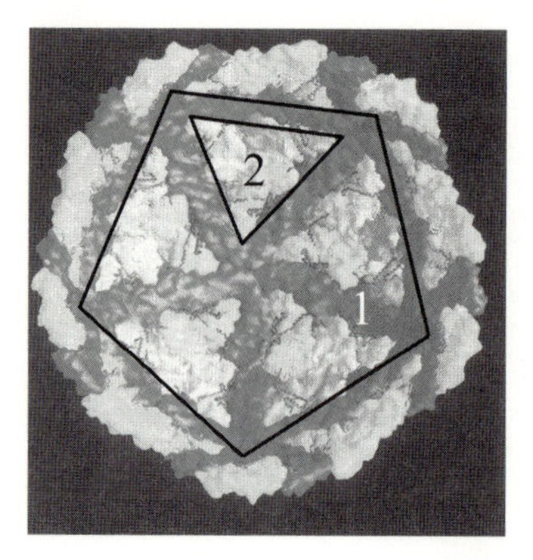

Fig. 6.5. Schematic view of the reovirus core, as seen from the interior. Five copies of λ_1 (1) radiate from an icosahedral fivefold axis and are arranged alternately to the members of a second set (2) thus forming a decamer. (adapted from [6.1])

a pattern with well defined and regularly spaced out bright spots, analogous to a ordinary crystal, though with *icosahedral* three-dimensional symmetry (Fig. 6.6). One characteristic property is that the density of the Bragg spots observed in each plane is considerably greater than the density we normally find for a periodic crystal. The reported diffraction pattern is surprising since one of the characteristic properties of icosahedral symmetry lies in having six fivefold rotation axes. Thus we observe the typical features of a diffraction pattern from a crystalline (thus periodic) structure in apparent contrast with a concomitant symmetry that cannot exist in a periodic structure. Systems with this kind of characteristic properties are called *quasicrystals*.

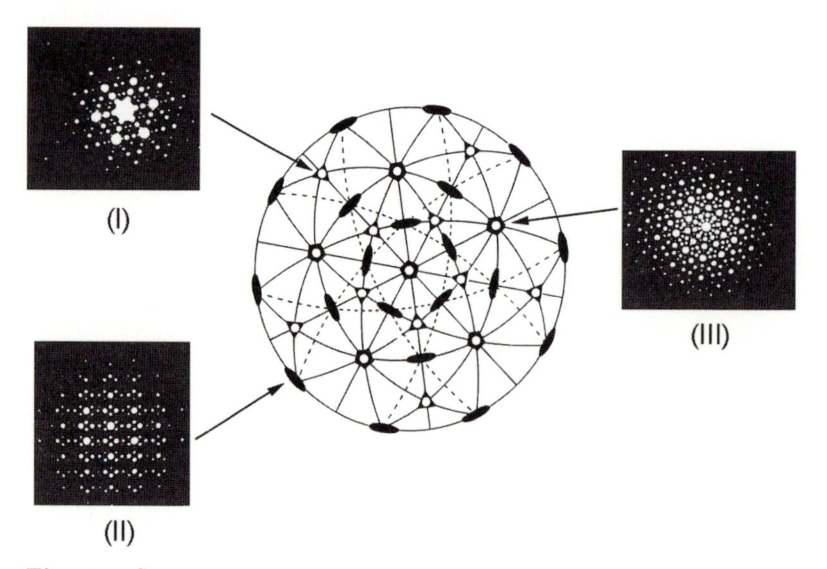

Fig. 6.6. Stereographic projection of the icosahedral point group and diffraction patterns (transmission electron microscopy) of fast quenched $Al_{84}Mn_{16}$, perpendicular to a threefold axis (I), a twofold axis (II), and a fivefold axis (III)

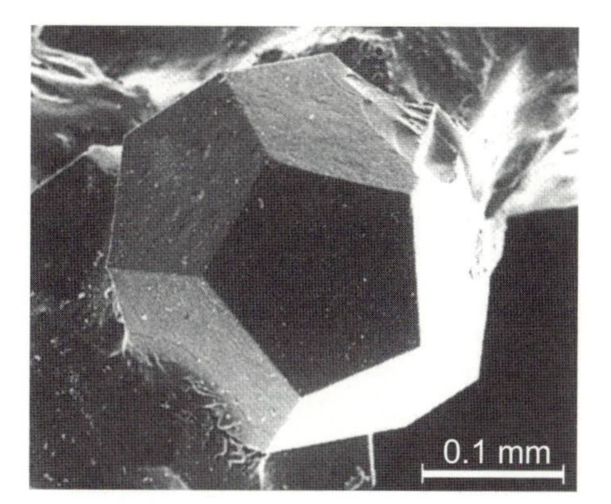

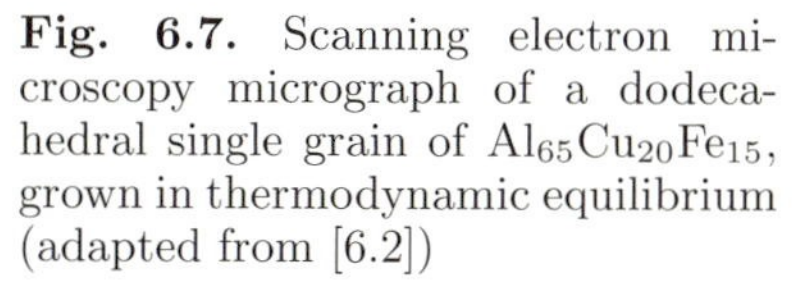

Fig. 6.7. Scanning electron microscopy micrograph of a dodecahedral single grain of $Al_{65}Cu_{20}Fe_{15}$, grown in thermodynamic equilibrium (adapted from [6.2])

The scanning electron micrographs of $Al_{84}Mn_{16}$ samples are characterised by having a morphology with dendritic nodules a few micrometers in cross-section. Each nodule is homogeneous and exhibits, in agreement with the characteristic features of the diffraction pattern, long range *orientational* order; on the other hand, the latter is associated with icosahedral symmetry. Figure 6.7 highlights the outer shape of the nodule, a regular pentagonal dodecahedron (compare the part (a) of Fig. 1.8).

Our interpretation of the above inconsistency has led us to generalise and complete the definition of a crystal, whereas the successive discovery of hundreds of other compounds with quasicrystalline phases and rotational symmetry with fivefold, eightfold, tenfold and twelvefold order allowed us to synthesise under thermodynamic equilibrium single-phase quasicrystalline samples. These kinds of quasicrystals are most suited to highlighting the structural, electrical, magnetic, mechanical and thermal properties of perfect quasicrystals. Thus, we can compare them with the samples obtained with non-equilibrium techniques and with the so-called *crystalline approximants*, namely those truly crystalline phases with composition very close to the quasicrystals.

We can understand the "pathology" of the structure of quasicrystals by following two research paths at the same time. The first implies considering the ideal systems, i.e. without any defects, that are aperiodic, though exhibiting long range order. Known as *incommensurate* phases, these systems have been observed, examined and synthesised since the 1960s. The second path refers to the search for a new and wider definition for a crystal. This requires us, among other things, to consider extending our notions on periodic functions (see Appendix).

6.2 The Enlarged Notion of Crystal

If the realisation of a crystalline state, with the associated long range order, is axiomatically tied to lattice periodicity, then the cases for $Al_{84}Mn_{16}$ and hundreds of other alloys become insoluble. On the other hand, over the last few years experimental structural studies on, and models for, these systems, as well as the incommensurate phases, have led to the emergence of a new and more global definition of the crystal. A solid that satisfies the diffraction conditions, in that it gives rise to an essentially *discrete* diffraction pattern, is to be deemed a crystal. This definition does not take into consideration either the symmetry of the system nor any *a priori* structural assumptions.

From a formal point of view, the general definition of a crystal requires us to examine two aspects in more depth: we have to topologically characterise the set of points that fulfil the diffraction conditions, and we have to develop model structures that fill the space, whose local properties are necessary and sufficient to determine the diffraction conditions. The very study of purposely built model systems aids us in analysing many characteristic properties of real quasiperiodic structures.

Let us take an ordinary crystal. The atomic positions $\boldsymbol{r}$ may be expressed as $(\boldsymbol{T}+\boldsymbol{x}_j)$, where $\boldsymbol{T}$ is a lattice translation and $\boldsymbol{x}_j$ the position of the j-th atom in the unit cell. If Q is a modulation wavevector, then the atomic position in a modulated phase is

$$\boldsymbol{r} = \boldsymbol{T} + \boldsymbol{x}_j + \boldsymbol{f}_j\left[\boldsymbol{Q}\cdot\ (\boldsymbol{T}+\boldsymbol{x}_j) + \phi_j\right] \tag{6.2}$$

where the function $\boldsymbol{f}_j(\boldsymbol{x}) = \boldsymbol{f}_j(\boldsymbol{x}+1)$ is periodic.

For simplicity, we shall take into account a one-dimensional chain of atoms, with interatomic distance x_1. The system is stabilised at low temperature if the charge density $\varrho(\boldsymbol{x})$ is spatially modulated. The deviation $\eta\varrho(\boldsymbol{x})$ in the charge density from its spatially homogeneous average value is well approximated as

$$\eta\varrho(\boldsymbol{x}) = \varrho\cos(2\pi x/\lambda)$$

where λ is the periodicity. In turn, this modulation induces a modulation in the atomic positions. If (x_1/λ) is a rational number, which can thus be expressed as the ratio between two prime integers (A/B), the structure is commensurate and a new unit cell with B atoms of the chain forms in the system. If, instead, (x_1/λ) is irrational, then the modulation is incommensurate with respect to the original lattice. In general, the modulated phase, where the atomic positions are given by (6.2), once it is subject to a scattering experiment, produces diffraction spots or peaks in positions

$$\boldsymbol{k} = m_1\boldsymbol{x}_1^* + m_2\boldsymbol{x}_2^* + m_3\boldsymbol{x}_3^* + m\boldsymbol{Q}. \tag{6.3}$$

In (6.3), m_1, m_2, m_3 are the indices for a reciprocal lattice vector, $\boldsymbol{k}'$.

When the components of modulation vector $\boldsymbol{Q}$ are *irrational* compared to the basis vectors of the reciprocal lattice $(\boldsymbol{x}_1^*, \boldsymbol{x}_2^*, \boldsymbol{x}_3^*)$, namely when $\boldsymbol{Q}$ is rationally independent, the observed diffraction spots do not refer to a lattice; rather, they refer to the Fourier module for a quasiperiodic function (see Appendix).

We can experimentally recognise an incommensurate crystal by the appearance in the diffraction pattern of the so-called satellite peaks. These peaks correspond to irrational multiples of reciprocal lattice vectors of the starting, non modulated, crystal.

Again with reference to the one-dimensional example, the spots in the scattered intensity correspond to a lattice whose vectors are

$$\boldsymbol{k} = (\pm A\boldsymbol{y}_1 \pm B\boldsymbol{y}_2)\ \hat{\boldsymbol{x}}$$

where A and B are integers, $\hat{\boldsymbol{x}}$ is the unit vector in direction x, and $\boldsymbol{y}_1 = (2\pi/x_1)\ \hat{\boldsymbol{x}}$, $\boldsymbol{y}_2 = (2\pi/\lambda)\ \hat{\boldsymbol{x}}$, where x_1/λ is irrational.

It is noteworthy that already in the simple example of a one-dimensional incommensurate structure to introduce the concept of reciprocal lattice implies there are two primitive translation vectors, $\boldsymbol{y}_1$ and $\boldsymbol{y}_2$. The ratio (y_1/y_2) is such that there are infinite pairs of integers A, B, which, for a fixed number ε, fulfil the condition $|A\ \boldsymbol{y}_1 - B\ \boldsymbol{y}_2| < \varepsilon$, which is true for any ε (see Appendix). Thus, the reciprocal lattice vectors for an incommensurate system make up a dense set in the reciprocal space. In general, the ensemble of wavevectors given by (6.3) is dense; as such it is reasonable for the intensity of the experimentally measured diffraction spots to tend towards zero quite rapidly as the values of the indices increase. If they do not tend towards zero it would be impossible to distinguish one spot from another because they would largely overlap. This means that we only observe a *finite* number of spots (Fig. 6.6), which makes it impossible to distinguish a quasiperiodic function from a semiperiodic function that is not quasiperiodic.

It is not easy to determine whether the $\boldsymbol{Q}$ components obtained from the experiment are irrational rather than rational with large denominators. Furthermore, we cannot define with absolute precision the position of the spots because of their intrinsic width, mainly due to the finite size of the sample and to instrumental effects. We do, however, assume that the physical properties of the system are independent of whether $\boldsymbol{Q}$ is irrational or not.

In an ordinary crystal, plane waves with arbitrary wavevector propagate, with the exception of those with a value $\boldsymbol{k}'$ for which the diffraction condition is fulfilled,

$$2\boldsymbol{k} \cdot \boldsymbol{k}' \pm |\boldsymbol{k}|^2 = 0. \tag{6.4}$$

The excluded states correspond to stationary waves with two components. These states remain extended but contribute to neither the propagation nor to energy transport. Since the $\boldsymbol{k}$ vectors form a dense set in quasicrystals, then any $\boldsymbol{k}'$ vector satisfies (6.4) and can thus contribute to diffraction. Besides this, we observe multiple diffraction, and the number of simple plane waves that contribute to the resulting stationary state grows with $|\boldsymbol{k}'|$.

Since we could suspect that propagation phenomena are inhibited in quasicrystals, it is convenient to consider the particle, e.g. the electron, associated with the plane wave with frequency ω and wavevector $\boldsymbol{k}'$. Let us refer to the dispersion relation $\omega(\boldsymbol{k}')$ of an ordinary crystal; this extends up to the Brillouin zone boundary. When $\boldsymbol{k}'$ obeys the Bragg law, a gap with width $\Delta\omega$ opens up in the THz region ($\Delta\omega \simeq 10^{12}$ Hz) of the phonon dispersion curve. The average lifetime of the particles consequently reduces to $\Delta t = (\Delta\omega)^{-1}$, namely 10^{-12} s; when group velocity is below 10^3 ms^{-1}, the resulting mean free path is less then 1 nm.

If we take any quasicrystal, the dispersion relation is a dense hierarchy of gaps, most of which are very narrow. When $\Delta\omega \simeq 10^{-4}$ THz, the particle may propagate over distances in the order of a few μm. As such, propagative and non-propagative states may coexist. Unlike their counterparts in the ordinary crystals, propagative states are much less effective and there are many more non-propagative states.

The size of the primitive unit cell in the physical space in a periodic crystal gives rise to a minimum interatomic distance; the above mentioned density property for wavevectors (which corresponds to the high density of Bragg spots observed in the quasicrystalline structures) implies that arbitrarily small vectors are allowed in the reciprocal lattice of an incommensurate structure. This condition poses the problem of the position of the atoms in the space, since these atoms cannot be found arbitrarily close to each other.

As surprising as it may seem, most of the crystalline phases are almost periodic. There are obviously also periodic structures, which are usually defined as "crystals", whereas the other phases are *incommensurate* crystalline phases. Among these phases, one important class is given by the modulated structures. The description of these crystals is obtained from an underlying structure with three-dimensional space group symmetry, which is associated with a periodic deviation, namely the modulation; the period of the latter, in the incommensurate case, is not compatible with the period of the lattice corresponding to the underlying space group.

If we observe modulation in atomic displacements, then the structure is *displacively* modulated, as shown in part (a) of Fig. 6.8. When the modulation regards the probability that crystallographic sites of the underlying structure may be occupied by atoms of a given species, we obtain an *occupationally* modulated structure, as schematised in part (b) of Fig. 6.8. A system that is made up of sub-systems whose underlying structures are mutually incommensurate, is called a *compositionally* modulated system and it is displayed in part (c) of Fig. 6.8.

Incommensurability may even occur in two-dimensional crystalline layers; it is observed quite frequently in atomic monolayers adsorbed onto crystalline substrates.

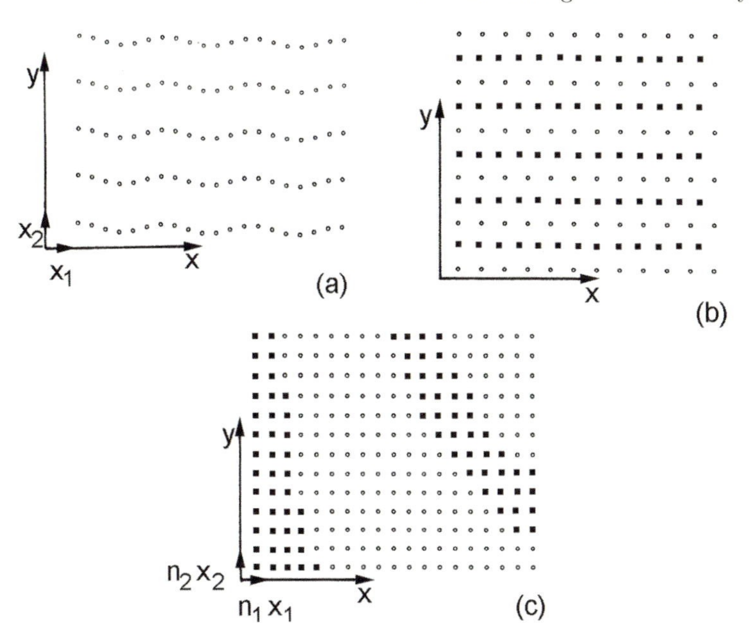

Fig. 6.8. Schematic illustration of incommensurate crystalline phases. (**a**) x_1 constant; x_2 constant, in the presence of transversal displacive modulation where $y = \cos(\pi/6) \times \cos(n_1x_1/5)$, with n_1 an integer. (**b**) Occupation modulation of lattice sites on a square lattice. Symbols (o) and (■) are respectively less than or equal to 0.5 and greater than 0.5 for the probability occupation function P of each lattice site. $P = \cos\left[\cos\left(5\pi/12\right) \cdot \left(n_1x_1\right)\left(1 + n_2x_2/9\right)\right]$. (**c**) Compositional modulation. Two sub-lattices are present, where the ratio between the lattice constant of the first (o) and that of the second (■) sub-lattice in x direction is the irrational number $(2/\sqrt{3})$: the two lattice constants are reciprocally incommensurate. Along direction y the ratio between the lattice constants for the two sub-lattices is unity

Lastly, it is currently possible to realise incommensurate artificial heterostructures. One significant example has been obtained by alternating layers of GaAs and AlAs, grown by molecular beam epitaxy; the artificial structure resulting from the sequence of layers is quasiperiodic. The scheme given in Fig. 6.9 shows alternating layers of fixed thickness: the first layer, L, consists of 1.7 nm of AlAs and 4.2 nm of GaAs, and alternates with a second layer, B, which consists of 1.7 nm of AlAs and 2 nm of GaAs.

Fig. 6.9. Fibonacci superlattice. B: AlAs (1.7 nm) + GaAs (2 nm). L: AlAs (1.7 nm) + GaAs (4.2 nm). The sequence of pairs $B - L$ is the Fibonacci chain

B	1
L	1
LB	2
LBL	3
LBLLB	5
LBLLBLBL	8
LBLLBLBLLBLLB	13
LBLLBLBLLBLLBLBLLBLBL	21

Fig. 6.10. Fibonacci chain. A particular local arrangement (the same as in Fig. 6.9) found in various generations is underlined. The Fibonacci numbers, on the right, give the number of B and L elements in each chain line, or generation

The L and B layers are subsequently added to the structure, according to a sequence that corresponds to the *Fibonacci chain*, F^n. Given an initial pair of LB layers, Fig. 6.10 shows the addition of the following layers according to the substitution rule $L \to LB$, $B \to L$. The result is a perfectly ordered, aperiodic, deterministic sequence of B and L layers.

F^n is a mathematical structure with peculiar properties; let us consider the Fibonacci n_n series, whose terms indicate the number of elements, whether they are B, or L, we find in the nth generation (line) of the chain. The Fibonacci n_n series is given as $n_0 = 0$, $n_1 = 1$, $n_2 = 1$, $n_3 = 2$, $n_4 = 3$, $n_5 = 5$, ... $n_n = n_{n-1} + n_{n-2}$. These numbers are reported on the right of each line of the Fibonacci chain (Fig. 6.10); they have the property that, as n increases, the ratio (n_{n+1}/n_n) converges onto the irrational number τ, called the golden ratio:

$$\tau = (1 + \sqrt{5})/2 = 1.61803398... \quad . \tag{6.5}$$

Each rational term (n_{n+1}/n_n) of the series is defined as the nth order *rational approximant* to τ.

The frequency, f, with which elements B and L respectively appear in the F^n chain is given by $f_n(B) = n_{n-2}$ and $f_n(L) = n_{n-1}$; in the limit of large n, $f_n(L)/f_n(B) = \tau$. Thus, periodic repetitions of groups of L and B elements are excluded. Even the τ^n elements, with increasing n, are a Fibonacci chain, namely $\tau^n = \tau^{n-1} + \tau^{n-2}$.

Coming back to our experimental realisation of the F^n chain with the above mentioned hetero-structure GaAs/AlAs, when the number of L and B layers increases, thus leading to a number of chain generations, we obtain a quasiperiodic sequence of L and B layers, whose diffraction pattern exhibits narrow peaks. These peaks are given two integer indexes instead of the single index, which is a feature of any one-dimensional periodic structure.

Since we have good control over the sequence growth, this system, just like several other similar systems, is a model for the quasiperiodic structures. The Fourier module for the artificial hetero-structure under examination has an order of four. We define as z the axis along which layers B and L grow; given that the structure is periodic in both x and y directions, two vectors are generated in these two directions, respectively $\boldsymbol{x}_1^*$ and $\boldsymbol{x}_2^*$, whereas the need to introduce a quasiperiodic function along the z axis requires another *two* vectors, $\boldsymbol{x}_3^*$ and $\boldsymbol{x}_4^*$, which are associated with mutually incommensurate periods.

6.3 Quasicrystals and Tilings

The quasicrystals form a special class of incommensurate *quasiperiodic* structures. The diffraction pattern exhibits well defined, narrow peaks and suggests that there is long range order in the system. The quasicrystals are associated with having perfect orientational order without translational periodicity. As the quasicrystals possess fivefold symmetry, which is associated with ideal orientational order, and quasiperiodicity, we are immediately led to assume that an analogy exists between these physical structures and the so-called Penrose tilings. This is a class of quasiperiodic tiling of the plane using tiles of two different shapes, where the vertices of the tiles satisfy the diffraction condition which, in turn, is associated with a Fourier transform with fivefold symmetry.

Very basically, we are in the presence of one of the three questions in Hilbert's 18th problem, namely, is there a tiling on whose tiles no symmetry group can transitively act? i.e. it does not act transitively on any partition of the n-dimensional Euclidean space, E^n, obtained using the above mentioned cell.

A tiling of E^n is a partition of the space into a countable family Υ of non-overlapping, closed sets C, called cells; $\Upsilon = \{C_1, C_2, ..., C_n ...\}$, such that

$$\text{int } C_i \cap C_j = \phi \qquad \text{for any } i \neq j \tag{6.6}$$

and

$$\cup_{i=1}^{\infty} C_i = E^n . \tag{6.7}$$

A number of different cells have been devised that meet the Hilbert requirements; in Fig. 6.11 an example of a two-dimensional solution ss given by Escher's "ghosts" is shown which, though they seem somewhat complex, constitute a modified version of plane tiling by equilateral triangles. The triangle side has two ghosts aligned in the sequence "head–tail–tail–head" and each triangle has three ghosts inside it. Since there is no rigid motion to cause any interchange between the ghosts inside the triangles and the ghosts on the sides, the symmetry group of this tiling does not transitively act on the tiling itself. Moreover, this is a unique tiling in that this is the only way to reciprocally assemble ghosts in such a way that the plane is completely covered without overlapping the tiles.

The Hilbert question may be put slightly differently; we could wonder if such a tile exists that can give rise to only aperiodic tilings. We are not aware

Fig. 6.11. The "ghosts" (M.C. Escher) are a complex modification of a plane tiling using equilateral triangles. The continuous line highlights a triangle and the crosses mark a triplet of ghosts inside the triangle. There is no rigid motion to allow the exchange of "inside" ghosts with those on the sides of a triangle

of any such tile at this time that, used on its own, could fulfil the required conditions. There are, however, tiles with such a shape that both periodic and aperiodic tilings can be generated. The initial assumption that any set of tiles that gives an aperiodic tiling can also give rise to a periodic tiling is, indeed, false. An *exclusively* aperiodic plane tiling could be realised using a set of six tiles with different shapes; this set was later reduced to four tiles and, later to two tiles. It was observed that such Penrose tiles gave rise to tilings with meaningful properties to understand the properties of quasicrystals.

The shape of the Penrose tiles can be chosen in a number of ways; for example, we can refer to two pairs of motifs which basically give rise to two "unit cells". The first motif, that is reported in Fig. 6.12, consists of two rhombi with equal length sides, whose acute angles are $\pi/5$ ("thin (T)") and $2\pi/5$ ("fat (F)") respectively.

The second pair of tiles (Fig. 6.13) is obtained from the first pair by finding the point along the principal diagonal of a "fat" rhombus that cuts the rhombus itself into two parts, whose ratio is given by the golden ratio τ; such a point is then connected to the vertices of the two obtuse angles. The two resulting motifs are called *dart* and *kite*, from their shapes. The side length of these two motifs is τ or 1, whereas the ratio between the area of the kite and the area of the dart is the golden ratio.

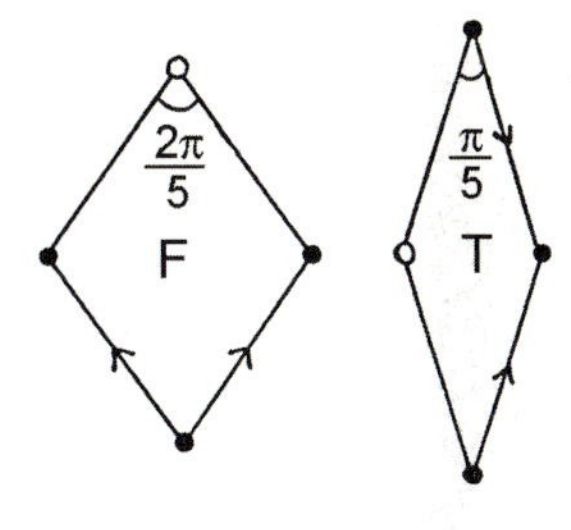

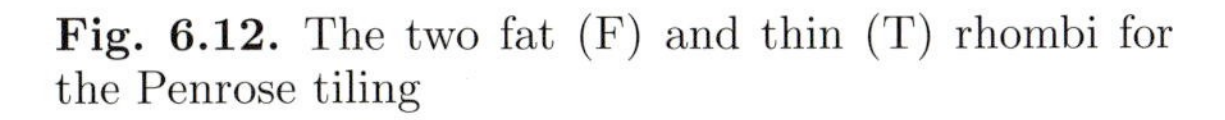

Fig. 6.12. The two fat (F) and thin (T) rhombi for the Penrose tiling

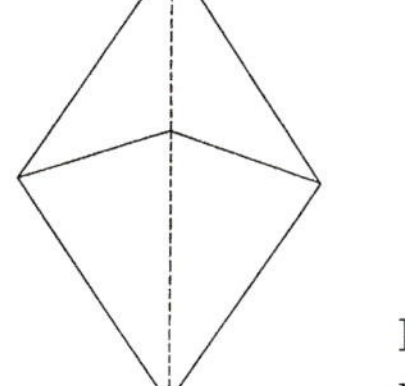

Fig. 6.13. "Dart" and "kite" construction starting from an F rhombus

Figure 6.14 shows how, by using rhombi, we can produce both periodic (part (a)) and aperiodic (part (b)) uniform plane tilings, depending on the rules we adopt to assemble the tiles together. We force the aperiodic structure by using specific matching rules between tiles, namely construction rules for the tiling. For example, for a lozenge-based tiling, we index the vertices of tiles with different numbers of arrows, then we arrange tiles together in such a way that they share equal-index sides. The procedure is schematised in Fig. 6.15.

The graphic rules for the reciprocal matching of the tiles are local and do not guarantee that, even when correctly applied, defective local arrangements will not occur. A defect is a void interstice that cannot be filled with a tile. Void formation may occur both when we simply consider correct matching of first neighbour pairs of tiles and when we try to consider the arrangement taken up by tiles that are second, third, fourth, neighbours to the tile we are disposing on the plane.

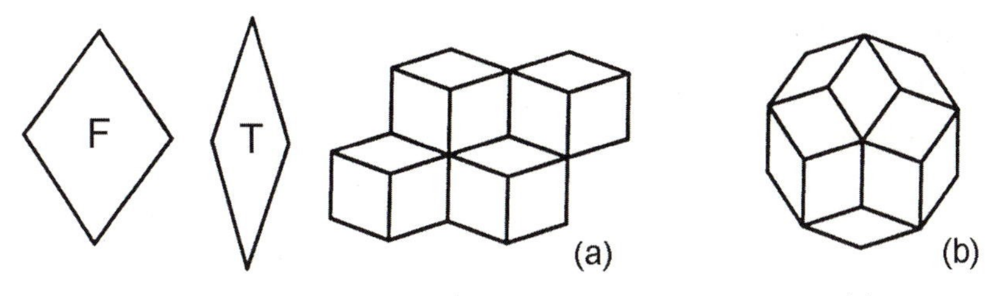

Fig. 6.14. Using rhombi F and T we obtain both periodic tilings (**a**), and aperiodic tilings (**b**)

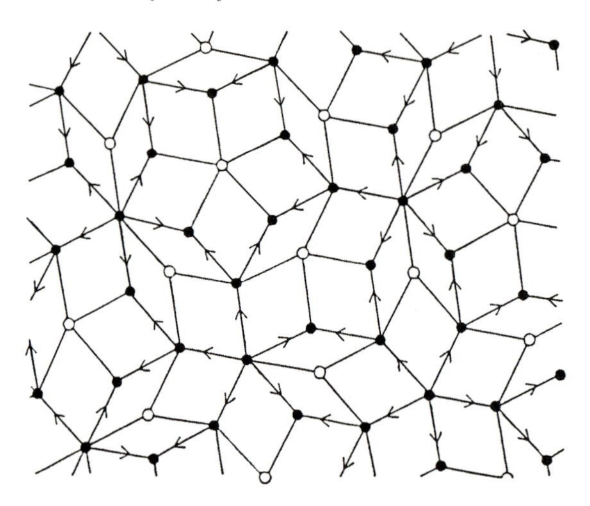

Fig. 6.15. Illustration of the local matching rules to construct an aperiodic Penrose tiling

The size of the already covered region which, at each stage of tiling growth, should be examined so that we can correctly add another tile, is *unlimited*; this is like implying that the error is intrinsic to aperiodicity.

A strict mathematical procedure has recently been introduced, whereby we define precise matching rules that can lead to both aperiodic and periodic structures. This procedure, called substitution, has allowed us to establish a clear connection between the two families of structures.

Even if we use the tiling matching rules, and having chosen a fixed set of tiles, we still cannot realise a single, unique aperiodic tiling; on the contrary, the number of Penrose tilings is uncountable. This is because these tilings are self-similar, namely they exhibit a characteristic behaviour with respect to a change of scale. Such a process is defined as the *inflation-deflation* procedure. As shown in Fig. 6.16, inflation is achieved by taking the tiles in a given tiling apart and rescaling them so that the size of the new tiles is the same size as the original tiles. Thus we realise a different tiling to the original

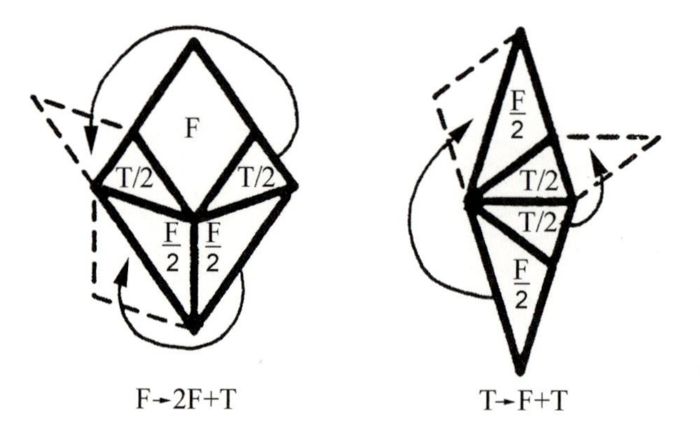

Fig. 6.16. A graphic example of inflation-deflation procedure applied to rhombi F and T; inflation scheme

one, yet formed by thin and fat rhombi with sizes that are greater than the original tiles.

Figure 6.10 clearly shows the inflation-deflation procedure for the one-dimensional Fibonacci chain which, in turn, is a self-similar structure. When we apply the substitution rules $L \rightarrow LB$, $B \rightarrow L$, the process becomes inflationary, whereas using the opposite rules we obtain the deflation process.

Coming back to our Penrose tilings, the inflation (deflation) procedure may be iterated *ad infinitum* to produce, at each stage, a generation of prototype rhombi that are larger (smaller) than the rhombi of the previous generation, thus giving rise to different plane tilings. Penrose tilings belonging to different generations are all distinct from each other since there is no Euclidean transformation, namely no distance-preserving affine transformation that could transform one tiling into another tiling.

Different generation tilings, however, are relatively homogeneous, and their local structures are repeated fairly regularly. This pattern similarity corresponds to the property that in the tiling each local surrounding, for any finite radius, is relatively dense. This means that in E^n every ball with radius greater than a fixed positive number contains at least one tiling vertex (an ordinary crystal is relatively dense).We observe that the Penrose matching rules allow only seven different rhombi arrangements at each vertex. Each of these local arrangements is relatively dense. This tiling property is called *local isomorphism*. Each local configuration, defined within a circle with radius R, e.g. centred on a vertex, re-appears within a distance of $2R$. Locally, two Penrose tilings are thus equal to each other, in that each local (finite) configuration observed in one of these tilings is observed in the other. This kind of local regularity is shown in the tiling in Fig. 6.15. Local isomorphism is a weaker form of repetitiveness, which in turn is a consequence of homogeneity. By comparison with quasicrystals and tilings, in ordinary crystals, repetitiveness is strong and results in long range order; in amorphous materials it is weak and produces short range order (see Sect. 4.7).

Local isomorphism is observed also in the Fibonacci chain shown in Fig. 6.10, where, for different adjacent generations, which correspond to different lines, a particular local configuration has been highlighted.

Given the local regularity of aperiodic structures, we immediately wonder whether this local regularity is the cause of the observed diffraction behaviour.

From the simulation of the Fourier transform for one- and two-dimensional aperiodic structures, we observe that the local isomorphism is not, in itself, sufficient to guarantee the condition required for diffraction. This condition seems to be governed by the properties of the substitution matrix applied in order to achieve the inflation-deflation procedure. For the Penrose tilings, this matrix is $S = \begin{pmatrix} 2 & 1 \\ 1 & 1 \end{pmatrix}$ and guarantees both local isomorphism and non-periodicity. In actual fact, the ratio between fat and thin rhombi is given by the golden ratio τ; we define a population vector as (F_n, T_n), whose components are the number of fat and thin rhombi we find in the nth generation

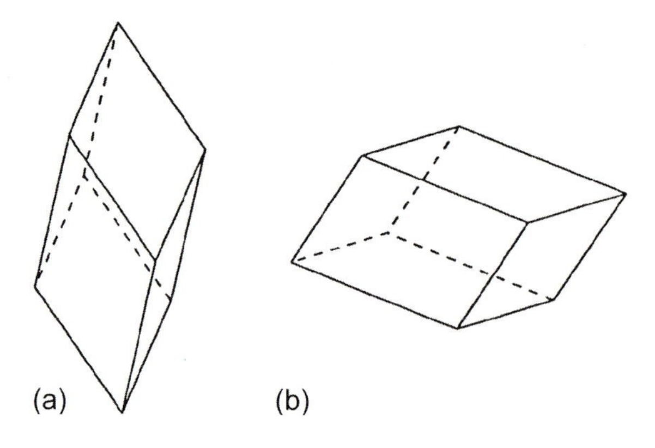

Fig. 6.17. Prolate (**a**) and oblate (**b**) rhombohedra needed to construct a three-dimensional Penrose lattice

tiling. Since (F_0, T_0) is the population vector for the arbitrarily chosen initial configuration, then

$$(F_n, T_n) = S^n(G_0, T_0). \tag{6.8}$$

If we let (F_0, T_0) be a linear combination of the S matrix eigenvectors, we can then study the trend in the population vectors as S is iterated. The eigenvectors are proportional to $(\tau, 1)$ and to $(-1, \tau)$, whereas the eigenvalues are τ^2 and $1/\tau^2$. The population vector is thus attracted to the eigenvector with an eigenvalue of τ^2, since the absolute value for the second possible eigenvalue is less than unity. It is exactly this Penrose tiling property that seems to be connected to the fulfilment of the diffraction condition.

The Penrose tilings have also been extended to three-dimensional structures; in this case the tiles are rhombohedra, respectively prolate and oblate, as shown in Fig. 6.17. Again, when we carry out a computer simulation of the Fourier transform for a three-dimensional Penrose tiling, as shown in Fig. 6.18, it is observed to qualitatively agree with the experimental electron diffraction patterns obtained from *real* quasicrystalline materials (Fig. 6.19).

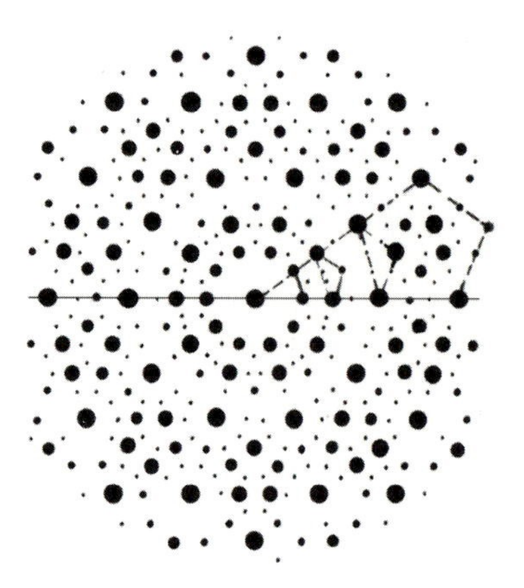

Fig. 6.18. Simulated Fourier spectrum perpendicular to a fivefold axis for a three-dimensional Penrose lattice. The dot arrangement in the plane is pentagonal; the distance between adjacent pairs of dots along each axis scales with τ

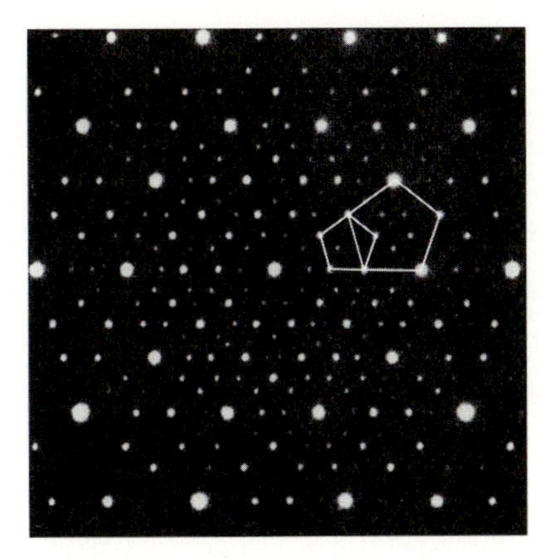

Fig. 6.19. Electron diffraction pattern for a fivefold axis of quasicrystalline $Al_{65}Cu_{20}Fe_{15}$; the lines highlight the pentagonal symmetry (adapted from [6.3])

6.4 Model Structures and Crystalline Approximants

When examining diffraction patterns obtained from real quasicrystals we observe the unambiguous presence of icosahedral symmetry in the reciprocal space, which lies at the origin of the intense reflections in the direction of the fivefold, threefold and twofold order rotation axes. The position of the high intensity diffraction spots observed along the binary axes (with even parity of the electron scattering pattern), may be obtained by τ inflation. We can reach the second spot in a row of spots along a diffraction line by multiplying τ by the distance from the initial spot to the centre. For threefold and fivefold axes (with odd parity), the inflation factor is τ^3, which reflects the symmetry of the primitive icosahedron in the reciprocal space. The quasicrystals exhibiting these features are called simple or primitive icosahedral quasicrystals.

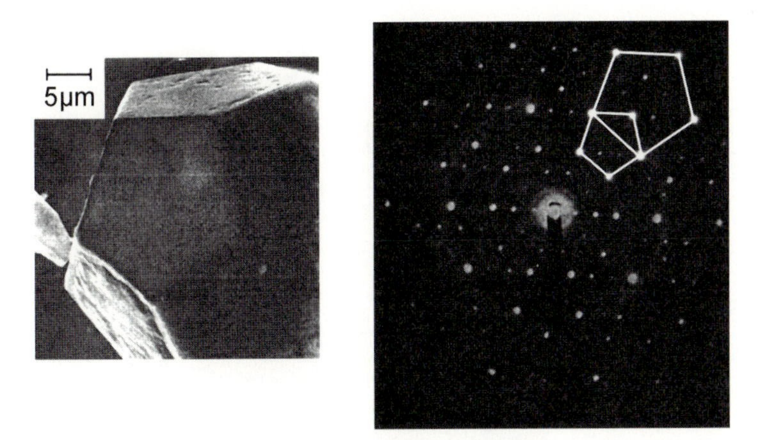

Fig. 6.20. Morphology (pentagonal dodecahedron) of a grain of $Zn_{50}Mg_{45}Y_5$ and pertinent electron diffraction pattern

Fig. 6.21. Morphology (rhombic triacontahedron) of a grain of $Al_{53.4}Li_{36.9}Cu_{9.7}$ and pertinent electron diffraction pattern

The morphology of the quasicrystalline phase depends on the growth conditions specific to the sample. When the quasicrystals form as a primary phase from the liquid, they often exhibit a faceted shape, which reveals the related point group symmetry. In the case of Zn–Mg–Y, just as for the prototype Al–Mn, the shape is a pentagonal dodecahedron, as shown in Fig. 6.20, whereas the morphology of the Al–Li–Cu alloy (Fig. 6.21) is a triacontahedron and that of Al–Pd–Mn in Fig. 6.22 is an icosidodecahedron.

Although all the simple quasicrystals exhibit the signature of τ^3 inflation along the odd parity directions, the details of the diffraction patterns from these phases, especially the intensities of the spots, differ significantly from material to material, suggesting that different atomic arrangements are present.

The inherent incommensurability of the icosahedral phase makes it difficult to reconstruct both the atomic structure and the scattered intensity distribution in the reciprocal space starting from a given set of diffraction data. If we try to index the diffraction patterns using the three Miller indices (hkl) used for periodic crystalline structures, we obtain *irrational* indices,

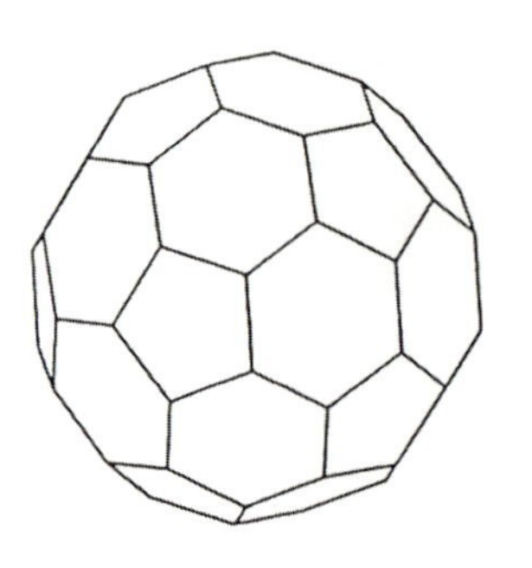

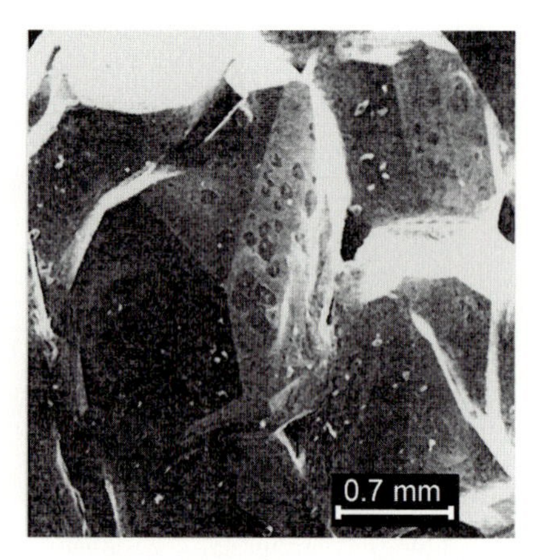

Fig. 6.22. Morphology (icosidodecahedron) of a grain of $Al_{68}Pd_{23}Mn_9$

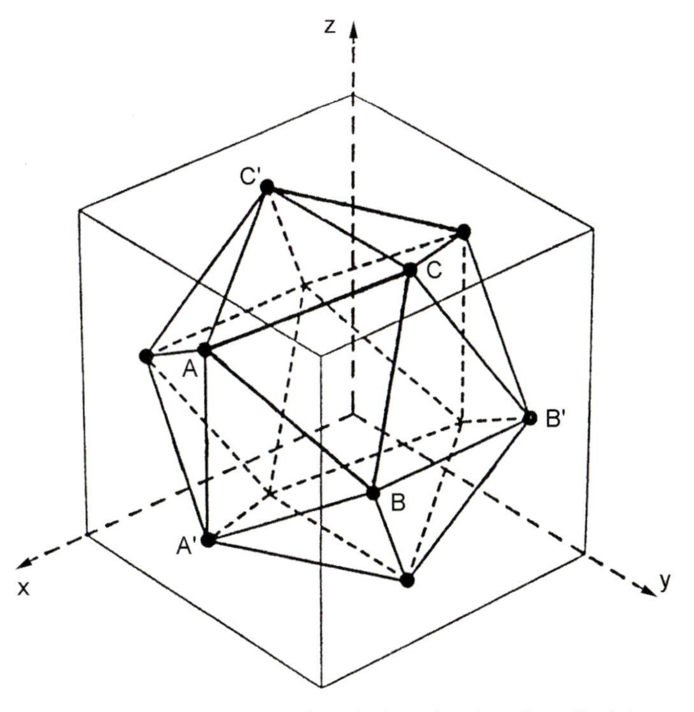

Fig. 6.23. Icosahedron with unit length edges inscribed in a cube; edges A, A', B, B', C, C', are parallel to axes z, x and y of the cube. Coordinates for points A, A', B, B', C, C' are, respectively, $(\tau, 0, \pm 1)$, $(\pm 1, \tau, 0)$ and $(0, \pm 1, \tau)$. The basis vectors $\boldsymbol{x}_1^*$, ... $\boldsymbol{x}_6^*$ used to index the diffraction spots produced by an icosahedral structure, by way of n_i integers, point out from the centre of the icosahedron to the six vertices A, A', B, B', C, C'

instead of integer indices. Assuming icosahedral symmetry, a good choice to assign the indices to the diffraction patterns requires using *six* linearly independent basis vectors to explore the whole reciprocal space. These vectors correspond to the vectors that point out from the centre to the vertices of the icosahedron inscribed in a six-dimensional hypercube as shown in Fig. 6.23. Each reciprocal lattice vector $\boldsymbol{k}_i$, and thus in particular each diffraction spot or peak, is given by six integer indices

$$\boldsymbol{k}_i = k_0 \left(n_1 \boldsymbol{x}_1^* + n_2 \boldsymbol{x}_2^* + n_3 \boldsymbol{x}_3^* + n_4 \boldsymbol{x}_4^* + n_5 \boldsymbol{x}_5^* + n_6 \boldsymbol{x}_6^* \right) \tag{6.9}$$

where k_0 is a constant that fixes the scale for the diffraction pattern. Owing to the inflation symmetry, we cannot choose a single value for k_0 *a priori*, unlike what we do for periodic crystals where the reference length scale is given by the length of the unit cell edge. For example, for a cubic crystal, the (100) reflection is located in correspondence to the wavevector $(2\pi/a)$ along the $(h00)$ axis, where a is the lattice constant of the material. For an ideal icosahedral quasicrystal, where a is the edge of the hypercube,

$$\boldsymbol{k}_i = \frac{2\pi}{a} \sum_i n_i \boldsymbol{x}_i^* \tag{6.10}$$

where $\boldsymbol{x}_i^*$ is the unit vector along the hypercube edge.

In the primitive icosahedral systems there are no restrictions to the succession of the indices, namely, each n_i can take on any integer, just like the simple cubic crystals; this structure is known as a *simple* icosahedral (SI) quasilattice.

A second kind of quasilattice has also been observed, which, for example, is associated with the alloys Al–Cu–Fe, Al–Cu–Ru, Al–Pd–Mn. This is a *face-centred* icosahedral (FCI) structure where the indices must have the same parity, namely the n_i must all be even, or odd, just like the indices for the face-centred cubic crystals.

The reciprocal space for a quasicrystal is dense. Equation (6.9) provides us with a six-dimensional periodic reciprocal lattice whose Fourier transform corresponds to a six-dimensional periodic distribution of mass density. This is not a new description since we currently use higher than three-dimensional space to analyse incommensurate phases, yet the quasicrystals are three-dimensional structures. Equation (1.28) holds for a three-dimensional periodic ordinary crystal, and the wavevectors, $\boldsymbol{k}$, that enter the Fourier expansion of the atomic density $\varrho(\boldsymbol{x})$ are vectors of the three-dimensional reciprocal lattice, according to (1.29); thus the atomic density is periodic in three dimensions.

In an icosahedral quasicrystal, the relation between the three-dimensional physical description of the system and its six-dimensional image is given by the *cut* and *projection* technique. In order to illustrate this relation to some extent it is worth referring to projections from two dimensions onto one dimension. This simple model also allows us to describe the structure of the one-dimensional incommensurate structures experimentally observed in Al–Cu–Co, Al–Ni–Si and Al–Cu–Mn. These materials with layered structure exhibit periodicity *within* each atomic plane. These planes, in turn, are stacked in agreement with the Fibonacci aperiodic sequence. The pertinent crystalline approximants are a series of CsCl-type structures where one of the two interpenetrating cubic lattices is made of lattice vacancies with repetition distances along the [111] axes such that they approximate a Fibonacci chain.

System modelling requires generating a one-dimensional aperiodic arrangement of atoms starting from a periodic square lattice, that simulates the hyperdimensional space. Given a pair of orthogonal axes, X and Y, for the two-dimensional lattice, we define a second pair, X^1 and Y^1, which is rotated at an angle of γ to the reference pair of axes. The physical space we are interested in is $\{X^1\}$, namely the axis X^1, as shown in Fig. 6.24.

The atoms on the square lattice are represented by segments with a fixed length, l, which stand for the atomic "surfaces" with a point-like central "nucleus". These segments lie perpendicular to $\{X^1\}$. The intersect of each segment with $\{X^1\}$ corresponds to the localisation of the point-like atoms in

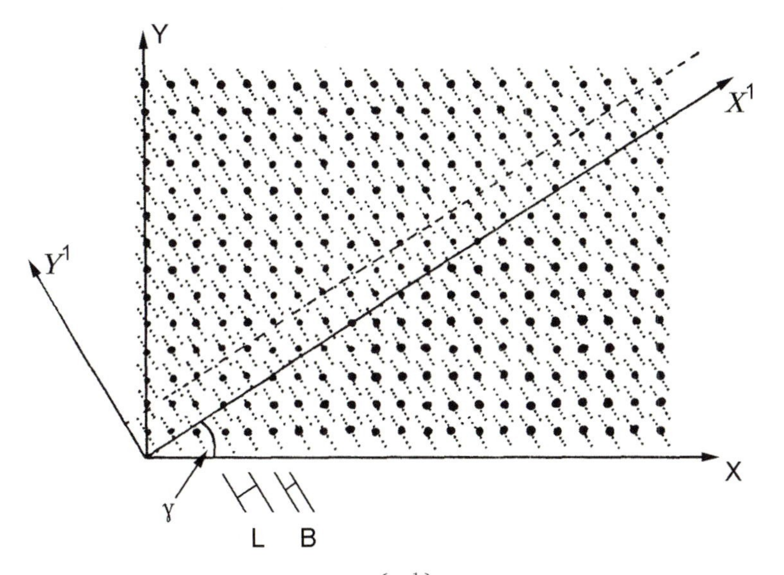

Fig. 6.24 One-dimensional cut $\{X^1\}$ through a two-dimensional square lattice. The atoms as represented by full dots, each with its surface ($\dot{}\cdot$) in the square lattice are indicated. Projections into the physical space $\{X^1\}$ of atoms within the two-dimensional strip between $\{X^1\}$ and the dashed line parallel to it, give a partition of $\{X^1\}$ into segments. If angle γ between X and $\{X^1\}$ is such that $\cot\gamma = \tau$, the sequence of the projected segments, respectively L and B is a Fibonacci chain. The figure shows, from the origin, the fifth, sixth and seventh generations as well as the first seven elements of the eighth generation of the chain obtained using the substitution rule $B \to L$; $L \to LB$ (see Fig. 6.10)

the physical space $\{X^1\}$. This is an example of m-dimensional cut through a periodic n-dimensional lattice, where n is greater than m. In this case, we make a one-dimensional cut through a two-dimensional periodic lattice. Then we project the atoms in the strip bounded by the X^1 axis and by the dashed straight line parallel to the X^1 axis (Fig. 6.24), which contains p bases for the square lattice adjacent to each other. If angle γ is irrational, then the sequence of interatomic distances δ between points that are projections onto $\{X^1\}$ of the atomic positions in the square lattice will also be irrational. In particular, if $\cot\,\delta = \tau$, the sequence coincides with a Fibonacci chain with (L) and (B) segments respectively.

If we use a graphic representation instead of the above procedure, then we overlook the details of the atomic basis and define a *collection domain.* Looking at Fig. 6.25 we see that all the points of the two-dimensional lattice in this domain are projected onto the physical space $\{X^1\}$. The strip with a width of F, which is parallel to $\{X^1\}$, and thus is rotated at an angle of γ', equal to γ, to the reference axes, is the two-dimensional collection domain. Let us consider a pair of any points with integer coordinates, that belong to

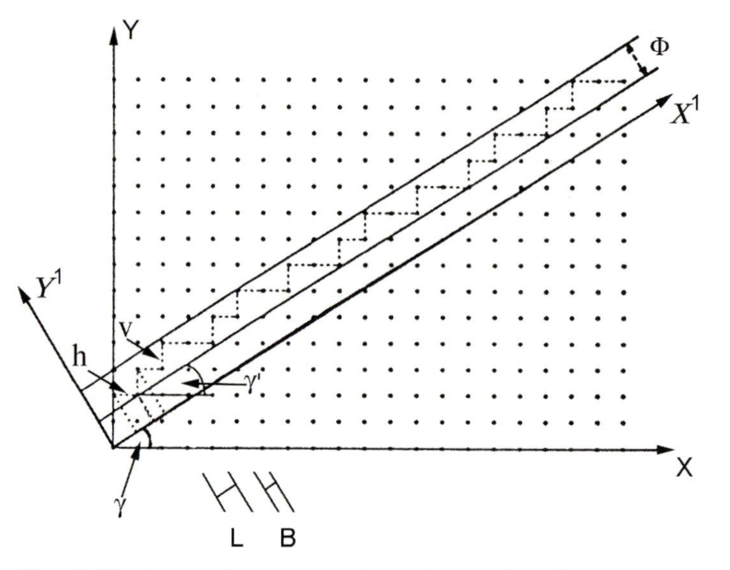

Fig. 6.25. Alternative construction scheme of a Fibonacci chain. The collection domain is the strip parallel to $\{X^1\}$, of width Φ. The atom sites contained in this domain are projected onto $\{X^1\}$; we consider the sequence of the segments on $\{X^1\}$, each resulting from the projection of adjacent atoms. When $\cot\gamma = \tau$, the projection onto $\{X^1\}$ of a pair of adjacent atoms lying parallel to X in the square lattice (h), gives us a segment L; the projection of a pair of atoms parallel to Y(v), gives a segment B. We generate a Fibonacci chain of L and B segments; the sixth and seventh generations are shown, as well as the first eleven elements of the eighth generation of the chain obtained using the substitution rule $B \to L$; $L \to LB$

the dashed stairs in Fig. 6.25. For both points, distance d from $\{X^1\}$ is given by

$$d = \frac{|am - b + q|}{\sqrt{m^2 + 1}}$$

where $m = \tan\gamma = 1/\lambda$ and, in our case, $q = 0$ since $\{X^1\}$ intersects the origin of the coordinate axes. The two points respectively have coordinates (a,b) and (c,d). So,

$$d_{(ab)} = \frac{a - b\lambda}{\sqrt{\lambda^2 + 1}}; \qquad d_{(cd)} = \frac{c - d\lambda}{\sqrt{\lambda^2 + 1}}$$

Only if λ is rational, namely if $\lambda = (a - c)/(b - d)$, will $\mathcal{D}(\delta)$, which gives the distance between the projections of the two points onto $\{X^1\}$, be periodic.

Again, we obtain one-dimensional *incommensurate* structures if $\cot(\gamma')$ is an irrational number. The atomic density along $\{X^1\}$ is obtained from the projection of the dashed segments, respectively vertical (v) and horizontal

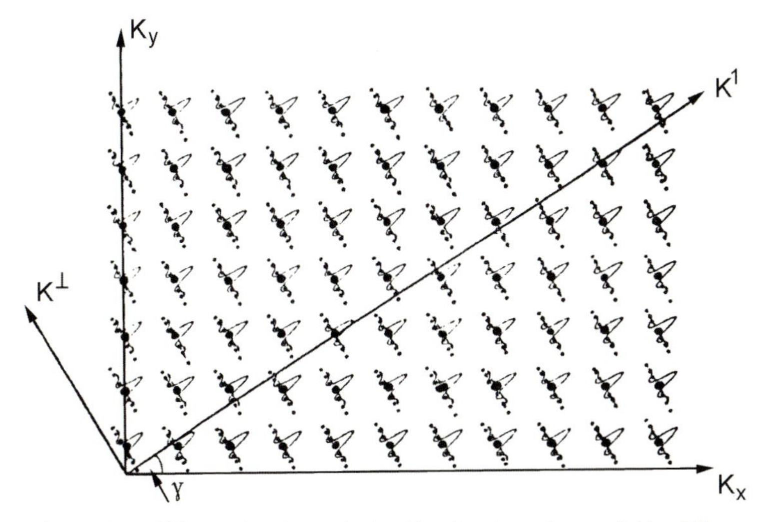

Fig. 6.26. Schematic view of the Fourier transform of the Fibonacci chain in Fig. 6.24. On the sites of a square two-dimensional lattice, reciprocal to the lattice in Fig. 6.24, we arrange the Fourier transform either of the atomic surfaces in Fig. 6.24 or of the collection domain in Fig. 6.25

(h), contained in the collection domain. In particular, where $\cot(\gamma') = \tau$, the atomic density is given by a Fibonacci chain with (L) and (B) interatomic spacings.

If we calculate the Fourier transform for the atomic basis, as shown in Fig. 6.26, and find the intersecting points between the two-dimensional structure factor and the K^1 axis, which is the reciprocal space for physical space X^1, we obtain the diffraction pattern produced by the chain under examination. The intersecting points are given by the vectors of the reciprocal lattice, $\boldsymbol{k}^1$. The Fourier transform of the basis, or of the collection domain, produces a set of delta functions along $\{K^1\}$ in coincidence with specific values for $\boldsymbol{k}^1$. The intensity of the delta functions is proportional to the square of the amplitude of the transform at the intersection point on the reciprocal space $\{K^1\}$

$$I = \left[\sin(\boldsymbol{k}^1 a)/\boldsymbol{k}^1 a\right]^2 \tag{6.11}$$

where a is the interatomic spacing for the two-dimensional lattice.

Let us suppose that the angle $\gamma = \cot^{-1}(\tau)$ between axis X of the two-dimensional lattice and the rotated axis $\{X^1\}$ is only slightly different from angle $\gamma' = \cot^{-1}(3/2)$, which gives us the orientation of the collection domain. When we project the points contained in the new collection domain onto $\{X^1\}$ the resulting structure is a periodic sequence of the very same segments in the Fibonacci chain, with repetition distance, namely unit cell, $(BLBLL)$. This structure, which is shown in Fig. 6.27 is called a *rational*

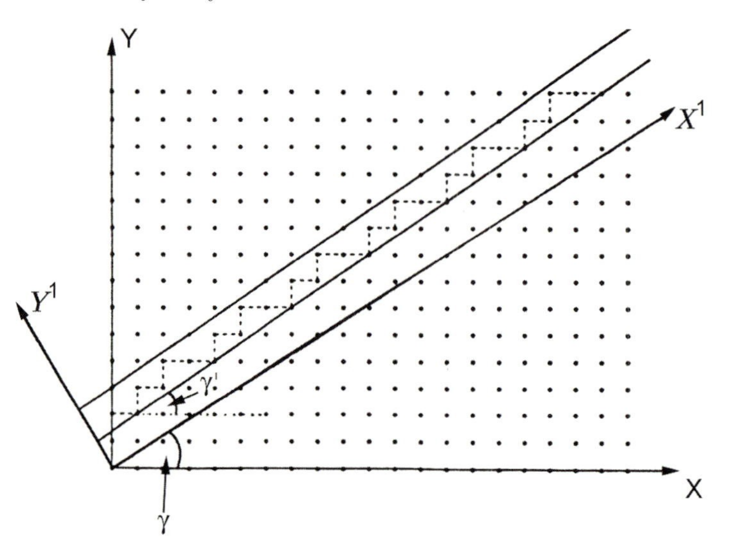

Fig. 6.27. Similar construction to Fig. 6.25, for the rational approximant (3/2) of the Fibonacci chain. The collection domain lies at an angle $\gamma'(\arctan\gamma' = 3/2)$ to $\{X^1\}$, slightly different to $\gamma(\arctan\gamma = \tau)$. The distances on $\{X^1\}$ between projections of adjacent atoms within the collection domain give a periodic sequence with unit cell $(BLBLL)$

approximant. The example refers to the rational approximant 3/2 of the Fibonacci chain. This approximant schematically explains the relation between crystalline approximants and quasicrystals.

We can build periodic approximants of the Fibonacci chain with ever greater unit cell sizes. All we have to do is give $\cot(\gamma')$ the value of a rational approximant (n_{n+1}/n_n) of τ.

Structure shifts along $\{X^1\}$ correspond to translations in the physical space. If we distort the unit cell, thus changing its size, we simulate deformation; as a consequence, the diffraction peaks broaden more and more as $\boldsymbol{k}^1$ increase.

If we cause a rigid translation of $\{X^1\}$ in direction Y^1, we obtain, once again, a Fibonacci chain where the arrangement of the L and B segments has changed. The diffraction pattern does not change, since the translation shifts the view-point of the structure some distance along $\{X^1\}$. This is a phase shift of the density waves, which is expected in incommensurate structures, but not in periodic systems. The shifts along Y^1 axis are associated with additional degrees of freedom; these collective modes are called *phasons.*

We can force the slope of the collection domain to fluctuate about an average value of τ^{-1} as shown in Fig. 6.28. These fluctuations are small amplitude and we obtain, again, a Fibonacci chain where defects are observed locally. In the pertinent diffraction pattern we still find Bragg spots at the same positions as for the ideal F^n chain, but a degree of *diffuse* scatter-

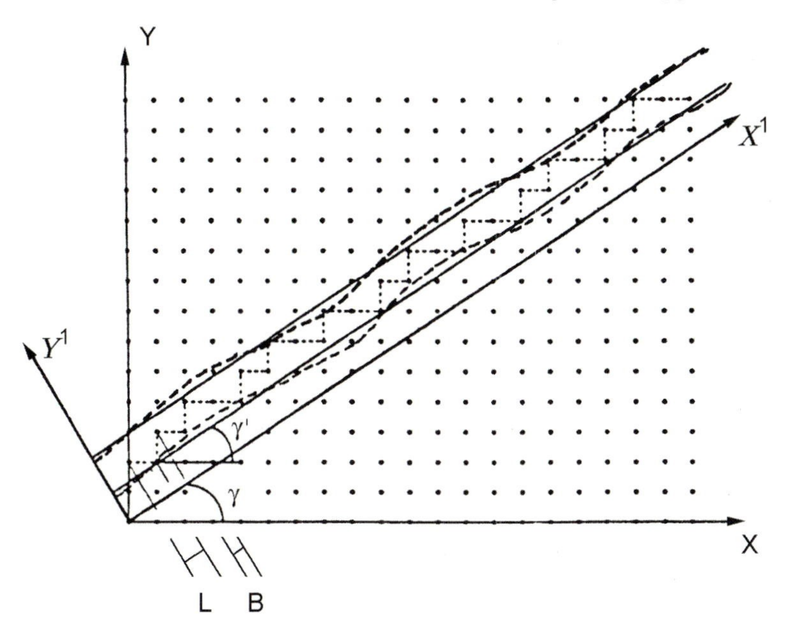

Fig. 6.28. The dashed curve is the collection domain whose slope varies with the amplitude of the bounded fluctuations about τ^{-1}: we introduce local defects into the Fibonacci chain, obtaining a model of random tiling

ing arises from these defects. This kind of structure lies at the roots of the quasicrystal models based on so-called random tilings.

We can also artificially corrugate the collection domain. In this case, we can immediately imagine that there is much more disorder inherent in the structure, as clarified by Fig. 6.29. Indeed, the forced fluctuations are unbounded; on the contrary, they grow along with $\{X^1\}$ values and can even give rise to random sequences of interatomic L and B distances. This kind of structural model for the quasicrystalline alloys is called *icosahedral glass*, and the kind of disorder we have introduced is called phason strain. The simulated diffraction patterns again exhibit rather narrow peaks whose shape, width and position depend on the specific features of the fluctuations.

From the experimental point of view, phason strain was observed in all the systems studied before 1988, and synthesised using rapid quenching methods, whereas it is absent in the so-called equilibrium quasicrystals with a high degree of structural perfection. Phason strain is associated with shifts in the experimental diffraction spots, or with spot broadening. It does not depend on $\boldsymbol{k}^1$, and it increases with the phason momentum, $\boldsymbol{k}^{\perp}$.

The close relationship between approximant periodic crystalline structures and quasicrystalline structures allow us to apply the cut and projection technique even to three-dimensional systems. The purely quasiperiodic three-dimensional quasicrystal is considered a section through the six-dimensional

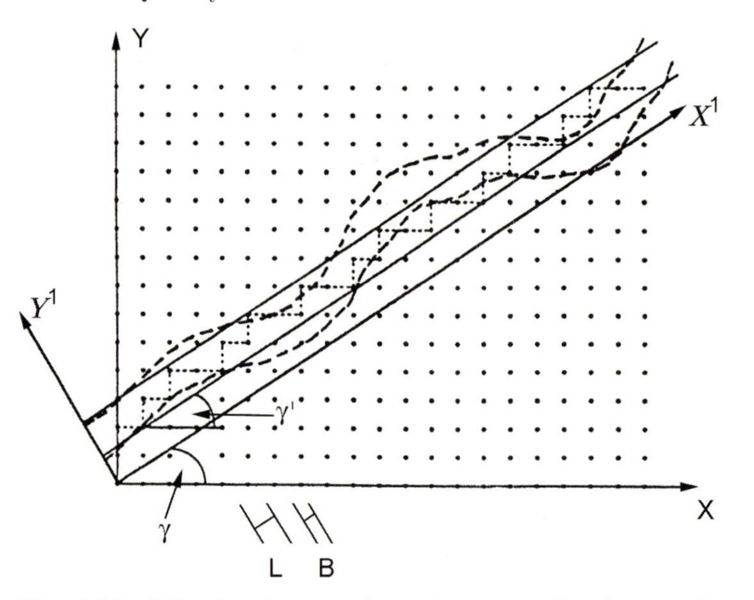

Fig. 6.29. Collection domain whose slope varies limitless: in the example, the amplitude of slope fluctuations increases with $\{X^1\}$. We introduce strong disorder into the Fibonacci chain. Model of quasicrystal with phason disorder, or icosahedral glass

periodic space, where a primitive hypercubic lattice is defined. Using the projection operation we can define the quasilattice constant for the quasicrystal. The link between such a constant and the diffraction pattern is not, however, as clear as it is for periodic crystals.

When we project a vector of the six-dimensional reciprocal space onto the real three-dimensional space, $\boldsymbol{R}$, with icosahedral symmetry, we obtain

$$\boldsymbol{r} = (2\pi/\boldsymbol{a}_j)\,\hat{\boldsymbol{P}} \sum_i n_i \boldsymbol{x}_i^* \tag{6.12}$$

where $\hat{\boldsymbol{P}}$ is the projection operator, as given by the matrix of rank three

$$\hat{\boldsymbol{P}} = \boldsymbol{1}/\sqrt{2} \begin{bmatrix} \sqrt{5} & 1 & 1 & 1 & 1 & 1 \\ 1 & \sqrt{5} & 1 & -1 & -1 & 1 \\ 1 & 1 & \sqrt{5} & 1 & -1 & -1 \\ 1 & -1 & 1 & \sqrt{5} & 1 & -1 \\ 1 & -1 & -1 & 1 & \sqrt{5} & 1 \\ 1 & 1 & -1 & -1 & 1 & \sqrt{5} \end{bmatrix} . \tag{6.13}$$

From (6.12), the basis vectors $\boldsymbol{x}_i^*$ in (6.9) may be rewritten in the three-dimensional space. If we choose a cubic coordinate system, based on three orthogonal binary rotation axes of the icosahedron as in Fig. 6.23, the vectors $\boldsymbol{x}_i^*$ are given by $(\pm 1, \pm\tau, 0)$ and their permutations, where τ is the golden ratio. All the $\boldsymbol{k}_i$ vectors have cubic coordinates $(h+h'\tau, k+k'\tau, l+l'\tau)$ where h, h', k, k', l, l' are integers.

Once we define the projection of the hypercube edge as the quasilattice constant, $\boldsymbol{a}_{qr}$, we obtain

$$\boldsymbol{a}_{qr} = \boldsymbol{a}/\sqrt{2}. \tag{6.14}$$

If we look at the value of the quasilattice constant for all known quasicrystals, the quasicrystals can be divided into two groups: for the Al–transition metal alloys $\boldsymbol{a}_{qr} \simeq 0.46$ nm, whereas for the quasicrystals obtained from Al–Mg–Zn and from Al–Li–Cu, $\boldsymbol{a}_{qr} \simeq 0.52$ nm. When we normalise to the atomic diameters of the alloy constituents the values are about 1.65–1.75 and about 2.00 respectively for the two groups.

The relative intensities of the spots in the diffraction patterns are caused by the atomic arrangements, which are repeated in the quasilattice; these intensities reflect the differences between the two groups of quasicrystals. With reference to the crystalline approximant structures, two distinct geometric structures have been identified in the arrangements of the atomic clusters forming the basic atomic motifs. These clusters are the Mackay icosahedron for the Al–transition metal group of alloys shown in Fig. 6.30 and the Pauling triacontahedron for the Al–Mg–Zn group displayed in Fig. 6.31. Both clusters contain three different atomic shells, each with icosahedral symmetry.

Modelling the atomic structure of icosahedral quasicrystals is based on these elementary structures which allow us to computer simulate the diffraction patterns, thus very closely reproducing the intensity differences in the experimental diffraction patterns.

Despite the above results, no quasicrystal structure is presently understood with the degree of accuracy and completeness we have reached in the analysis of crystalline structures. However, we expect considerable ad-

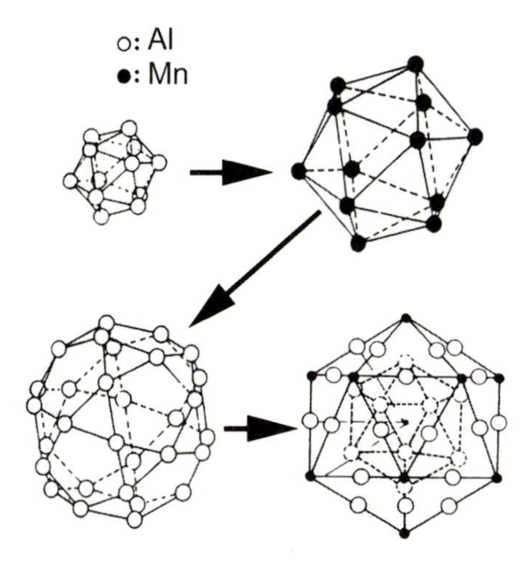

Fig. 6.30. Construction of a Mackay icosahedron cluster for $Al_{84}Mn_{16}$; open dots: Al; full dots: Mn (adapted from [6.4])

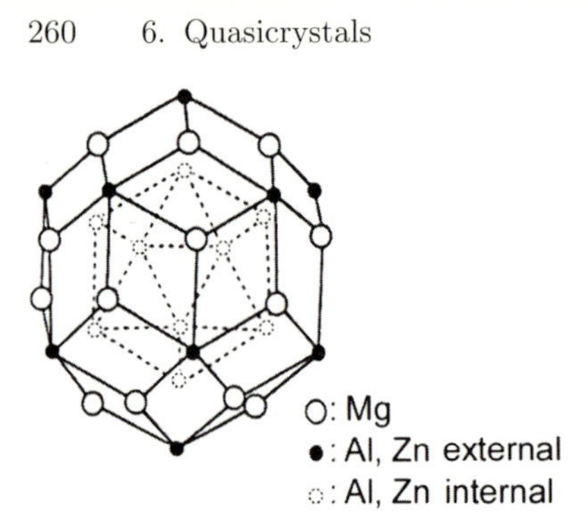

Fig. 6.31. Construction of a Pauling triacontahedron cluster (see Fig. 6.21) for $Mg_{39.5}(Al,Zn)_{60.5}$; open dots: Mg; full dots: Al, or Zn (external); dashed dots: Al, or Zn (inner)

vances in this field from the study of the recently discovered stable *binary* quasicrystalline alloy $Cd_{85}Yb_{15}$. Indeed, in a systematic exploration of the Cd–Yb phase diagram, icosahedral $Cd_{85}Yb_{15}$ was found; this phase congruently melts at 909 K. The obtained single crystals have a big orthogonal unit cell and selected-area electron diffraction patterns show that their lattice is icosahedral, primitive. The phase is found both in the solidified and in the fully annealed state, thus it is thermodynamically stable. The reason why no other rare earth forms stable quasicrystals with Cd lies in the combination of Yb size and valency. Yb can be divalent (radius 0.194 nm) and trivalent (radius 0.174 nm). It is divalent Yb to form the quasicrystal; the only other divalent and trivalent rare earth is Eu (radius of divalent Eu, 0.204 nm) that does not form any binary quasicrystal with cadmium. As the radii of most rare earths lie in the range 0.175–0.185 nm, a condition to obtain a binary quasicrystal with Cd is that the radius of the partner lies between 0.185 and 0.204 nm. Indeed, calcium, though not a rare earth, fulfils this condition and it has been found to form stable binary quasicrystals with Cd.

The crystalline approximant of $Cd_{85}Yb_{15}$ is the cubic phase $Cd_{85.7}Yb_{14.3}$ (space group Im3) with lattice parameter $a = 1.564$ nm. This structure is modelled by packing into a bcc skeleton clusters whose atomic shells have icosahedral symmetry. Unlike the above discussed Mackay icosahedron and Pauling triacontahedron, four atoms, arranged in a tetrahedron at the centre of each cluster, break the icosahedral symmetry, as shown in Fig. 6.32.

In an energetic picture, a stable quasicrystal corresponds, like an ordinary crystal, to a symmetric atom arrangement associated to the lowest energy. The role of the tetrahedron is to provide the asymmetry required by energy stabilisation. Here the structurally simple quasi-unit cell, through spatial repetition according to local matching rules, realises the lowest energy state that coincides with the highest packing density.

Alternatively to the energy picture, the presence of the tetrahedron supports the idea that quasicrystals are entropy-stabilised. Here the advantage with respect to the energy picture and the associated strict matching rules between elementary structural units is that in the entropic framework the exact

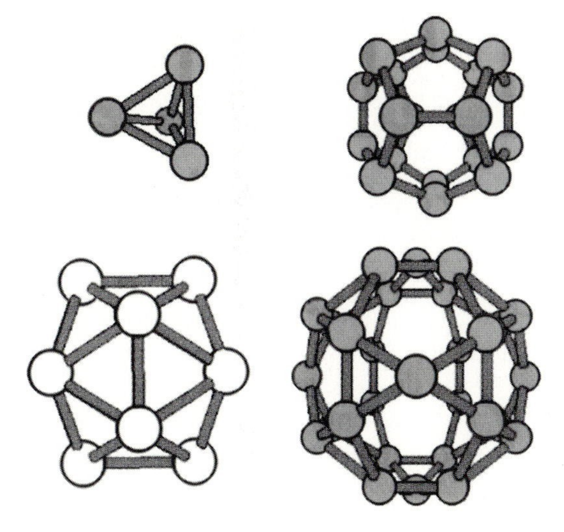

Fig. 6.32. Construction of the 66-atom icosahedral cluster in the quasicrystalline $Cd_{85}Yb_{15}$ alloy. The first shell is a tetrahedron of Cd atoms around the cluster centre. The second shell is a Cd dodecahedron (20 atoms). The third shell is an icosahedron of 12 Yb atoms. The fourth shell is a Cd icosidodecahedron, obtained by putting 30 Cd atoms on the edges of the Yb icosahedron (adapted from [6.5])

tetrahedron orientation is irrelevant when modelling quasicrystal structure because all orientations are equivalent.

Also, the observation that $Cd_{85}Yb_{15}$ is a congruent melter supports the entropy picture. Indeed, when cooling down the high temperature liquid CdYb, the first phase that forms is the quasicrystal, which can be in equilibrium with the melt; this indicates that the entropies of the two phases are comparable. Formation of the crystalline approximant requires a solid state reaction at a lower temperature, involving the quasicrystal and the other crystal phase.

Besides CdYb, an icosahedral phase has been found to form as a primary precipitation phase in the crystallisation process of the binary amorphous $Zr_{70}Pd_{30}$ alloy. Although the icosahedral phase is metastable and converts to an equilibrium crystalline phase upon annealing, its discovery has made it possible to predict the existence of several binary icosahedral alloys in other systems.

6.5 Structural Properties and Stability of Real Quasicrystals

We can often understand the features of the real quasicrystalline systems synthesised using non-equilibrium methods in terms of the icosahedral glass model. From a pictorial point of view, in two dimensions the difference between the computer simulated structures using a quasiperiodic model and using an icosahedral glass, shown in Fig. 6.33, is clear. In part (a) of the figure, the tiling with pentagons connected to each other through their edges using Penrose matching rules completely and aperiodically covers the plane;

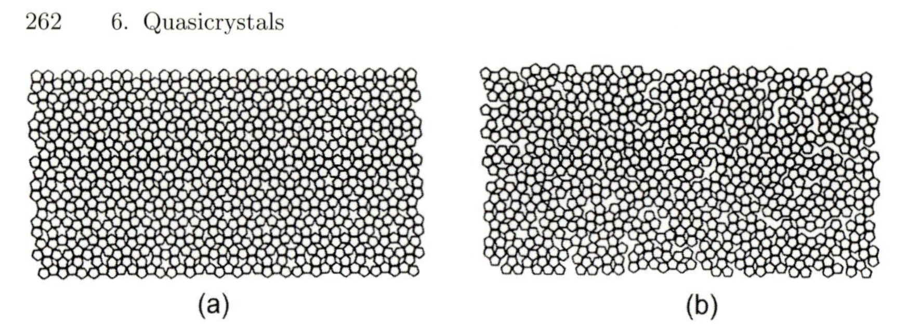

Fig. 6.33. (**a**) Quasiperiodic arrangement of pentagons connected together by the edges using local matching rules: the result is a Penrose tiling. (**b**) Random arrangement of pentagons connected together by their edges. A two-dimensional version of the icosahedral glass is obtained

in part (b), the same pentagons are randomly connected together through their edges; the inherent disorder is considerable, as evidenced by several large voids.

When we extend the icosahedral glass (part (b) of Fig. 6.33) into three dimensions, we obtain a system with atomic clusters where each cluster has icosahedral symmetry. The clusters are reciprocally connected in such a way to exhibit the same orientation, yet there is a certain degree of randomness in the way the clusters are interconnected. From Fig. 6.34 it is evident that cluster growth is governed by purely local rules, thus avoiding the formation of quasiperiodic patterns in the resulting structure. From the semantic point of view, this is not a glassy system, in the strict sense of the term, since we observe long range orientational order; yet, a high degree of disorder is intrinsic to the structure of this model system.

The development of the icosahedral glass model matches the observations of finite peak widths in the high resolution X-ray diffraction patterns for Al–Mn; from these patterns the coherence length for the samples is around

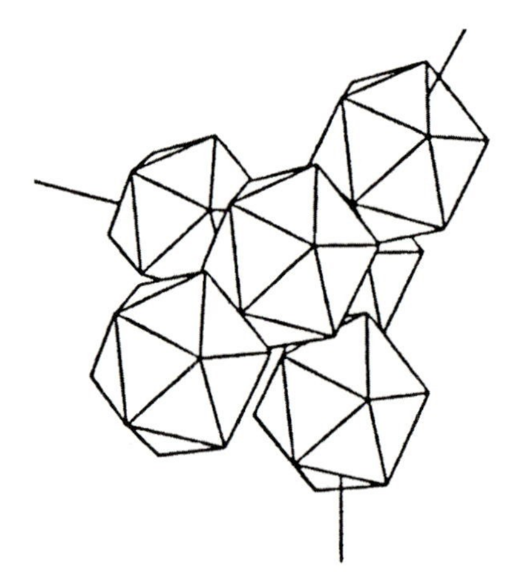

Fig. 6.34. Icosahedra connected together along the threefold axes form the basic structural element of the three-dimensional icosahedral glass model

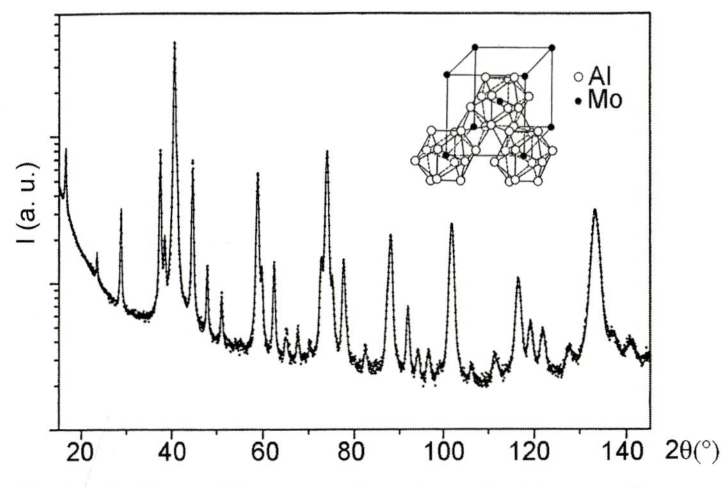

Fig. 6.35. X-ray diffraction pattern from $Al_{12}Mo$ crystalline approximant. The structure is made of a bcc lattice of Mo atoms (full dots), each located in the centre of an icosahedron of Al atoms (open dots) interconnected through octahedra. (adapted from [6.6])

30 nm, even though icosahedral orientational order extends over much longer distances. Thus, these quasicrystalline systems exhibit icosahedral orientational correlations that extend over much greater distances than the translational correlation length, analogous to the liquid crystals.

Even the crystalline approximant phases, where α–AlMnSi, $Al_{12}Mo$ and R–AlLiCu are prototypes, are formed by clusters of atoms with a substantially icosahedral shape all packed together in a body-centred cubic lattice, exemplified in Fig. 6.35. For the approximants, the size of the atomic clusters required to obtain agreement between the random packing model and the experimental data is just the cluster size in the crystalline phase. This result suggests that the unit cells in the crystalline and icosahedral phases are identical.

It is immediate to represent the growth process of a quasicrystal using a very simple scheme. When the liquid alloy is cooled, stable icosahedral clusters form. If the material is further cooled slowly enough, the clusters will grow and form a periodic crystalline arrangement of clusters arranged in a bcc lattice with coordination eight. If, though, the cooling conditions are off-equilibrium, each cluster will coordinate with a lower number of neighbours; consequently, interconnection defects form in a quasiperiodic structure.

This conceptual scheme is quite realistic for those systems prepared with non-equilibrium techniques; however, some experimental results are incompatible with this scheme. Indeed, quasicrystalline alloys have been obtained by solid state diffusion reactions, where the icosahedral phase nucleates in the absence of a pre-existing liquid, or amorphous phase with its associated short range order. Moreover, systems have been observed where the icosahedral phase is a stable equilibrium phase, such as AlLiCu, AlCuFe and TiZrNi. In particular, as regards the TiZrNi system, the alloy $Ti_{45}Zr_{38}Ni_{17}$, obtained by cooling down the melt, includes the Laves hexagonal C14 phase and a solid solution α; after annealing in a vacuum at 843 K for 64 hours, the material exhibits one single *stable* phase with a $Ti_{41.5}Zr_{41.5}Ni_{17}$ composition and primitive icosahedral structure similar to AlLiCu.

One important feature of the icosahedral glass model is given by the predicted dependence of the broadening of the simulated diffraction peaks on the phason momentum $|\boldsymbol{k}^{\perp}|$. In the first models this dependence was more than linear, which gave rise to much broader diffraction peaks than those experimentally observed at high $|\boldsymbol{k}^{\perp}|$ values. However, when structural rearrangement is introduced into the model by slightly heating the system, it improves the agreement with the experimental results. This indicates that the widths of the diffraction peaks are approximately linearly dependent on the phason momentum.

Although the icosahedral glass model for quasicrystalline systems allows quite a considerable degree of disorder in the structure, it does offer us a simple representation of the structure with a good qualitative reproduction of the features of the diffraction patterns. This suggests that a real icosahedral structure is probably midway between the extremes of a purely random model and the almost perfect periodic order associated with the Penrose tiling.

Investigations into quasicrystals that have been amorphised and, on the contrary, investigations into quasicrystals obtained from the glassy phase have both demonstrated that the same icosahedral structural units are exhibited in liquid, amorphous and quasicrystalline systems; these very same units are found also in the crystalline approximants. Ordinary metallic glasses cannot be transformed into quasicrystals; thus, we wonder if the local order in the amorphous materials from which we can synthesise quasicrystals, which we call qc. amorphous, already include some of the typical features that favour the establishment of long range orientational order during the transition from the metallic glass to the quasicrystal.

In Fig. 6.36 the structure factor for the qc. amorphous $Al_{84}Mn_{16}$ is compared with that of the metallic glass $Co_{80}P_{20}$. We notice that both exhibit a pre-peak, a main peak at $k_1 \simeq 30\ \mathrm{nm}^{-1}$ and a second peak. The main peak in $Al_{84}Mn_{16}$ is caused by the broadening and overlapping of the contributions from the most intense lines in the quasicrystalline diffraction pattern. Once peak positions are normalised to the position of k_1, the position of the second peak of $Al_{84}Mn_{16}$ is given by $r_2 = (k_2/k_1) \simeq 1.66$, whereas for $Co_{80}P_{20}$ it is given by $r_2 \simeq 1.72$. The value for $Al_{84}Mn_{16}$ is typical of other qc. amorphous system, such as $Al_{84}V_{16}$ and $Al_{70}Fe_{13}Si_{17}$, whereas the position of the second peak in other ordinary metallic glasses, such as $Pd_{80}Si_{20}$ and $Fe_{80}B_{20}$ stays in the interval $r_2 = 1.70 - 1.75$. The difference is quite meaningful.

The simulation of chemical medium range order between like atoms does not influence the position of the peaks, even though such ordering does pro-

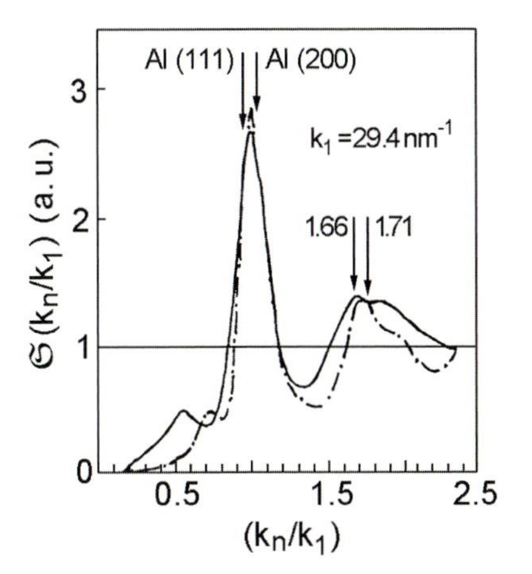

Fig. 6.36. Comparison between experimental structure factors $\mathfrak{S}(k_n/k_1)$. —: $\mathfrak{S}(k_n/k_1)$ for amorphous $Al_{84}Mn_{16}$; k_1 is the position of the principal peak of $\mathfrak{S}(k)$, taken as the unit wavevector. $- \cdot -$: $\mathfrak{S}(k_n/k_1)$ for amorphous $Co_{80}P_{20}$ (adapted from [6.7])

duce an increase in the intensity of some peaks, in particular in the pre-peak at $r_0 = (k_0/k_1) = 0.55$.

Now, the general features of the diffraction pattern generated by an amorphous structure depend on the contributions arising from first, second and, at the most, third neighbour correlations (see Sect. 4.3); thus, r_2 is presumably given by small scale topological features, including a few atoms. The simulation of the trend in the structure factor $\mathfrak{S}(k)$ for three different clusters, where each cluster is formed by thirteen atoms, is given in Fig. 6.37. In the first case, the atoms form an ideal icosahedron with twelve spheres at the vertices in rigid contact with each other and with the thirteenth atom, 5% smaller, in the central position ($r_2 = 1.65$). The second structure is a distorted icosahedron formed by like atoms ($r_2 = 1.72$). Lastly, an fcc cluster of identical atoms is simulated ($r_2 = 1.84$). The values for r_2 for the two icosahedral arrangements match the experimental results for $Al_{84}Mn_{16}$

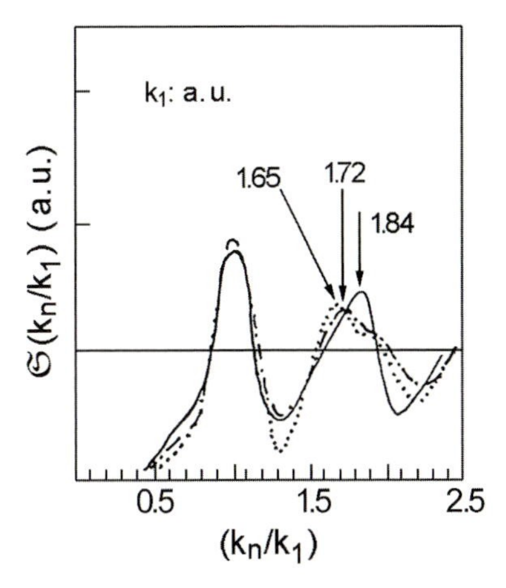

Fig. 6.37. $\mathfrak{S}(k_n/k_1)$ calculated for thirteen-atom clusters with various geometries $\cdots$: ideal icosahedron, where the central atom is 5% smaller than the surface atoms; $- \cdot - \cdot -$: distorted icosahedron, obtained using the dense, random packing of hard spheres model; —: fcc packing (adapted from [6.7])

and for $Co_{80}P_{20}$, thus suggesting that there are regular icosahedra in the qc. amorphous alloys.

Although the non-relaxed icosahedral glass model cannot correctly reproduce the width of the diffraction peaks observed in real quasicrystals, it remains a valid tool in the search for mutual correlations between amorphous and quasicrystalline structures. One specific X-ray diffraction investigation has been performed on the $Al_{75}Cu_{15}V_{10}$, $Al_{53}Mn_{20}Si_{27}$ and $Al_{65}Cu_{20}Fe_{15}$ alloys; the starting materials were amorphous materials and they were annealed to grow the icosahedral phase (i).

Compared to i-AlCuFe, the patterns for the first two alloys exhibit fewer, considerably broader peaks that do not change under further heat treatment. The correlation length, estimated by the peak width, is around 30 nm both for i-AlCuV, and for i-AlMnSi. Whereas we do not observe any relation between the phason momentum $|\boldsymbol{k}^{\perp}|$ and the diffraction peak width for i-AlCuFe, we do notice progressive peak broadening as a function of $|\boldsymbol{k}^{\perp}|$ for i-AlCuV and i-AlMnSi. The dependence is nearly linear, in agreement with the predictions of the structural model of a partially relaxed icosahedral glass.

After keeping the glassy phase at high temperature (693 K) for an hour, by high resolution electron microscopy, and keeping the beam parallel to a fivefold axis of the i-AlCuV sample, we obtain an image that highlights defective areas in the grown quasicrystalline structure, as shown in Fig. 6.38. In particular, the exhibited fringes are an index of incongruity between the spatial

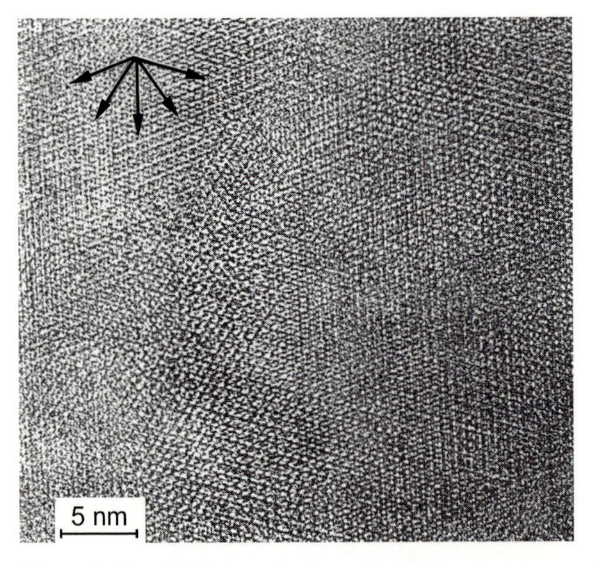

Fig. 6.38. High resolution electron microscopy image of quasicrystalline $Al_{75}Cu_{15}V_{10}$. The incident beam is kept parallel to a fivefold axis of the system. The fringes are attributed to reciprocally incongruent spatial arrangements in various directions (adapted from [6.8])

order that develops along different directions and reflect a random distribution of phason disorder. The simulation of the image lattice, which produces the fringes in Fig. 6.38, produces a structure formed by pentagons connected together at their edges in a random fashion. Some elementary structures we typically find in Penrose tilings ("diamond" and "boat") appear with unusual frequency. However, this kind of tiling is specifically characterised by, above all, defects, that are highlighted in black in Fig. 6.39; this is a typical feature of the icosahedral glass model. Thus, in real quasicrystalline alloys, obtained in non-equilibrium conditions, the presence of structural defects can be represented by a icosahedral glass model, possibly relaxed. This is true also of quasicrystals grown from a non-equilibrium system, such as an amorphous phase.

Though most experimental efforts to understand the atomic structure of the quasicrystals were focused on reciprocal space studies, scanning tunneling microscopy (STM) has been used to observe the real space structure of a fivefold surface in the prototypical $Al_{68}Pd_{23}Mn_9$ icosahedral quasicrystal.

The STM images exhibit atomically flat terraces that are much like the terraces of ordinary periodic crystals. However, the step heights between terraces exhibit only two incommensurable values, $L = 0.678$ nm and $B = 0.422$ nm; the L/B ratio is 1.61, very close to the golden ratio τ. The flat terraces are separated by steps in the sequence $\underline{LLBLLBLBLL}$, which is part of the seventh generation of the Fibonacci chain F^n, $B\underline{LLBLLBLBLL}BL$, obtained iteratively by the substitution rule $B \rightarrow L$, $L \rightarrow BL$. These observations do indeed confirm the theoretical prediction that any icosahedral quasicrystal will exhibit a finely faceted surface with flat terraces perpendicular to a fivefold axis of the quasicrystal. The terraces are separated from

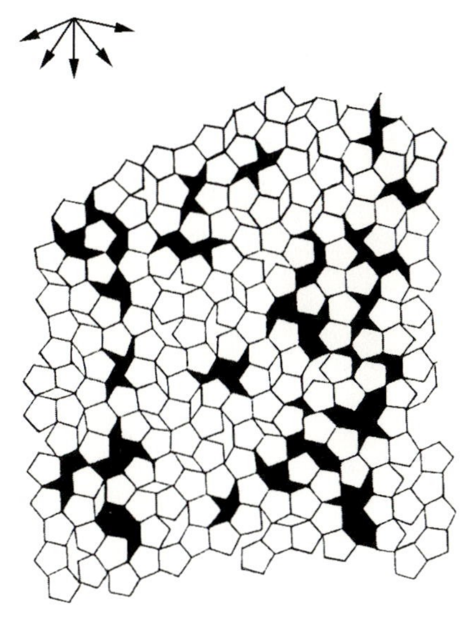

Fig. 6.39. Simulation of the image lattice that produces the structure in Fig. 6.38. The simulation is obtained using pentagons connected together by their edges in a random way, resulting in an icosahedral glass (adapted from [6.8])

each other by steps with only two possible different heights, where the ratio between the higher and lower step height is τ, and the step sequence makes a Fibonacci chain.

The experimental proof that the normals to the terrace surfaces are parallel to a fivefold axis of the quasicrystal is given by the terrace shown in Fig. 6.40 observed using high resolution STM. We notice a highly regular distribution of pentagonal-shaped "holes" the same size and orientation along the whole terrace; these holes give rise to a set of parallel lines of holes.

The distribution of these pentagonal objects reflects the quasicrystalline nature of the samples. If we draw a set of lines that form a grid parallel to the pentagon sides we observe only two possible distances between pairs of these lines, $L = 1.181$ nm and $B = 0.738$ nm; the ratio between them is very close to τ and their spatial sequence reproduces a Fibonacci chain.

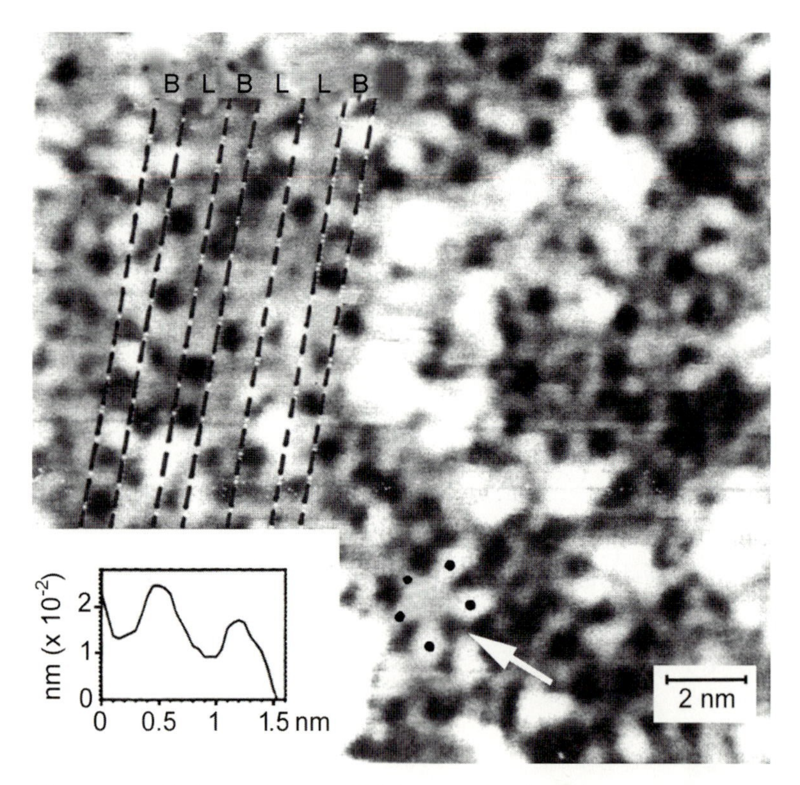

Fig. 6.40. High resolution STM image of a fivefold surface of quasicrystalline $Al_{68}Pd_{23}Mn_9$; symmetry elements with fivefold order are highlighted. A pentagon-shaped star is shown by the arrow. The separations between the dotted lines aligned with the edges of pentagonal spots constitute a non-periodic sequence, with alternating L and B distances that reproduce part of the Fibonacci chain. $L = 1.18$ nm; $B = 0.738$ nm; the ratio (L/B) is almost equal to the golden ratio τ (adapted from [6.9])

The morphological features of a threefold surface of an icosahedral AlPdMn alloy are similar to those of a fivefold surface; again, a terrace-step configuration is found. Each terrace displays a fine structure with long range order compatible with the threefold symmetry. This fine structure is rougher than that observed in fivefold surfaces, and we find holes of typical depth in the tenth of a nanometer range. As to the steps, they can be straight and they bound triangles, or portions of triangles, whose edges are invariably parallel.

When studying the quasicrystalline systems we obviously want to know what mechanisms give the quasicrystals their considerable structural stability.

The specific features of the diffraction patterns obtained from quasicrystals prove the existence of fivefold orientational order and a dense distribution of diffraction spots, or peaks. Each of these peaks is very narrow, being delta-like in samples prepared using equilibrium methods where the phason disorder is strongly limited. This feature suggests that the electrons in quasicrystals undergo strong diffusion throughout the entire reciprocal space.

The electrical resistivity typical of these structures is very high; in the extreme cases of icosahedral samples of AlPdRe with high structural perfection, at low temperature, (below 1 K), the resistivity is around 30Ω cm, thus being 9–10 orders of magnitude greater than in pure aluminium. Resistivity grows as the temperature rises until, at room temperature, it is about 200 times the value at 4 K. The result is anomalous if compared to the ordinary metals, insulators and semiconductors. This kind of behaviour can be associated with the packing hierarchy for icosahedral clusters and with the intensity of the Coulomb interaction, that give rise to repeated localisation of the binding electrons. As a result, a significant electron-quasilattice interaction occurs indicating that the cohesive mechanism in quasicrystals is electronic.

In the same sense, we have to interpret the measurements of electronic specific heat coefficient, γ_{el}, which give us an estimate for the density of the electronic states around the Fermi level. The rather small values (0.1–0.3 $\mathrm{mJ\,K^{-2}mol^{-1}}$), measured for all stable quasicrystals, lie between thirty and ten percent of the value for the free electron model. It is likely that in these systems a mechanism exists that opens up a gap in the electronic density of states at the Fermi level.

The existence of localised versus extended electron states in quasicrystals, corresponding respectively to a discrete, or continuous, eigenvalue spectrum, has not been clarified completely. The problem has recently been investigated by angle resolved photoemission in the decagonal quasicrystal $Al_{71.8}Ni_{14.8}Co_{13.4}$. Here crystalline order exists along the tenfold axis, while the planes with tenfold symmetry, lying parallel to the sample surface and normal to the tenfold axis, show quasi-periodicity. Both emission angles and kinetic energies (i.e. momenta $\boldsymbol{k}$) of valence electrons scattered by soft X-rays from the near surface regions were measured. Looking at the behaviour of electrons in the decagonal planes, the effect of quasicrystallinity was tested, while looking at normal incidence to the planes the effect of crystalline order was assessed.

Both $s-p$ and d states exhibit band-like behaviour, with the symmetry of the quasiperiodic lattice, as determined by low energy electron diffraction. Moreover, the Fermi level is crossed by dispersing d bands, similarly to what is observed in ordinary crystals. The measured broad bandwidths and effective masses are comparable to free electrons suggesting that at least some states, both within and out of the quasicrystalline planes, are extended in real space. However, the weakness of the observed features, from both $s-p$ and d states, is probably due to the damping of such states, which is typical of localisation.

Indeed, quasicrystals could have *critical* states, with a power-law decay of their amplitudes (in amorphous metals the decay is exponential). Such states should be localised to a cluster of size R, which can resonantly tunnel to locally isomorphous clusters, within a $2R$ distance (see Sect. 6.3). Thus the properties of critical eigenstates could depend on a dispersion relation determined by the potential profile that results from the local environment, with strongly dampened amplitudes.

Among ordinary metallic crystalline materials the electron phases are alloys whose crystalline structure changes at specific values for the number of average valence electrons per atom, $\langle Z \rangle$. In the framework of the nearly free electron (NFE) model, the compositional limits for the various electron phases are determined by the condition of geometrical contact between the Fermi surface and the first Brillouin zone.

We can assume that the stability of quasicrystals is also attributable to an analogous mechanism, and apply the NFE model to quasicrystalline alloys. From an analysis of the X-ray, or neutron, diffraction patterns and supposing that the Fermi surface and the pseudo-Brillouin zone have spherical symmetry, we obtain

$$\frac{4}{3}\pi \boldsymbol{k}_{\mathrm{F,NFE}}^{3} = \frac{\left[(2\pi)^{3}/2\right]\langle Z \rangle}{(4/3)\,\pi \langle r \rangle^{3} \eta} . \tag{6.15}$$

The volume of the Fermi surface is given by the ratio between the volume of the occupied electronic states and the volume of the pseudo-Brillouin zone; $\langle r \rangle$ is the average atomic radius of the alloy, thus for a $A_{1-x}B_x$ system, $\langle r \rangle = r_{\mathrm{A}}(1-x) + r_{\mathrm{B}}x$ and η is the packing efficiency, namely 0.688 for ideal icosahedral packing.

The Fermi wavevector values calculated with the NFE model, $k_{\mathrm{F,NFE}}$, can be plotted against k_{F} values deduced from the position of the most intense peak in the experimental diffraction pattern. For a meaningful set of quasicrystalline alloys, both icosahedral and decagonal, we observe the linear relationship shown in Fig. 6.41.

This allows us to extend the sequence of electron phases so that they also encompass the quasicrystals. The peculiar quasicrystalline structure is

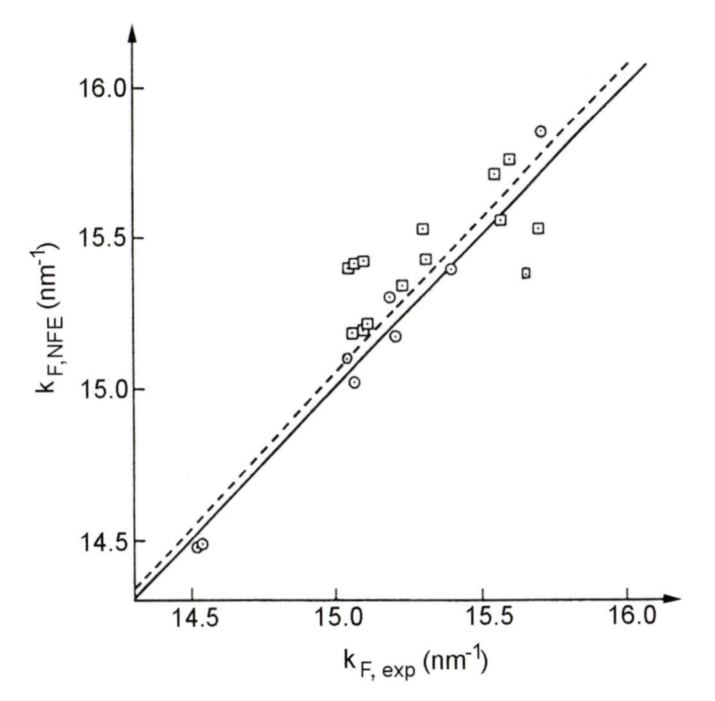

Fig. 6.41. Relation between experimental values of the Fermi wavevector, k_{F} and values calculated with the nearly free electron (NFE) model, $k_{F,\mathrm{NFE}}$. Dots and squares refer to alloys of the Al–transition metal family and Ga–Mg–Zn family, respectively (adapted from [6.10])

bounded by an average lower $\langle Z \rangle$ value of 1.45 and an upper value of 2.6. Table 6.1 gives the structure, the average concentration of valence electrons per atom, $\langle Z \rangle$, and the upper $\langle Z \rangle$ limit for the electron phase sequence including crystalline, quasicrystalline and glassy states.

Electron phase	Structure	$\langle Z \rangle$	Limit $\langle Z \rangle$
α	fcc	1.00 – 1.41	1.362
β	bcc	1.35 – 1.60	1.480
ζ	hcp (c/a=1.633)	1.22 – 1.83	–
γ	complex cubic	1.54 – 1.70	1.620
δ	complex cubic	1.55 – 2.00	–
μ	β–Mn	1.40 – 1.54	–
ϵ	hcp (c/a=1.570)	1.65 – 1.89	1.700
η	hcp (c/a=1.750)	1.92 – 2.00	–
qc	icosahedral	1.45 – 2.60	–
amorphous	disordered	> 1.80	–

Table 6.1. Crystalline structure, average concentration of conduction electrons per atom, $\langle Z \rangle$ and limiting $\langle Z \rangle$ values for various electron phases (adapted from [6.10])

It is significant that the values for the parameter $\langle Z \rangle$ associated with various crystalline phases, quasicrystalline phases and the glassy phase, make up a hierarchy.

The idea that an essentially electronic mechanism is responsible for the stability of quasicrystals was recently supported by a detailed analysis of structure stability and electronic transport properties performed on several films of $Al_{1-x}(Cu_2Fe)_x$, both amorphous and quasicrystalline icosahedral, with Al contents ranging from 30 to 80 at.% and the Cu_2Fe ratio kept constant.

In the amorphous films an electronic induced structural peak at $k = k_p$, shifting as a function of composition just below, but parallel to, $2k_F$ is observed (see Sect. 3.3). Correspondingly, the deviations of electronic transport properties from the NFE model are largest. As to the icosahedral phase, with ideal atom packing at the composition $Al_{62.5}Cu_{25}Fe_{12.5}$, it shows two main structural peaks. The position of the first peak coincides with that of the electronic induced peak in the amorphous alloy with the same composition. Assuming that the $2k_F$ value coincides in both phases, the second peak of the icosahedral phase falls even closer to $2k_F$, thus optimising the structural stability.

The initially amorphous films were annealed in several steps, until crystallisation into the icosahedral phase occurred at about 700 K. The thermopower $S(T)$ of the alloy was measured down to 10 K after each annealing step, as shown in Fig. 6.42 for two representative samples. In the amorphous phase both films show proportionality between thermopower and tempera-

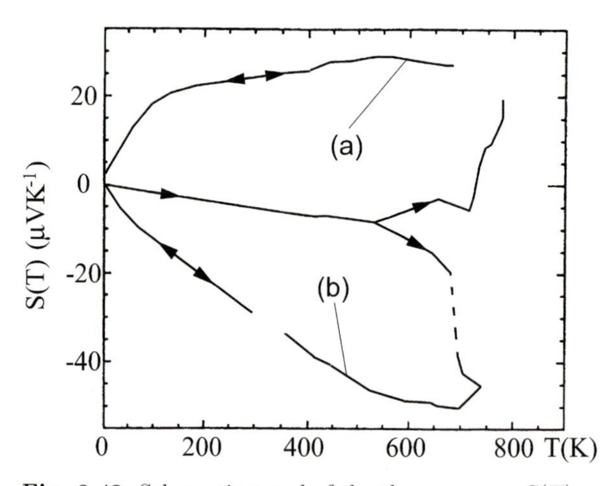

Fig. 6.42. Schematic trend of the thermopower $S(T)$ as a function of temperature for a representative quasicrystalline film with composition $Al_{62.5}Cu_{25}Fe_{12.5}$ (curve a) and $Al_{63.5}Cu_{24}Fe_{12.5}$ (curve b). Arrows indicate the verse of phase change; double arrows refer to reversible changes of the icosahedral phase. No data are available in the regions where the curves are interrupted (adapted from [6.11])

ture, in agreement with the free electron model. By contrast, dramatic differences in magnitude and even in sign, both positive and negative, of $S(T)$ were observed in icosahedral films. $S(T)$ is strongly affected by band structure effects; as a general trend it is observed that those samples with largest electrical resistivity (measured independently), hence those that have best transformed into the icosahedral phase, display the largest positive thermopower. Thus positive $S(T)$ values are a fingerprint of best quasicristallinity.

In amorphous systems, electrons can be treated as spherical waves, the atoms being located at Friedel minima of the pair potential (see Sect. 3.3). Owing to scattering, those electrons whose wavelength matches the sphere diameter, $\lambda_{\mathrm{F}} = (2\pi/2k_{\mathrm{F}}) = (2\pi/k_{\mathrm{p}})$, will be localised because multiple backscattering is extremely effective. Gaps, or pseudo-gaps in the electronic density of states will open at the Fermi energy E_{F}. Let us now suppose that starting from the amorphous state, each annealing step allows for a better atom rearrangement in the Friedel minima; this leads to enhanced backscattering, to a deeper pseudo-gap and to lower conductivity. In the icosahedral phase atom location at the shells becomes ordered (orientationally) and angular correlations arise. The ultimate effect of such an ordering process is even stronger coherent backscattering of electron waves, which in turn enhances interference effects and localisation. Thus, the observed low conductivity of perfect quasicrystals depends on multiple backscattering and a coherent interference effect.

The strong similarity between atomic structure and electronic transport properties of amorphous and icosahedral phases again supports the idea that both phases are stabilised by a Hume–Rothery mechanism, the essential difference between the two phases being the orientational order in the icosahedral phase.

A. Appendix

The following is a collection of results pertaining to the theory of the almost periodic functions, and particularly the quasiperiodic functions [A.1] [A.2], presented in axiomatic form. The aim is to define certain properties of the Fourier spectrum for those classes of function that are useful in comparing the diffraction patterns obtained from quasicrystalline materials.

Let $\Re$ be the real line, X a complete metric space and $\mu = \mu(x_1, x_2)$ a metric on X. Let $f(t) : \Re \to X$ be a continuous function with values in X; the range of f is the set $\mathcal{D}_f \{x \in X : x = f(t), t \in \Re\}$.

We say that a set of real numbers $E \subset \Re$ is *relatively dense* if there is a number $\lambda > 0$ such that any interval $(\alpha, \alpha + \lambda) \subset \Re$, of length λ, contains one number from E.

A number τ is called an *ε-almost period* of $f : \Re \to X$ if

$$\sup_{t \in \Re} \mu\left(f(t+\tau), f(t)\right) \leq \varepsilon. \tag{A.1}$$

We say that a continuous function $f : \Re \to X$ is *almost periodic* if, for each $\varepsilon > 0$, it has a relatively dense set of ε-almost periods; this means that there is a number $\lambda = \lambda(\varepsilon) > 0$ such that each interval $(\alpha, \alpha+\lambda) \subset \Re$ contains at least one number $\tau = \tau_\varepsilon$ that satisfies (A.1).

When we analyse (A.1) we notice that an almost periodic function will allow an infinite number of τ translations such that in each point the translated function differs by less than ε from the original, non-translated, function. This guarantees that the sum of two almost periodic functions is in turn almost periodic.

Let, for example,

$$f(t) = \sin(2\pi t) + \sin(2\pi\tau t). \tag{A.2}$$

This function is non-periodic when τ is irrational; however, let us consider the multiples of τ with a module of one; these are a countable number of irrationals such that, when each is translated by a suitable integer they are contained in the interval $[0, 1]$. As such, they *densely* cover this interval. For this reason, for each positive ε number there is an integer N_ε such that

$$|\sin(2\pi\tau N_\varepsilon)| < \varepsilon. \tag{A.3}$$

Each periodic function is also almost periodic. For, if f is periodic with period T, then all numbers $nT(n = \pm 1, \pm 2 ...)$ are also periods of f, and thus are almost periods of f for each $\varepsilon > 0$.

The set of numbers nT is, in turn, relatively dense.

The period of a periodic function gives us a unique definition of the set of its Fourier exponents. If T positive is the period, then all the Fourier exponents are multiples of $2\pi/T$.

The almost periods of an almost periodic function and the relative Fourier exponents are closely related together; for each d positive and for all the natural numbers N there is an $\varepsilon = \varepsilon(\delta; N) > 0$ such that each ε-almost period of an almost periodic function $f(t) : \Re \rightarrow X$ fulfils the system of inequalities

$$|\exp(\mathrm{i}\lambda_n \tau) - 1| \leq \delta \quad (n = 1, 2, ...N). \tag{A.4}$$

Conversely, for each positive ε there is a positive $\delta = \delta(\varepsilon)$ and a natural number $N = N(\varepsilon)$ such that each real number τ that fulfils the system of inequalities (A.4) is an ε-almost period of an almost periodic function $f(t)$. As such, an almost periodic function $f(t)$ may be expanded in a Fourier series

$$f(t) = \sum_n f(n) \exp[2\pi k_n t]. \tag{A.5}$$

The Fourier spectrum for an almost periodic function thus consists of a *countable* sum of *delta* peaks.

The family of almost periodic functions is closely tied to the family of periodic functions of several variables, including a countable number of variables and, in particular, to the family of the limit periodic functions. A function $\tilde{f}(t_1, t_2, ...t_n) : \Re^n \rightarrow X$ is a limit periodic function of the n variables $t_1, t_2 ...t_n$ if it is the uniform limit on $\Re^n$, of a sequence $f_k(t_1, t_2, ..., t_n)$ $(k = 1, 2, ..., n)$ of continuous periodic functions.

A function $\tilde{f}(t_1, t_2, ...)$ of a countable number of variables is called limit periodic if a sequence of continuous periodic functions $f_k(t_1, t_2, ..., t_{m_k})$ exists (such that $\lim_{k \rightarrow \infty} m_k = \infty$) which uniformly converges to $\tilde{f}(t_1, t_2, ...)$.

Let $F(t_1, t_2, ...t_n)$ be a continuous periodic function and $T_1, T_2, ..., T_n$ its periods; we assume that $T_1^{-1}, T_2^{-1}, ..., T_n^{-1}$ are linearly independent and we define the diagonal function $f(t) = F(t, t, ..., t)$.

The set of values for the diagonal function $f(t)$ is dense everywhere in the set of values of $F(t_1, t_2, ...t_n)$; this means that for all points $\left(t_1^{(0)}, t_2^{(0)}, ..., t_n^{(0)}\right) \in \Re^n$ and for every $\varepsilon > 0$ there is a $\xi \in \Re$ such that

$$\left\| F\left(t_1^{(0)}, t_2^{(0)}, ..., t_n^{(0)}\right) - f(\xi) \right\| < \varepsilon. \tag{A.6}$$

Each almost periodic function $f(t)$ is the diagonal function of a limit periodic function of a finite number, or a countable number of variables. This is true also if $F(t_1, t_2, ...t_n)$ is a periodic continuous function, with periods $T_1, T_2, ..., T_n$, and the diagonal function $f(t) = F(t, t, ..., t)$ is almost periodic.

A non-empty set of numbers is called a *module* if it is a group under the operation of addition. For example, the set of all real numbers is a module; the set of all the real integers is a module. We shall here only consider countable modules of real numbers. Given a finite, or countable set of real numbers $\alpha_1, \alpha_2, ...\alpha_l, ...$ we can obtain a module containing all of these numbers, considering the numbers $m_1\alpha_1 + m_2\alpha_2 + ... + m_l\alpha_l$, with $m_1, m_2...m_l$ integers and l arbitrary and finite. This set of numbers is the smallest module containing the set $\{\alpha_1, \alpha_2, ...\alpha_l\}$ and is given as $\mathcal{M}\{\alpha_1, \alpha_2, ...\alpha_l\}$

A sequence of numbers $t_m \in \Re$ is called *f-increasing* if $f(t + t_m) \to f(t)$ uniformly, that is if t_m is an ε_m-almost period where $\varepsilon_m \to 0$.

For every almost periodic function $f : \Re \to X$ a countable module $\mathcal{M}_f$ exists, where a sequence $\{t_m\}$ is f-increasing if and only if

$$\exp(\mathrm{i}\lambda t_m) \to 1, \qquad \text{where } \lambda \in \mathcal{M}_f. \tag{A.7}$$

It follows from this result that if two almost periodic functions $f(t)$ and $g(t)$ have equal modules ($\mathcal{M}_f = \mathcal{M}_g$) then they have a single unique set of increasing sequences. Thus, for two almost periodic functions $f(t)$ and $g(t)$ to have the same increasing sequences $\mathcal{M}_f$ must and only has to be equal to $\mathcal{M}_g$.

The kind of basis of Fourier exponents of the almost periodic function $f(t)$ in the module $\mathcal{M}_f$ gives us the characteristic properties of the limit periodic function $\tilde{f}(t_1, t_2, ...)$ where $f(t)$ is the diagonal function. In particular, if the basis in $\mathcal{M}_f$ is integer, that is each Fourier exponent of $f(t)$ is a linear combination of $\gamma_1, \gamma_2...$ with integer coefficients, then $\tilde{f}(t_1, t_2, ..., t_n)$ is periodic with periods $2\pi/\gamma_1, 2\pi/\gamma_2, ...$ If the basis is finite, though not necessarily integer, then $\tilde{f}(t_1, t_2, ..., t_n)$ is a limit periodic function of a finite number n of variables. If the basis is both integer and finite, then $\tilde{f}(t_1, t_2, ..., t_n)$ is periodic in each variable. This class of almost periodic functions is called conditionally periodic functions, with periods $2\pi/\gamma_1, 2\pi/\gamma_2, ..., 2\pi/\gamma_n$, or *quasiperiodic* functions. Thus, a quasiperiodic function is uniformly approximated by a finite sum of exponential functions, for a fixed value m, with argument $[2\pi\mathrm{i}(n_1\gamma_1 + n_2\gamma_2 + ... + n_m\gamma_m)t]$.

One immediate example is once again given by the sum of two sine functions with irrational periods

$$f(t) = A\sin(2\pi t) + B\sin(2\pi\nu t) \tag{A.8}$$

where ν is an irrational number. In general terms, the quasiperiodic function $f(t)$ is written as

$$f(t) = \sum_{n_1...n_m} f(n_1, ..., n_m)\exp[2\pi\mathrm{i}(n_1\gamma_1 + n_2\gamma_2 + ... + n_m\gamma_m)t]. \tag{A.9}$$

This function is usually not periodic; however it can be obtained from a function $g(t)$ of m variables, where g is periodic in each variable

$$g(t_1, ..., t_m) = \sum_{n_1...n_m} f(n_1, ...n_m) exp[2\pi i(n_1 t_1 + n_2 t_2 + ... + n_m t_m)]. \tag{A.10}$$

Equation (A.10) turns into (A.9) if $t_i = \gamma_i t$.

In the case of functions of several variables, the formal generalisation is immediate: a function $f(\boldsymbol{x})$ in an N-dimensional space is quasiperiodic if its Fourier transform is given as

$$f(\boldsymbol{k}) = \sum_{h_1...h_m} F(h_1, ..., h_m)\delta(\boldsymbol{k} - h_1\boldsymbol{x}_1^* - ... - h_m\boldsymbol{x}_m^*). \tag{A.11}$$

This way the Fourier wavevectors belong to the set $\boldsymbol{M}^*$

$$\boldsymbol{M}^* = \sum_{i=1}^{n} h_i \boldsymbol{x}_i^* \tag{A.12}$$

where the n vectors $\boldsymbol{x}_i$ are chosen so that they are rationally independent. In general, a given set of n vectors $\boldsymbol{x}_1...\boldsymbol{x}_n$ is rationally independent if the equality $\sum_{i=1}^{n} n_i \boldsymbol{x}_i = 0$ holds only when all the n_i rational coefficients are null.

The set of wavevectors in the Fourier transform, with the $\boldsymbol{M}^*$ structure (A.12), is called the *Fourier module*; the number of basis vectors gives the rank of $\boldsymbol{M}^*$ and its dimension is the same as that of the space spanned by the basis vectors. The non-quasiperiodic, but almost periodic functions are the limit of the quasiperiodic functions when the rank of the Fourier module diverges. At the other end of the spectrum, the reciprocal lattice of an ordinary crystal is a Fourier module where the rank, namely three, is equal to the dimension, since we obtain periodicity along each of the directions given by the basis vector triplet x_1^*, x_2^*, x_3^*.

References

Chapter 1

Further Reading

Ashcroft, N.W., Mermin, N.D.: *Solid State Physics* (Holt, Rinehart and Winston, New York, 1976)
Coxeter, H.S.M.: *Regular Polytopes*, 3rd ed. (Dover, New York, 1973)
Hilbert, D., Cohn–Vossen, S.: *Geometry and the Imagination*, 3rd ed. (Chelsea, New York, 1990)

Chapter 2

[2.1] *Webster's III New Intl. Dictionary* (G.&C. Merriam Co., Springfield, MA, 1971)
[2.2] C.M. Escher: *Grafica e Disegni* (B. Taschen Verlag, Köln, 1992)
[2.3] D. Lee, J. Cheng, M. Yuan, C.N.J. Wagner, A.J. Ardell: J. Appl. Phys. **64**, 4772 (1988)

Further Reading

Dunlop, A., Legrand, P., Lesueur, D., Lorenzelli, N., Morillo, J., Barbu, A., Bouffard, S.: Europhys. Lett. **15**, 765 (1991)
Goodstein, D.L.: *States of Matter* (Prentice Hall, Englewood Cliffs, N.J., 1975)
Haberland, H. (Ed.): *Clusters of Atoms and Molecules* (Springer Verlag, Berlin, 1994)
Haruyama, O., Kuroda, A., Asaki, N.: J. Non-Cryst. Solids **150**, 483 (1992)
Ossi, P.M.: Phil. Mag. **B76**, 541 (1997)
Poole, P.H., Grande, T., Sciortino, F., Stanley, H.E., Angell, C.A.: Comput. Mater. Sci. **4**, 373 (1995)
Price, D.L., Saboungi, M.-L., Reijers, H.T.J., Kearley, D., White, R.: Phys. Rev. Lett. **66**, 1894 (1991)
Richardson, J.M., Price, D.L., Saboungi, M.-L.: Phys. Rev. Lett. **76**, 1852 (1996)
Shechtman, D., Blech, I., Gratias, D., Cahn, J.W.: Phys. Rev. Lett. **53**, 1951 (1984)
Steeb, S., Lamparter, P.: J. Non-Cryst. Solids **156–158**, 24 (1993)
Ziman, J.M.: *Models of Disorder* (Cambridge University Press, Cambridge, U.K., 1979)

Chapter 3

[3.1] W.H. Zachariasen: J. Am. Chem. Soc. **54**, 3841 (1932)
[3.2] J. Schroers, R. Busch, A. Masuhr, W.L. Johnson: Appl. Phys. Lett. **74**, 2806 (1999)
[3.3] P.M. Ossi: Phil. Mag. **B79**, 2129 (1999)

Further Reading

Bakke, E., Busch, R., Johnson, W.L.: Appl. Phys. Lett. **67**, 3260 (1995)
Barratt, J.-L., Latz, A.: J. Phys. Condens. Matter **2**, 4289 (1990)
Brodin, A., Borjesson, L., Engberg, D., Torell, L.M., Sokolov, A.P.: Phys. Rev. **B53**, 11511 (1996)
Chryssikos, G.D., Kamitsos, E.I., Patsis, A.P., Bitsis, M.S., Karakassides, M.A.: J. Non-Cryst. Solids **131–133**, 1089 (1991)
Götze, W.: in *Amorphous and Liquid Materials*, ed. by E. Lüscher, G. Jacucci, G. Fritsch (Martinus Nijhoff, Dordrecht, 1986) p. 34
Hafner, J.: J. Non-Cryst. Solids **117–118**, 18 (1990)
Kauzmann, W.: Chem. Rev. **43**, 219 (1948)
Knaak, W., Mezei, F., Farago, B.: Europhys. Lett. **7**, 429 (1988)
Kob, W., Andersen, H.C.: Phys. Rev. Lett. **73** 1376 (1994)
Michaelsen, C., Gente, C., Bormann, R.: J. Appl. Phys. **81**, 6024 (1997)
Nagel, S.R., Tauc, J.: Phys. Rev. Lett. **35**, 380 (1975)
Ossi, P.M., Pastorelli, R.: Surf. Coat. Technol. **125**, 61 (2000)
Peker, A., Johnson, W.L.: Appl. Phys. Lett. **63**, 2342 (1990)
Pick, R.M., Dreyfus, C., Aouadi, A.: in *Proc. Workshop on Non-Equilibrium Phenomena in Supercooled Fluids, Glasses and Amorphous Materials* (World Scientific, Singapore, 1996), p. 148
Sagel, A., Sieber, H., Fecht, H.-J., Perepezko, J.H.: Phil. Mag. Lett. **77**, 109 (1998)
Souletie, J.: J. Phys. France **51**, 883 (1990)

Chapter 4

[4.1] R.J. Temkin, W. Paul, G.A.N. Connell: Adv. Phys. **22**, 581 (1973)
[4.2] T. Fukunaga, K. Suzuki: Sci. Rep. Res. Inst. Tôhoku Univ. **A29** 153 (1981)
[4.3] P. Lamparter: Zeits. Naturforsch. **50a**, 329 (1995)
[4.4] P. Kizler: Phys. Rev. Lett. **67**, 3555 (1991)
[4.5] J.M. Dubois, G. Le Caër: in *The Structure of Non-Crystalline Materials*, ed. by P.H. Gaskell, J.M. Parker, E.A. Davis (Taylor and Francis, London, 1983) p. 206
[4.6] R. Capelletti, A. Miotello, P.M. Ossi: J. Appl. Phys. **81**, 1 (1997)
[4.7] A. Bonizzi, R. Checchetto, A. Miotello, P.M. Ossi: Europhys. Lett. **44**, 627 (1998)
[4.8] P.M. Ossi, A. Miotello: Appl. Organometall. Chem. **15**, 430 (2001)
[4.9] G. Lucovsky, T.M. Hayes: in *Amorphous Semiconductors*, ed. by M.H. Brodsky (Springer Verlag, Berlin, 1979) p. 215
[4.10] P.M. Ossi: Rivista del Nuovo Cimento, **15**, n. 5, 1 (1992)
[4.11] L.C. Chen, F. Spaepen: Nature **336**, 366 (1988)
[4.12] J.M. Dubois, F. Montoya, C. Back: Mater. Sci. Engin. **A178**, 285 (1994)
[4.13] Y. Hirotsu: Mater. Sci. Engin. **A179–180**, 97 (1994)

[4.14] L. Cervinka: J. Non-Cryst. Solids **156–158**, 94 (1993)
[4.15] J.L. Finney: Proc. R. Soc. London, Ser. A, **319**, 495 (1970)
[4.16] G.S. Cargill (III): J. Appl. Phys. **41**, 12 (1970)
[4.17] C.H. Bennett: J. Appl. Phys. **43**, 2727 (1972)

Further Reading

Cervinka, L.: J. Non-Cryst. Solids **232–234**, 1 (1998)
Chartier, P., Mimault, J., Girardeau, T., Jaouen, M., Tourillon, G.: J. All. Comp. **194**, 77 (1993)
Dubois, J.M., Gaskell, P.H., Le Caër, G.: Proc. R. Soc. London **A402**, 323 (1985)
Egelstaff, P.A.: J. Non-Cryst. Solids **156–158**, 1 (1993)
Frank, F.C.: Proc. R. Soc. London **A215**, 43 (1952)
Gaskell, P.H., Wallis, D.J.: Phys. Rev. Lett. **76**, 66 (1996)
Helliwell, J.R.: Acta Crystall. **A54**, 738 (1998)
Kizler, P.: J. Non-Cryst. Solids **150**, 342 (1992)
Lamparter, P., Sperl, W., Steeb, S., Blétry, J.: Zeits. Naturforsch. **37a**, 1223 (1982)
Rehr, J.J., Albers, R.C.: Rev. Mod. Phys. **72**, 621 (2000)
Sadoc, J.F., Dixmier, J., Guinier, A.: J. Non-Cryst. Solids **12**, 46 (1973)
Suzuki, K., Fukunaga, T., Shibata, K., Otomo, T., Mizuseki, H.: in *Thermodynamics of Alloy Formation*, ed. by Y.A. Chang, F. Sommer (TMS, Warrendale, PA, 1997) p. 125
Uhlherr, A., Elliott, S.R.: J. Phys. Condens. Matt. **6**, L99 (1994)
Wagner, C.N.J.: in *Amorphous Metallic Alloys*, ed. by F.E. Luborsky (Butterworths, London, 1983) p. 58

Chapter 5

[5.1] H. Wu, S.R. Desai, Lai-Sheng Wang: Phys. Rev. Lett. **76**, 212 (1996)
[5.2] I.A. Harris, R.S. Kidwell, J.A. Northby: Phys. Rev. Lett. **53**, 2390 (1984)
[5.3] O. Echt, K. Sattler, E. Recknagel: Phys. Rev. Lett. **47**, 1121 (1981)
[5.4] G. Torchet, J. Farges, M.F. de Feraudy, B. Raoult: in *The Chemical Physics of Atomic and Molecular Clusters*, ed. by G. Scoles, Proc. Intl. School of Physics "E. Fermi", Course CVII (North Holland, Amsterdam, 1990) p. 513
[5.5] B. W. van de Waal: Phys. Rev. Lett. **76**, 1083 (1996)
[5.6] W.D. Knight, K. Clemenger, W.A. de Heer, W.A. Saunders: Phys. Rev. **B31**, 2539 (1985)
[5.7] K. Clemenger: Phys. Rev. **B32**, 1359 (1985)
[5.8] E.C. Honea, A. Ogura, C.A. Murray, K. Raghavachari, W.O. Sprenger, M.J. Farrold, W.L. Brown: Nature **366**, 42 (1993)
[5.9] J.M. Hunter, J.L. Fye, M.F. Jarrold, J.E. Brower: Phys. Rev. Lett. **73**, 2063 (1994)
[5.10] E. Rohlfing, D. Cox, A. Kaldor: J. Chem. Phys. **81**, 3322 (1984)

Further Reading

Benedeck, G., Colombo, L., Gaito, G., Serva, S.: in *The Physics of Diamond*, ed. by A. Paoletti, A. Tucciarone, Proc. Intl. School of Physics "E. Fermi", Course CXXXV (IOS Press, Amsterdam, 1997) p. 575

Benedeck, G., Martin, T.P, Pacchioni, G. (Eds.): *Elemental and Molecular Clusters*, Proc. 13th Intl. School, Erice, Italy (Springer, Berlin, 1988)

Dresselhaus, M.S., Dresselhaus, G., Eklund, P.C.: J. Mater. Res. **8**, 2054 (1993)

Fultz, B., Frase, H.N.: in *Ultrafine Grained Materials*, ed. by R.S. Mishra, S.L. Semiatin, C. Suryanarayana, N.N. Thadhani, T.C. Lowe (TMS, Warrendale, PA, 2000) p. 3

Gleiter, H.: J. Appl. Cryst. **24**, 79 (1991)

Gleiter, H.: Mater. Sci. Forum **189–190**, 67 (1995)

Heer, W.A. de, Knight, W.D., Chou, M.Y., Cohen, M.L.: in *Solid State Physics*, ed. by H. Ehrenreich, D. Turnbull, **40** (Academic Press, New York, 1987), p. 93

Heiney, P.A.: J. Phys. Chem. Solids **53**, 1333 (1992)

Hirai, H., Kondo, K., Yoshizawa, N., Shiraishi, M.: Appl. Phys. Lett. **64**, 1797 (1994)

Jarrold, M.F., Constant, V.A.: Phys. Rev. Lett. **67**, 2994 (1991)

Johnson, R.D., Bethune, D.S., Yannoni, C.S.: Acc. Chem. Res. **25**, 169 (1992)

Krishnan, A., Dujardin, E., Treacy, M.M.J., Hugdahl, J., Lynum, S., Ebbesen, T.W.: Nature **388**, 451 (1997)

Martin, T.P., Bergmann, T., Göhlich, H., Lange, T.: in *Cluster Models for Surface and Bulk Phenomena*, ed. by G. Pacchioni, P.S. Bagus, F. Parmigiani (Plenum Press, New York, 1992) p. 3

Pacchioni, G., Rosch, N.: Acc. Chem. Res. **28**, 390 (1995)

Richard, P., Gervois, A., Oger, L., Troadec, J.P.: Europhys. Lett. **48**, 415 (1999)

Scoles, G. (Ed.): *The Chemical Physics of Atomic and Molecular Clusters*, Proc. Intl. School of Physics "E. Fermi", Course CVII (North Holland, Amsterdam, 1990)

Siegel, R.W.: Ann. Rev. Mater. Sci. **21**, 559 (1991)

Weissmuller, J., Schubert, P., Franz, H., Birringer, R., Gleiter, H.: in *The Physics of Non-Crystalline Solids*, ed. by L. David Pye, W.C. La Course, H.J. Stevens (Taylor & Francis, London, 1992) p. 26

Yacaman, M.J., Herrera, R., Zorrilla, C., Tehuacanero, S., Avalos, M.: Mater. Res. Soc. Symp. Proc. **206** (1991) p.183

Chapter 6

[6.1] K.M. Reinisch, M.L. Nibert, S.C. Harrison: Nature **404**, 960 (2000)

[6.2] M. Audier, P. Guyot: in *Quasicrystals and Incommensurate Structures in Condensed Matter*, ed. by M.J. Yacaman, D. Romeu, V.Castano, A. Gomez (World Scientific, Singapore, 1990) p. 288

[6.3] A.P. Tsai, A. Inoue, T. Masumoto: Jpn. J. Appl. Phys. **26**, 1505 (1987)

[6.4] M. Audier, P. Guyot: in *Extended Icosahedral Structures*, ed. by M.V. Jaric, D. Gratias (Academic Press, S. Diego, 1989) p. 1

[6.5] A.P. Tsai, J.Q. Guo, E. Abe, H. Takakura, T.J. Sato, Nature **408**, 537 (2000)

[6.6] S. Enzo, G. Mulas, F. Delogu, R. Frattini, J. Metast. and Nanocryst. Mater. **2**, 417 (1999)

[6.7] J. Dixmier, A. Chenoufi, B. Bouchet-Fabre, Europhys. Lett. **13**, 511 (1990)

[6.8] A.P. Tsai, K. Iraga, A. Inoue, T. Masumoto, K. Satoh, K. Tsuda, M. Tanaka: Mater. Sci. Eng. **A181–182**, 750 (1994)
[6.9] T.M. Schaub, D.E. Bürgler, H.-J. Güntherodt, J.B. Suck: Phys. Rev. Lett. **73**, 1255 (1994)
[6.10] P.M. Ossi: J. All. Comp. **186**, 153 (1992)
[6.11] C. Roth, G. Schwalbe, R. Knöfler, F. Zavaliche, O. Madel, R. Haberkern, P. Häussler: J. Non-Cryst. Solids **252**, 869 (1999)

Further Reading

Abe, E., Sato, T.J., Tsai, A.P.: Phil. Mag. Lett. **77**, 205 (1998)
Bellingeri, P., Ossi, P.M.: J. All. Comp. **316**, 39 (2001)
Hubert, H., Devouard, B., Garvie, L.A.J., O'Keefe, M., Buseck, P.R., Petuskey, W.T., McMillan, P.F.: Nature **391**, 376 (1998)
Merlin, R., Bajema, K., Clarke, R., Juang, F.Y., Bhattacharya, P.K.: Phys. Rev. Lett. **55**, 1768 (1985)
Mizushina, I., Watanabe, M., Murakoshi, A., Hotta, M., Kashiwagi, M., Yoshiki, M.: Appl. Phys. Lett. **63**, 373 (1993)
Penrose, R.: in *Introduction to the Mathematics of Quasicrystals*, ed. by M.V. Jaric (Academic Press, S. Diego, 1989) p. 53
Penrose, R.: Math. Intelligencer **2**, 32 (1979)
Rothenberg, E., Theis, W., Horn, K., Gille, P.: Nature **406**, 602 (2000)
Rouxel, D., Cai, T.-H., Jenks, C.J., Lograsso, T.A., Ross, A., Thiel, P.A.: Surf. Sci. **461**, L521 (2000)
Shechtman, D., Blech, I., Gratias, D., Cahn, J.W.: Phys. Rev. Lett. **53**, 1951 (1984)
Stephens, P.W., Goldman, A.I.: Phys. Rev. Lett. **56**, 1168 (1986)
Stephens, P.W., Goldman, A.I.: Phys. Rev. Lett. **57**, 2331 (1986)
Takeuchi, S.: Mater. Sci. Forum **150–151**, 35 (1994)
Yurjev, G.S., Sheromov, M.A.: Nucl. Inst. Meth. **A359**, 181 (1995)
Zhang, T., Inoue, A., Matsushida, M., Saida, J.: J. Mater. Res. **16**, 20 (2001)

Appendix

[A.1] C. Corduneanu: *Almost Periodic Functions* (Interscience, New York, 1968)
[A.2] L. Amerio, G. Prouse: *Almost Periodic Functions and Functional Equations* (Van Nostrand Reinhold, New York, 1971)

Index

Printing (Computer to Film): Saladruck Berlin
Binding: Stürtz AG, Würzburg